BIOZONE

D0608371

THE LIVING
EARTH

NGSS   INTEGRATING BIOLOGY AND EARTH SCIENCE

THE LIVING EARTH

Meet the Writing Team

Tracey Greenwood
I have been writing resources for students since 1993. I have a Ph.D in biology, specialising in lake ecology and I have taught both graduate and undergraduate biology.

Tracey
Senior Author

Kent Pryor
I have a BSc from Massey University majoring in zoology and ecology and taught secondary school biology and chemistry for 9 years before joining BIOZONE as an author in 2009.

Kent
Author

Lissa Bainbridge-Smith
I worked in industry in a research and development capacity for 8 years before joining BIOZONE in 2006. I have an M.Sc from Waikato University.

Lissa
Author

Richard Allan
I have had 11 years experience teaching senior secondary school biology. I have a Masters degree in biology and founded BIOZONE in the 1980s after developing resources for my own students.

Richard
Founder & CEO

ISBN 978-1-927309-55-1

First Edition

Third printing with clarifications and change from inverse notation to use of the solidus.

Copyright © 2018 Richard Allan
Published by **BIOZONE International Ltd**

Printed by REPLIKA PRESS PVT LTD using paper produced from renewable and waste materials

Next Generation Science Standards (NGSS) is a registered trademark of Achieve. Neither Achieve nor the lead states and partners that developed the Next Generation Science Standards were involved in the production of this product and do not endorse it.

Purchases of this book may be made direct from the publisher:

BIOZONE Corporation
USA and Canada
FREE phone: 1-855-246-4555
FREE fax: 1-855-935-3555
Email: sales@thebiozone.com
Web: www.thebiozone.com

Cover photograph

The tropical jungles of Southeast Asia support a wide variety of animals and plants, including the orangutan, the Asian elephant, and the Sumatran rhinoceros. They are the oldest tracts of tropical rainforest dating to about 70 million years old. As with rainforests elsewhere in the world the jungles of Southeast Asia are under pressure from human activities such as logging and slash and burn land clearance.

PHOTO: © quickshooting/www.stock.adobe.com

Thanks to:

The staff at BIOZONE, including Mike Campbell for design and graphics support, Paolo Curray for IT support, Allan Young for office handling and logistics, and the BIOZONE sales team.

All rights reserved. No part of this publication may be reproduced, stored in a retrieval system, or transmitted in any form or by any means, electrical, mechanical, **photocopying**, recording or otherwise, without the permission of BIOZONE International Ltd.
This book may not be **re-sold**. The conditions of sale specifically prohibit the photocopying of exercises, worksheets, and diagrams from this book for any reason.

Contents

CODES: Activity is marked: ● to be done   ✓ when completed

Contents

IS6 Ecosystem Stability and the Response to Climate Change

Basic Skills for Life Science Students

CODES: **Activity** is marked: • to be done ☑ when completed

The Living Earth: A Flow of Ideas

This concept map shows the broad areas of content covered within each Instructional Segment of The Living Earth.
Anchoring phenomena are indicated in red boxes and dashed red arrows show conceptual connections between topics.
Green boxes show some of the crossing cutting concepts linking some of the topics. Can you find other connections?

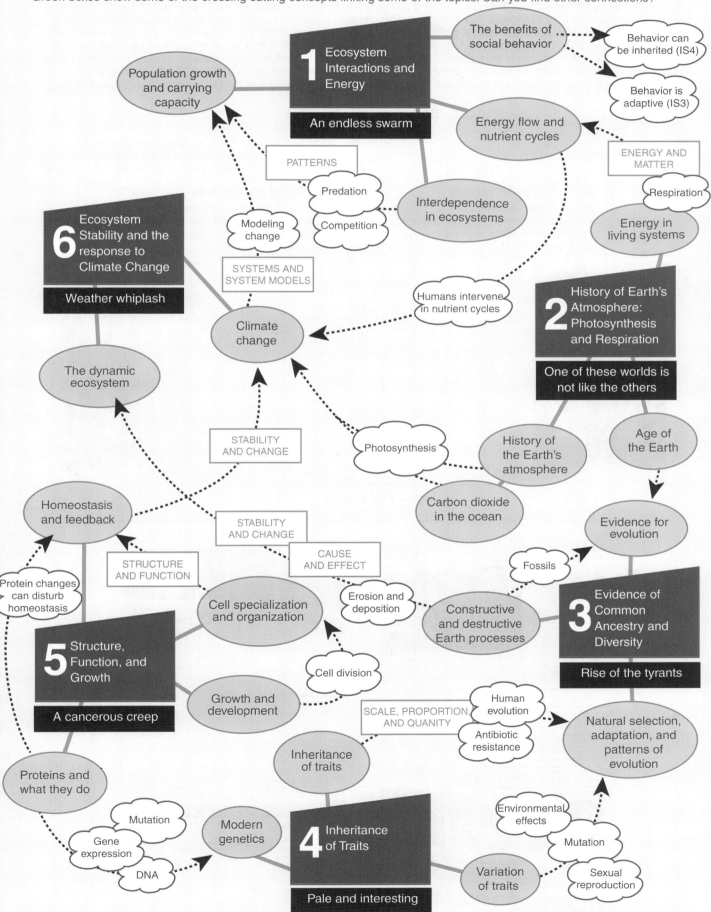

Using This Book

Activities make up most of this book. These are presented as integrated instructional sequences over multiple pages, allowing you to build a deeper understanding of phenomena as you progress through each chapter. Each chapter begins with an **Anchoring Phenomenon**. This is something you would have seen or experienced (e.g. whiplash weather) but may not necessarily be able to explain. The anchoring phenomenon is revisited at the close of the chapter. Most of the remaining activities in a chapter are designed to lead you an understanding of that phenomenon. Each activity begins with a task to **engage** your thinking, asking you to review your current understanding of a phenomenon and setting the scene for the content to follow. The activity then allows you to **explore** related content through modeling, experimentation, or data analysis. You can then **explain** phenomena described through models, simulations, data, descriptions, or photographs. Many activities will also require you to **elaborate** (expand) on the ideas covered and then to **evaluate** your understanding.

Structure of a chapter

Chapter introduction
Identifies the activities relating to the guiding questions.

Summative assessment
This can be used as a formal assessment of the NGSS performance expectations addressed in the chapter.

Anchoring phenomenon
The first activity is always an anchoring phenomenon. It introduces a phenomenon that is explained by the rest of the activities in the chapter.

Anchoring phenomenon revisited
Once you have completed the activities in the chapter should be able to explain various aspects of the anchoring phenomenon more fully.

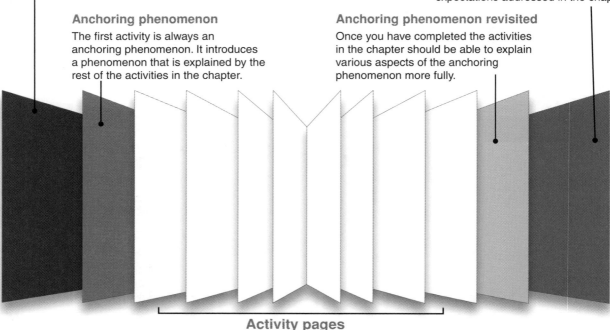

Activity pages

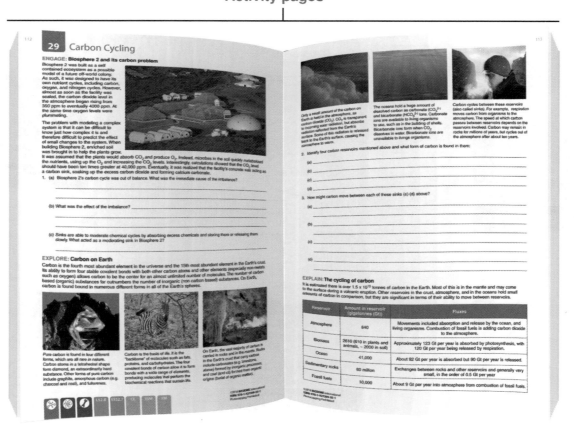

This identifies the Instructional Segment to which this chapter applies.

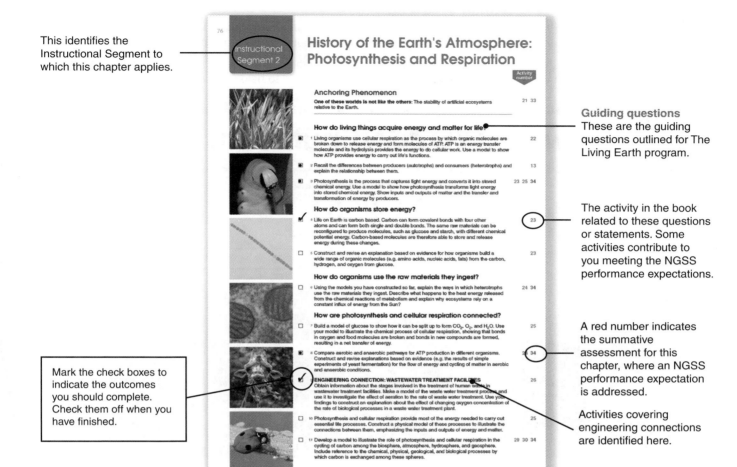

Guiding questions
These are the guiding questions outlined for The Living Earth program.

The activity in the book related to these questions or statements. Some activities contribute to you meeting the NGSS performance expectations.

Mark the check boxes to indicate the outcomes you should complete. Check them off when you have finished.

A red number indicates the summative assessment for this chapter, where an NGSS performance expectation is addressed.

Activities covering engineering connections are identified here.

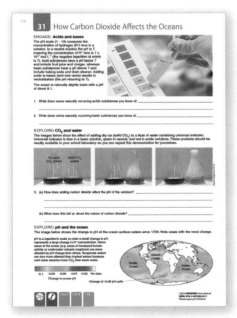

Group discussions, informative articles, simple practical activities, and modeling, are used to **engage** with and **explore** material related to the activity.

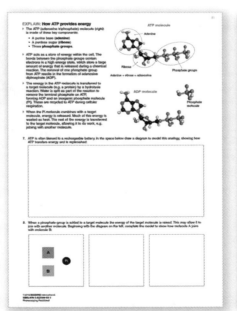

You will need to use your knowledge and the evidence presented to **explain** trends in data or how a particular system works.

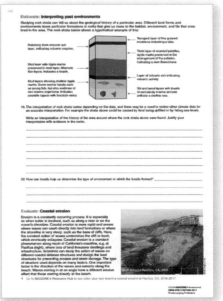

You can **elaborate** on the skills and understanding gained during the activity and **evaluate** your understanding using further modeling and data analysis.

©2018 **BIOZONE** International
Photocopying Prohibited

Using BIOZONE's Resource Hub

▶ BIOZONE's Resource Hub provides links to online content that supports the activities in the book. From this page, you can also check for any errata or clarifications to the book or model answers since printing.

▶ The external websites are, for the most part, narrowly focused animations and video clips directly relevant to some aspect of the activity on which they are cited. They provide great support to help your understanding.

www.BIOZONEhub.com

Then enter the code in the text field   **TLE1-9551**

Search for an activity here.

Q Search activity number, title, keyword...

The Living Earth

The Living Earth is an innovative new resource for the new California Science Framework (grades 9-12). It integrates NGSS Life Science and Earth and Space Sciences to deliver a phenomenon-based curriculum that meets the diverse needs of California's teachers and students. With an integrated 3D approach and flexibility as its central theme, this program delivers multiple learning experiences for students via the printed book and its supporting online resources. The Resource Hub provides student access to videos, animations, 3D models, explanatory content, spreadsheets, databases, and research opportunities. Teachers can also access much of the relevant source material used to develop content. There is much to explore!

Instructional segment (IS) and chapter title.

IS 1

Ecosystem Interactions and Energy

- 1 An Endless Swarm
- 2 The Earth's Systems
- 3 Abiotic Factors Influence Distribution
- 4 The Ecological Niche
- 5 Populations Have Varied Distribution
- 6 Population Growth
- 7 Modeling Population Growth
- 8 The Carrying Capacity of an Ecosystem
- 9 Species Interactions Can Regulate Populations
- 10 Predation Can Control Some Populations

- 11 Organisms Compete for Limited Resources
- 12 Human Activity Alters Populations
- 13 Producers, Consumers, and Food Webs
- 14 Energy in Ecosystems
- 15 Nutrient Cycles
- 16 Humans Intervene in Nutrient Cycles
- 17 Group Behavior Improves Survival
- 18 Individuals in Groups Often Cooperate
- 19 An Endless Swarm Revisited
- 20 Summative Assessment

Click on an activity title to go directly to the resources available for that activity.

View resources →

IS 2

History of Earth's Atmosphere: Photosythesis

©2018 **BIOZONE** International
Photocopying Prohibited

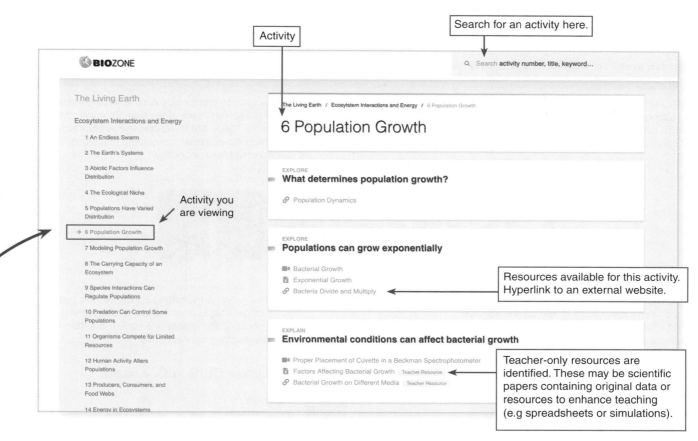

Activity

Search for an activity here.

Activity you are viewing

Resources available for this activity. Hyperlink to an external website.

Teacher-only resources are identified. These may be scientific papers containing original data or resources to enhance teaching (e.g spreadsheets or simulations).

The Resource Hub icons

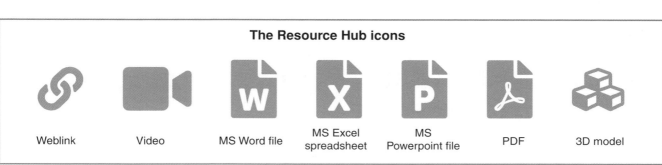

Weblink | Video | MS Word file | MS Excel spreadsheet | MS Powerpoint file | PDF | 3D model

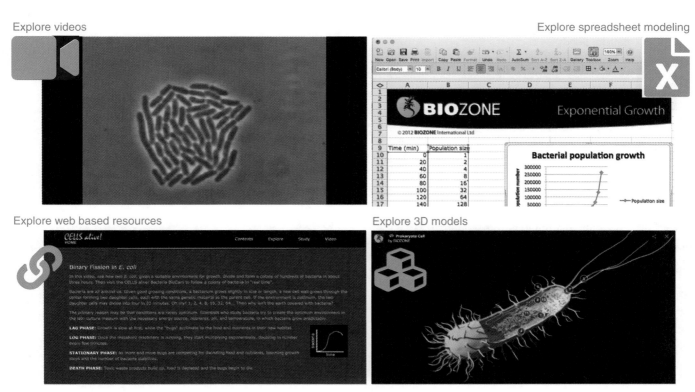

Explore videos

Explore spreadsheet modeling

Explore web based resources

Explore 3D models

©2018 **BIOZONE** International
Photocopying Prohibited

x

Using the Tab System

The tab system is a useful way to quickly identify the crosscutting concepts, disciplinary core ideas, and science and engineering practices embedded within each activity. The tabs also indicate whether or not the activity is supported online.

The hub tab indicates that the activity is supported online at the BIOZONE RESOURCE HUB. Online support may include videos, articles, and computer models.

The **orange** disciplinary core idea tabs indicate the disciplinary core ideas that are covered in the activity. These are described in the introduction to each chapter, under the guiding questions. The code itself is unimportant.

| | | | LS2.B | ESS2.7 | CE | SSM |

The **blue** science and engineering practices tabs use picture codes to identify the science and engineering practices (SEPs) relevant to the activity. There may be several SEPs per activity.

The **green** crosscutting concepts tabs indicate activities that share the same crosscutting concepts. Many activities have more than one crosscutting concept code.

Science and Engineering Practices

 Asking questions (for science) and defining problems (for engineering)
Asking scientific questions about observations or content in texts helps to define problems and draw valid conclusions.

 Developing and using models
Models can be used to represent a system or a part of a system. Using models can help to visualize a structure, process, or design and understand how it works. Models can also be used to improve a design.

 Planning and carrying out investigations
Planning and carrying out investigations is an important part of independent research. Investigations allow ideas and models to be tested and refined.

 Analyzing and interpreting data
Once data is collected it must be analyzed to reveal any patterns or relationships. Tables and graphs are just two of the many ways to display and analyze data for trends.

 Using mathematics and computational thinking
Mathematics is a tool for understanding scientific data. Converting or transforming data helps to see relationships more easily while statistical analysis can help determine the significance of the results.

 Constructing explanations (for science) and designing solutions (for engineering)
Constructing explanations for observations and phenomena is a dynamic process and may involve drawing on existing knowledge as well as generating new ideas.

 Engaging in argument from evidence
Scientific argument based on evidence is how new ideas gain acceptance in science. Logical reasoning based on evidence is required when considering the merit of new claims or explanations of phenomena.

 Obtaining, evaluating, and communicating information
Evaluating information for scientific accuracy or bias is important in determining its validity and reliability. Communicating information includes reports, graphics, oral presentation, and models.

Cross Cutting Concepts

 Patterns
We see patterns everywhere in science. These guide how we organize and classify events and organisms and prompt us to ask questions about the factors that create and influence them.

 Cause and effect
A major part of science is investigating and explaining causal relationships. The mechanisms by which they occur can be tested in one context and used to explain and predict events in new contexts.

 Scale, proportion, and quantity
Different things are relevant at different scales. Changes in scale, proportion, or quantity affect the structure or performance of a system.

 Systems and system models
Making a model of a system (e.g. physical, mathematical) provides a way to understand and test ideas.

 Energy and matter
Energy flows and matter cycles. Tracking these fluxes helps us understand how systems function.

 Structure and function
The structure of an object or living thing determines many of its properties and functions.

 Stability and change
Science often deals with constructing explanations of how things change or how they remain stable.

©2018 **BIOZONE** International
Photocopying Prohibited

Instructional Segment 1

Ecosystem Interactions and Energy

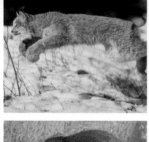

Anchoring Phenomenon

An endless swarm: The high density and swarming of migratory locusts. 1 19

What factors affect the size of populations within an ecosystem?

☐ 1 Identify the various **abiotic** and **biotic** components of **ecosystems**. Analyze and interpret data to describe how these different components influence one another. 3 4

☐ 2 In what way is the Earth a system of systems? Describe the general groupings of Earth materials and processes (**atmosphere**, **hydrosphere**, **biosphere**, **geosphere**, **anthrosphere**) each of which is shaped by its own processes and interactions with other systems. Develop a model to show how the spheres interact. 2

☐ 3 What is a population? Describe different patterns of population growth and explain the role of **carrying capacity** in limiting population growth. Use mathematical and computational thinking and modeling to predict the effect of chosen interdependent factors on the size of a population over time. 5-8

☐ 4 Conduct investigations to test how different parameters change population size. Analyze your findings and describe the population changes mathematically. Use mathematical representations to support and revise evidence-based explanations about factors affecting populations and diversity in ecosystems of different scale. How well does an ecosystem model at one scale relate to a model at another scale? 6-8 20

☐ 5 Categorize factors influencing population growth as **density dependent** (DD) or **density independent** (DI) and describe how they are different. Analyze and interpret data to explain how DD and DI factors affect the flow of energy and matter and that this is how they affect population size. 8-11

☐ 6 Describe the ways organisms obtain and store energy. Explain how this energy is transferred in ecosystems through **food chains** and **food webs**. Use the conceptual model of an energy pyramid and calculate energy fluxes to explain the energy available at each successive **trophic level** in an ecosystem. 13 14

☐ 7 Use predictive models of predator-prey population cycles to support claims about the relative amounts of energy at different trophic levels. 14

☐ 8 Explain how nutrients (matter) cycle within and between ecosystems including between abiotic and biotic components. Use mathematical representations to show that matter and energy are conserved as matter cycles and energy flows through ecosystems. 15 20

☐ 9 How do populations behave as a system with many interacting parts (members)? Evaluate the evidence for the role of group behavior in the survival and reproductive success of individuals and populations. 17 18 20

What are common threats to remaining natural ecosystems and biodiversity? How can these threats be reduced?

☐ 10 Explain how humans might cause density dependent and density independent changes to ecosystems by altering the availability of resources and changing the landscape (including through climate change). Describe how these changes might affect the size and diversity of populations. 12

☐ 11 Obtain information to summarize the various positive and negative ways in which humans influence ecosystem resources and disrupt the usual nutrient cycles. 16

1 An Endless Swarm

ANCHORING PHENOMENON: The high density and swarming of migratory locusts

A swarm of locusts is one of nature's most incredible animal events. So astonishing and destructive are these swarms they are recorded in numerous historical accounts, including those of Greek and Roman historians. Locust 'plagues' have historically been particularly catastrophic in North Africa, where they have long been associated with famine.

Under certain conditions particular species of normally solitary shorthorned grasshoppers may form vast swarms (dense aggregations) that migrate across the country eating everything in their path. Swarms have been known to contain billions of locusts (the swarming form of grasshoppers), last multiple generations, and many years.

Locusts are the swarming or gregarious form of certain grasshopper species.

Locust swarms may contain up to 80 million individuals per km²

1. Identify a species in your local area that:

   (a) Swarms: _____

   (b) Migrates: _____

2. Discuss in groups or as a class what factors in the environment might cause a normally solitary species to suddenly form a voracious giant swarm. Summarize your ideas here:

   _____

   _____

   _____

   _____

3. Swarming occurs regularly, which suggests the behavior has advantages to the individuals in the swarm. Discuss within your group what these advantages might be and summarize them below:

   _____

   _____

   _____

4. Discuss how human activities might be involved with or affected by swarming locusts: _____

   _____

   _____

   _____

©2018 **BIOZONE** International
ISBN: 978-1-927309-55-1
Photocopying Prohibited

2 The Earth's Systems

Hydrosphere: all liquid and surface water. Ice is sometimes called the cryosphere.

Atmosphere: all the gases (air), including water vapor.

Geosphere: the Earth itself.

Biosphere: all living things.

Anthrosphere: The part of the environment made or modified by humans.

ENGAGE: The Earth is made up of spheres

The model above shows the Earth's four spheres. These interact as a complex system that maintains life on Earth. The anthrosphere is part of the biosphere but is sometimes classified as a fifth sphere because of the impact that humans have on all other systems.

1. Observe the environment around you and identify elements of each of the five spheres:

   (a) Biosphere: _____

   (b) Anthrosphere: _____

   (c) Hydrosphere: _____

   (d) Geosphere: _____

   (e) Atmosphere: _____

2. Interactions between spheres involve movement of energy and/or matter between them. Look at the photo below. Identify the spheres present and briefly describe any interactions that could be occurring between them.

   _____

   _____

   _____

   _____

   _____

   _____

   _____

   _____

   _____

3. Anthrospheric changes can have significant impacts on the other four spheres. Using your local environment as inspiration, describe a way in which human activity can affect the other spheres:

   _____

   _____

   _____

©2018 **BIOZONE** International
ISBN: 978-1-927309-55-1
Photocopying Prohibited

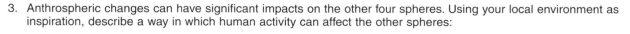

SSM   LS2.A

EXPLORE: Ecosystems have many components

Ecosystems are natural units made up of the living organisms (biotic factors) and the physical conditions (abiotic factors) in an area. **Abiotic factors** include non-living factors associated with the geosphere, hydrosphere, and atmosphere (below). **Biotic factors** include all the living organisms and their activities.

The interactions of living organisms with each other and with the physical environment help determine the features of an ecosystem. The components of an ecosystem are linked to each other (and to other ecosystems) through nutrient cycles and energy flows.

Biotic factors

These are all the living organisms in the environment, including their interactions.
• Plants
• Animals
• Microorganisms (e.g. bacteria)
• Fungi
• Protists (e.g. algae and protozoans)

Atmosphere (air)

• Wind speed
• Wind direction
• Humidity
• Light intensity/quality
• Precipitation
• Temperature

Hydrosphere (water)

• Dissolved nutrients
• pH
• Salinity
• Dissolved oxygen
• Precipitation
• Temperature

Geosphere (rock/soil)

• Nutrient availability
• Soil moisture
• pH
• Composition
• Temperature
• Depth

4. (a) Which spheres are represented in the savanna ecosystem model above? _____

_____

_____

(b) Activities in one sphere can affect other spheres and may cause changes at the ecosystem level. Develop a model, e.g. a diagram or mind map, of interactions within and between the biotic and abiotic components of an ecosystem.

3 Abiotic Factors Influence Distribution

ENGAGE: Distribution of the common sea star

The common sea star (*Asterias rubens*) is a widely distributed marine invertebrate, found throughout the Atlantic, often in very high numbers. It can be found at a wide range of depths between 0-400 m and experiences large variations in abiotic factors.

Scientists collected adult specimens from two populations in the White Sea (off the Northwest coast of Russia) and the Barentz Sea (off the Northern coasts of Norway and Russia). They exposed them to a range of salinities (amount of dissolved salt in parts per thousand) within a five-compartment chamber (right) and recorded the number of animals found in different salinities. The animal was placed in the center of the chamber with each arm experiencing water of different salinity. The animal then crawled into the compartment with the preferred salinity. All other factors were kept constant. The results are shown below.

Sea star choice chamber. Each compartment contains water of a different salinity.

Salinity (‰)	Frequency of choice (%)	
	White Sea	Barentz Sea
15.0	0	0
17.5	3	0
20.0	12	1.2
22.5	36	7.5
25.0	42	3.4
27.5	31	6.2
30.0	18	30.2
32.5	9	39.6
35.0	8	42.1
37.5	0	29.6
40.0	0	14
42.5	0	9.8

NEED HELP?
See Activity 97

1. (a) Plot the two sets of data from the table above on the grid provided.

   (b) What do the plots show? _____

   _____

   _____

   (c) What was the preferred salinity for each of the sea star populations? _____

   _____

   (d) What do these results suggest about the salinity of the two areas of collection? _____

   _____

   _____

   (e) Describe the abiotic conditions the common sea star as a species can tolerate: _____

   _____

   _____

   _____

©2018 **BIOZONE** International
ISBN: 978-1-927309-55-1
Photocopying Prohibited

CE LS2.A

2. Based on the White Sea sea star population, draw a general diagram (model) to show how the numbers of individuals in a population change over an abiotic gradient. Label it to show the optimal (preferred), marginal, and unavailable habitat. Mark the tolerance range (the range for an abiotic factor outside of which no individuals can survive):

EXPLORE: Estuarine habitats

Estuary: high tide

Estuary: low tide

An estuary is a semi-enclosed coastal body of water, which has a free connection with the ocean and where marine and freshwater environments meet and mix. Estuarine water is brackish (it has more salt than fresh water but not as much as seawater) but salinity varies with tidal flows. Estuaries provide habitat for young fish and migratory bird populations. They are dynamic environments with large fluctuations in the abiotic conditions as the tide rises or falls to cover or expose tidal flats. Important abiotic factors include pH, salinity, temperature, and dissolved oxygen.

The estuarine habitat of the striped shore crab

The striped shore crab (*Pachygrapsus crassipes*), right, is a widespread species along the west coast of North America, inhabiting a region extending high into the intertidal zone where it is exposed to air for about half of each day. They live in hard mud and rocky substrates where they can burrow or hide. They cannot live in soft sand as they may become suffocated. They forage in and out of the water, feeding mostly at night on algae, limpets, and smaller crabs.

Peter D. Tillman CC2.0

3. (a) Thinking about estuarine environments (above), what are some of the challenges faced by the striped shore crab living there?

_____

_____

_____

_____

(b) Suggest what physiological, structural, or behavioral features might be important to the striped shore crab's survival?

_____

_____

_____

_____

©2018 **BIOZONE** International
ISBN: 978-1-927309-55-1
Photocopying Prohibited

SNAPSHOT: ELKHORN SLOUGH, CALIFORNIA

EXPLAIN: How do abiotic factors affect organisms?

▶ Elkhorn Slough National Estuarine Research Reserve (above) is a large (688 ha) tidal salt marsh and estuary located half way between Santa Cruz and Monterey. The estuary extends 11 km inland from the coast and provides habitat for over 700 species including plants, invertebrates, birds, marine mammals, and fish.

▶ The reserve is made up of several different areas, including South Marsh. Habitats range from oak woodlands and coastal chaparral to marshes and wetlands.

▶ The reserve is owned and managed by the California Department of Fish and Wildlife. Along with researchers from the National Oceanic and Atmospheric Administration (NOAA), they monitor the health of the reserve and carry out research in on-site field laboratories.

▶ Some of the research involves monitoring abiotic factors and the effect of their changes on the plants and animals within the reserve.

▶ Environmental tolerance factors for two organisms found at South Marsh are shown below. Chinook salmon is a migratory fish species, which moves into coastal streams to spawn. The Olympia oyster is a resident filter-feeding bivalve mollusk.

▶ Selected physical data for South Marsh over two years (2016-2017) is presented on the next page.

CA EP&Cs I: The ecosystem services provided by natural systems are essential to human life (I b)

Wetlands like the Elkhorn Slough provide essential services to humans and the environment.

▶ The physical and biotic environment of the wetland acts as a natural filter for water before it enters the sea.

▶ The high productivity of wetlands also means they are able to remove and store large amounts of carbon dioxide from the atmosphere, slowing global warming.

▶ Monitoring protected coastal areas allows better management of resources to benefit both humans and wildlife.

Olympia oyster (*Ostrea lurida*)

- Salinity of 12-25 ppt (parts per thousand) is optimal for growth. Death occurs at salinities below 5 ppt or above 25 ppt.
  Brackish water is 5-30 ppt, seawater is ~35 ppt.
- More likely to spawn when salinity is over 20 ppt.
- Water temperature of 18°C for 4 hours is required for spawning.
- Need a dissolved oxygen (DO) of 4 mg/L or greater.
- Optimum temperature is 16°-19°C but temperatures up to 27°C are tolerated.
- pH range of 7.5-8.5 is required for optimal growth.

Spawning coloration

Chinook salmon (*Oncorhynchus tshawytscha*)

- Salinity > 15 ppt. Tolerance depends on stage.
- Optimal temperature for adults is 14.5-17°C but they tolerate 3-20°C.
- Optimum temperature for fertilization and fry development is below 9-10°C and should not exceed 13.5-14.5°C.
- Spawn at temperatures below 14.5°C.
- Newly hatched salmon need a minimum DO of >10 mg/L. Adults prefer a DO of >7 mg/L.
- pH range 4.0-9.0 is required for survival. Optimum pH is narrow at 7.5-7.8.

©2018 **BIOZONE** International
ISBN: 978-1-927309-55-1
Photocopying Prohibited

Water temperature and dissolved oxygen, South Marsh, Elkhorn Slough, 2016 - 2017

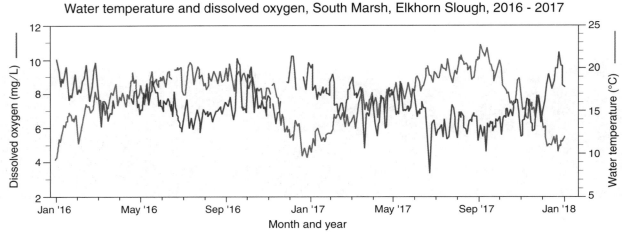

Water pH, South Marsh, Elkhorn Slough, 2016 - 2017

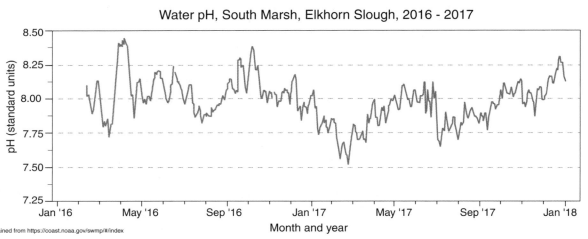

Data obtained from https://coast.noaa.gov/swmp/#/index

4. Using the information about the Chinook salmon and Olympia oyster on the previous page to help you, describe how optimum and tolerance ranges for physical conditions differ:

_____

_____

_____

_____

5. Study the graph of water temperature and dissolved oxygen. Is there any relationship between the two? _____

_____

_____

6. Keeping in mind the environmental tolerance parameters required by the Chinook salmon, comment on how well the species would have tolerated the conditions displayed in the two graphs above, and explain your reasoning. Comment on how the conditions would have affected the different biological stages of the fish where possible:

_____

_____

_____

_____

_____

_____

_____

_____

_____

©2018 **BIOZONE** International
ISBN: 978-1-927309-55-1
Photocopying Prohibited

ELABORATE: Alien invaders and a system out of balance

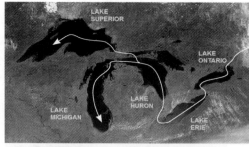

Trout

Sea lamprey

Alewife

USGS

USFW

▶ The alewife (*Alosa pseudoharengus*) is a migratory species of herring found along the Atlantic Coast of North America. Like salmon, the adults move from the sea into freshwater streams to breed. Alewife have gained access to the four upper Great Lakes using the Welland Canal to bypass the natural barrier of the Niagara Falls (top photo).

▶ In the Great Lakes, they are a nuisance species, and have displaced many of the once commercially important native Great lakes fish species. Alewife populations in Lakes Huron and Michigan became so abundant in the 1960s that they made up most of the lakes' biomass. During these periods of very high abundance, unexplained massive alewife die-offs occurred, polluting shorelines and causing a public nuisance. Unfortunately, the obvious native predator, the lake trout, had already been decimated as a result of another alien, the sea lamprey (seen feeding on a lake trout in the center left image). This prompted the introduction of salmon to control alewife numbers.

▶ In the years since, an important salmon fishery has developed around alewife as a forage fish. Now, the alewife population is in decline, but the native forage fish species that the alewife displaced may not now be able to recover.

▶ Adult alewife (image left) and their juveniles need a dissolved oxygen (DO) level >3.6 mg/L. The eggs and larvae need a DO >5 mg/L. Given its importance as a bait and forage fish, alewife introductions to California have been considered in the years prior to 1997. A related species, the American shad, *Alosa sapidissima*, was successfully introduced to the Sacramento River in 1871 and now forms an important recreational fishery.

7. Based on information provided, the physical data for South Marsh on the opposite page, and the resources available through **BIOZONE's Resource Hub**, decide whether an alewife fishery is possible or desirable in Elkhorn Slough. As a group, argue a case either for or against its introduction to this region. What features of California's waterways could influence the success and risk of an introduction? What similarities are there to the Great Lakes scenario? What species would it compete with and potentially displace? Summarize your arguments below:

_____

_____

_____

_____

_____

_____

_____

_____

_____

_____

_____

_____

_____

_____

_____

_____

_____

EVALUATE: Communicate your findings

8. As a group, present your arguments (as outlined above to the class), e.g. as a poster or oral presentation.

©2018 **BIOZONE** International
ISBN: 978-1-927309-55-1
Photocopying Prohibited

4 The Ecological Niche

ENGAGE: The niche is the functional role of an organism

The **ecological niche** (or niche) of an organism describes its functional position in its environment. The full range of environmental conditions under which an organism can exist describes its **fundamental niche**.

▸ The fundamental niche is influenced by the physical environment and the organism's adaptations for exploiting it.

▸ The presence of other organisms may constrain the breadth of an organism's niche so that the organism exploits only part of the niche 'space' available to it. The niche an organism actually occupies is called its **realized niche**.

The physical conditions influence the habitat. A factor may be well suited to the organism, or present it with problems to be overcome.

Adaptations enable the organism to exploit the resources of the habitat. The adaptations take the form of structural, physiological and behavioral characteristics of the organism.

Physical conditions
- Substrate
- Humidity
- Sunlight
- Temperature
- Salinity
- pH
- Exposure
- Altitude
- Depth

Resources offered by the habitat

- Food sources • Shelter • Mating sites
- Nesting sites • Predator avoidance

Adaptations for:
- Locomotion
- Activity pattern
- Tolerance to physical conditions
- Predator avoidance
- Defense
- Reproduction
- Feeding
- Competition

Resource availability is affected by the presence of other organisms and interactions with them: competition, predation, parasitism, and disease.

The habitat provides opportunities and resources for the organism. The organism may or may not have the adaptations to exploit them fully.

1. (a) Name an organism in your area and identify what type of environment it is commonly found in: _____

_____

_____

_____

(b) List some adaptations its has that enable it to exploit certain resources or parts of the environment in which it lives (e.g. nocturnal, camouflage):

_____

_____

_____

_____

(c) What do you know of your organism's niche? i.e. its functional role in the environment. Describe what you know below:

_____

_____

_____

_____

_____

_____

©2018 **BIOZONE** International
ISBN: 978-1-927309-55-1
Photocopying Prohibited

LS2.A   CE

EXPLORE: Organisms can't always exploit all of their fundamental niche

On the Scottish coast, two species of barnacles, *Semibalanus balanoides* and *Chthalamus stellatus*, coexist in the same general environment. The barnacles naturally show a layered distribution, with *Semibalanus* concentrated on the lower region of the shore, and *Chthalamus* on the upper shore. When *Semibalanus* were experimentally removed from the lower strata, *Chthalamus* spread into that area. However, when *Chthalamus* were removed from the upper strata, *Semibalanus* failed to establish any further up the shore than usual.

High tide mark

A

Inset enlarged, right

Low tide mark

B   C

Settling *Semibalanus* larvae die from desiccation at low tide

Chthamalus adults

Settling *Chthamalus* larvae are crowded out by *Semibalanus*

Semibalanus adults

A   *Chthalamus* range when *Semibalanus* present

B   *Chthalamus* range when *Semibalanus* absent

C   *Semibalanus* range when *Chthalamus* present *or* absent

2. (a) Which of the barnacles appeared to exploit its entire fundamental niche? _____

   (b) What physical factor limited the range of this barnacle? _____

3. (a) Describe the range of the fundamental niche for *Chthalamus*: _____

   _____

   _____

   (b) Was this range realized (was it fully exploited)? _____

   (c) Explain your answer: _____

   _____

   _____

   _____

   _____

   (d) Based on this case study, what can you say about how the presence of other organisms might affect the distribution or population size of a species?

   _____

   _____

   _____

   _____

   _____

   _____

EXPLAIN: Making a prediction about niche

Can two species with the same fundamental niche coexist (live together) in the same environment? This question has been studied in many different situations. One of the more well known is the *Paramecium* experiment carried out by G.F. Gause. First he grew two different separate populations of *Paramecium* with the same resource needs and recorded the growth of the populations over time. The graphs below show the outcome of this first experiment.

Paramecium grown in isolation

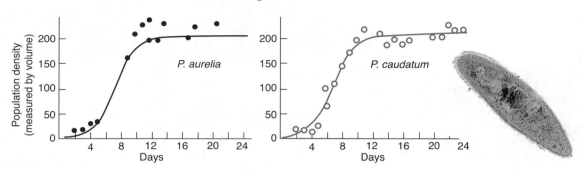

In a second experiment Gause grew the two species together and recorded the growth of the populations over time.

4. (a) Make a prediction about the outcome of this second experiment. Consider that the *Paramecium* species require the same resources (including type of food, depth of water, temperature, etc.)

_____

_____

   (b) Go **BIOZONE's resource hub** and read the page on Gause's experiment and the outcome of growing *P. aurelia* and *P. caudatum* together. Was your prediction correct? Can you explain the experiment's outcome?

_____

_____

Gause's law

The outcome of the second experiment led Gause to formulate the **competitive exclusion principle** (Gause's law) which states that two species that compete for exactly the same resources cannot coexist. Competition between species for the same resources narrows the niche of each species, producing the **realized niche** for each.

Specialization of particular characteristics (such as beak size) to better exploit a narrower niche is called character displacement and it often happens when species exploit similar resources.

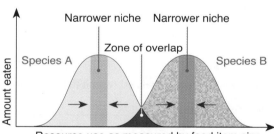

The phenomenon is well recorded in Darwin's finches, where different species have broadly similar and overlapping diets, but exploit some food resources more effectively because of their differing beak sizes. Among the Galápagos ground finches *Geospiza*, right, the medium ground finch (*G. fortis*) can exploit larger, harder seeds than the small ground finch (*G. fuliginosa*).

5. Explain why two species, competing for the same resources, cannot coexist. What evidence is there to support this?

_____

_____

_____

_____

©2018 **BIOZONE** International
ISBN: 978-1-927309-55-1
Photocopying Prohibited

5 Populations Have Varied Distributions

ENGAGE: Populations

The photo (left) shows wolf pack in Yellowstone Park. This pack may roam over a large area, but tend to stay within the geographical boundaries of the park. As of December 2016, there were 108 wolves in 11 packs in Yellowstone Park.

1. Based on this information, work in pairs to construct a definition of a population:

_____

_____

_____

_____

_____

2. Now that you have defined what a population is, study your own environment and note down examples of populations you can easily observe. What do you notice about the populations? Do the different populations all contain similar numbers of organisms? How are the individuals within a population dispersed relative to each other?

_____

_____

_____

_____

_____

_____

EXPLORE: Population density varies

Your observations may have revealed that different populations can exist naturally at different densities. Density is expressed as number of individuals per unit area (for land organisms) or per unit volume (for aquatic organism). The density of populations is affected by the availability of resources. Population density is higher where resources are plentiful and lower where they are scarce or highly variable.

3. Use the information below to construct models showing the difference between high density and low density populations.

In low density populations, individuals are spaced well apart. There are only a few individuals per unit area. Highly territorial or solitary animal species, such as tigers, leopards, and bears, occur at low densities.

In high density populations, there are many individuals per unit area. This can be a natural feature of highly social species (e.g. ants and termites) or species that reproduce asexually to form large colonies, e.g. corals.

©2018 **BIOZONE** International
ISBN: 978-1-927309-55-1
Photocopying Prohibited

LS2.A

EXPLORE: Population distribution

▸ Population **distribution** describes how organisms are distributed in the environment relative to each other.

▸ Three distribution patterns are usually recognized: random, clumped (or aggregated), and uniform. In the examples described below, the circles represent individuals of the same species.

Random distribution

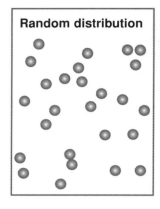

Dune grass

Oyster bed

In random distributions, the position of one individual is independent of all other individuals. Random distribution is uncommon but can occur in uniformly structured environments where unpredictable factors determine distribution, e.g. oyster larvae settling on rocks after being carried by ocean currents or dandelion seeds germinating after being dispersed by wind or currents.

Clumped distribution

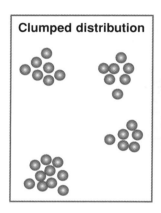

Musk oxen herd

Pronghorn herd

In clumped (or aggregated) distributions, individuals are grouped in patches (often around a resource). Clumped distributions are the most common type of distribution pattern in nature and are typical of herding and other highly social species and in environments where resources are patchy.

Uniform distribution

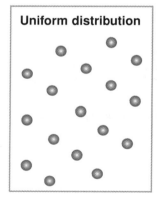

Gannet colony

Allocasuarina in Australia

In uniform (regular) distributions individuals are evenly spaced and the distance between neighboring individuals is maximized. Uniform distributions occur in territorial species, e.g. breeding colonies of seabirds, but also occurs in plants that produce chemicals to inhibit the growth of nearby plants.

4. Name the type of distribution would you expect to see when:

   (a) Resources are not evenly spread out: _____

   (b) Resources are evenly spread out: _____

   (c) Animals have greater hunting success as a group: _____

   (d) Animals are territorial: _____

5. How do you think resource availability (e.g. food, water, space) affect distribution? _____

   _____

   _____

   _____

   _____

©2018 **BIOZONE** International
ISBN: 978-1-927309-55-1
Photocopying Prohibited

6 | Population Growth

ENGAGE: Populations tend to increase in number

We have all seen or heard about plagues of pest species, such as flies, cockroaches, or mice (below). In order for a population to grow in size, more individuals must be added to the population than are removed.

1. (a) Describe an example where you observed a rapid increase in a population's numbers:

   _____

   _____

   _____

   _____

   _____

   (b) Did the population keep growing rapidly? _____

   _____

   _____

Mouse population explosion

Grant Singleton CSIRO cc 3.0

EXPLORE: What factors regulate population growth?

▶ The number of individuals in a population (designated **N**) is simply determined by quantifying the gains and losses to the population. Births, deaths, immigrations (movements into the population) and emigrations (movements out of the population) are events that collectively determine the number of individuals in a population.

▶ Scientists usually measure the rate of these events. These rates are influenced by environmental factors, such as the availability of resources, and by the biology of the species itself (e.g. its intrinsic capacity to increase or biotic potential). Species are highly variable in this respect, some populations grow very rapidly and some very slowly.

Births (B) →   → Emigration (E)

Immigration (I) →   → Deaths (D)

2. Using the terms B, D, I and E, develop an equation to describe how D, B, E, and I determine population growth:

   _____

3. Now construct equations to express the following:

   (a) A population in equilibrium: _____

   (b) A declining population: _____

   (c) An increasing population: _____

4. A population started with 100 individuals. Over a year, population data were collected. Calculate birth rates, death rates, and net migration rate, and use your equation to calculate the rate of population change for the data below (as percentages):

   (a) Births = 14: Birth rate = _____ (b) Net migration = +2: Net migration rate = _____

   (c) Deaths = 20: Death rate = _____ (d) Rate of population change = _____

   (e) State whether the population is increasing or declining: _____

©2018 **BIOZONE** International
ISBN: 978-1-927309-55-1
Photocopying Prohibited

 SC SSM LS2.A

EXPLORE: Populations can grow exponentially

In populations of bacteria, population growth can be very rapid. Some bacterial populations are able to double every 20 minutes. The doubling time here is effectively also the generation time, the average time between two consecutive generations in a population. In the example below, a bacterial cell was placed into a liquid growth medium and the number of cells present were counted every 20 minutes for six hours.

5. Complete the table below by doubling the number of bacteria for every 20 minute interval.

6. Graph the results on the semi-log grid below using suitable labels for each axis.

NEED HELP? See Activity 93

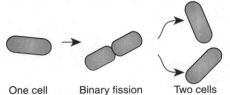

One cell   Binary fission   Two cells

Time (min)	Population size
0	1
20	2
40	4
60	8
80	
100	
120	
140	
160	
180	
200	
220	
240	
260	
280	
300	
320	
340	
360	

7. Describe the shape of the graph: _____

8. (a) Predict the number of cells present after 380 minutes: _____

   (b) Plot this value on the graph above.

9. (a) Why do you think a semi-log (log-linear) grid is used to plot microbial growth? _____

   _____

   _____

   _____

   (b) What would be the shape of the plot if both axes were on a linear scale? _____

   _____

©2018 **BIOZONE** International
ISBN: 978-1-927309-55-1
Photocopying Prohibited

10. The growth of the bacterial population on the previous page was rapid, doubling every 20 minutes. Do you think this growth can be sustained indefinitely? If not, what might eventually limit the growth of the bacterial population?

_____

_____

_____

EXPLAIN: Environmental conditions can affect bacterial growth

Researchers wanted to study the effect of nutrient levels on the growth of two *E.coli* cultures. They did this by placing one culture in a growth medium containing limited nutrients (labeled minimal medium) and another into a growth medium containing a complex mixture of nutrients (labeled complex medium). The growth of each culture was measured over time using a spectrophotometer at 660 nm. As the bacterial cultures grew, the solutions became more turbid (cloudier) and the absorbance reading increased.

E.coli dividing

1 μm

11. (a) Predict which culture will show the greatest growth:

_____

_____

(b) Why did you make this prediction? _____

_____

12. (a) The growth results for *E.coli* are shown in the table right. Plot the results for *E.coli* growth on the two media below:

E.coli growth results

Incubation time (min)	Absorbance at 660 nm	
	Minimal medium	Complex medium
0	0.021	0.014
30	0.022	0.015
60	0.025	0.019
90	0.034	0.033
120	0.051	0.065
150	0.078	0.124
180	0.118	0.238
210	0.179	0.460
240	0.273	0.698
270	0.420	0.910
300	0.598	1.070

(b) Describe the differences you can see in the shapes of the curves: _____

_____

_____

_____

(c) What do you think may be causing these differences? _____

_____

_____

_____

©2018 **BIOZONE** International
ISBN: 978-1-927309-55-1
Photocopying Prohibited

EXPLAIN: Exponential growth cannot usually be sustained

As you saw in your plot of bacterial growth, the change in population's numbers over time can be represented as a population growth curve. On a linear scale, exponential growth, which is density-independent, produces a J-shaped growth curve (below). While some populations may initially exhibit this type of growth, it is rarely sustained in nature.

Recall the relationship between the factors determining population growth. If we want to compare populations of different sizes, it is useful to express population parameters such as rates of birth, death, and population growth on a **per capita** (per individual) basis. The maximum per capita rate of population increase under ideal conditions (or **biotic potential**) is designated r_{max}. We calculate this using a simple equation (in words to the right, below):

$r_{max} = B - D / N$	r_{max} = Births − Deaths ÷ Population number

Exponential growth (right) is then expressed as:

$dN/dt = r_{max}N$	Population growth rate at time t $=$ per capita rate of increase X population number

The value of r_{max} is constant for any one species as a feature of its biology. However, in natural populations r, the per capita rate of increase, is usually lower than this because resource limitation prevents organisms achieving their biotic potential. Organisms will show exponential growth if r is positive and constant so the equation is often given simply as **$dN/dt = rN$** (right).

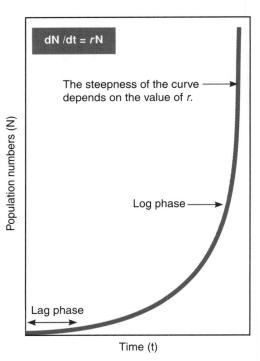

$dN/dt = rN$

The steepness of the curve depends on the value of r.

Log phase

Lag phase

Population numbers (N)

Time (t)

13. (a) Explain why there is a lag in the curve: _____

_____

(b) What factors do you think could limit continued exponential growth of the population? _____

_____

_____

EXPLORE: Environmental constraints limit population growth and size

▸ Populations do not generally continue to grow unchecked. Their numbers generally stabilize around a maximum that can be supported by the environment. This maximum number is called the **carrying capacity** (K).

▸ As the population number approaches K, the resources required for continued growth become limiting and there is increasing environmental resistance to further expansion. As a result, the rate of population growth slows.

▸ If the change in numbers is plotted over time, the curve is S shaped (sigmoidal), and the pattern of growth is called logistic. **Logistic** (density dependent) **growth** is typical of populations that live at or near carrying capacity.

▸ Logistic growth is expressed mathematically as: $dN/dt = rN(K-N/K)$

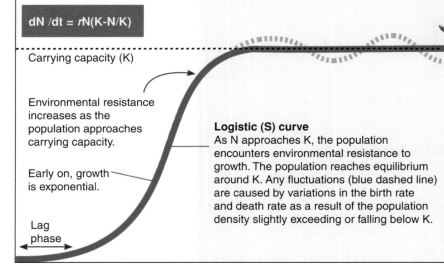

$dN/dt = rN(K-N/K)$

Carrying capacity (K)

Environmental resistance increases as the population approaches carrying capacity.

Early on, growth is exponential.

Lag phase

Population numbers (N)

Time (t)

Logistic (S) curve
As N approaches K, the population encounters environmental resistance to growth. The population reaches equilibrium around K. Any fluctuations (blue dashed line) are caused by variations in the birth rate and death rate as a result of the population density slightly exceeding or falling below K.

Many animal populations show logistic growth and their populations exist at or near carrying capacity. Populations may fluctuate around K but these fluctuations tend to become less pronounced over time. In such species, e.g. mammals, competition and efficient use of resources tend to be very important.

©2018 **BIOZONE** International
ISBN: 978-1-927309-55-1
Photocopying Prohibited

EXPLAIN: Why do most populations show logistic growth?

The Fruita Historical Area of Capitol Reef National Park in Utah preserves a pioneer Mormon settlement. Mormon settlers established orchards to provide the community with fruit. Today the orchard is maintained by the National Park Service. Park rangers noticed that yellow bellied marmots (*Marmota flaviventris*) had been climbing the trees to eat the fruit. Concerned about potential damage to the trees, they began monitoring the population in order to recommend a management program. The data they collected between 1985-1996 is presented in the table below.

14. (a) Plot the data from the study on the grid (right):

(b) Estimate the carrying capacity for this population:

_____

_____

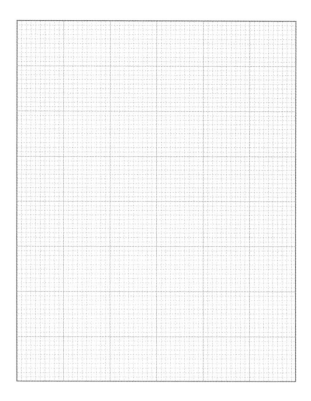

Number of marmots caught in the Fruita historical area of Capitol Reef National Park, Utah

NEED HELP? See Activity 97

Year	Population size
1985	47
1986	72
1987	95
1988	147
1989	112
1990	124
1991	132
1992	123
1993	113
1994	125
1995	117
1996	132

Data: https://www.saylor.org/site/wp-content/uploads/2011/12/SAYLOR.ORG-BIO313-RESOURCE_MANAGEMENT_ACTION_PLAN.pdf

ELABORATE: Logistic growth can be described mathematically

Plotting a logistic growth curve on a spreadsheet can help in understanding the effect of population size on the growth rate and how the logistic equation applies. For a hypothetical population of 2, r is 0.15 and K 100. The following formulae can be entered into the spreadsheet:

	A	B	C	D	E	F	G
1	r	t (period)	N	K	K-N/K	dN/dt	
2	0.15	0	2	100	=($D$2-C2)/$D$2	=$A$2*C2*E2	
3		=B2+1	=C2+F2				
4				Population at t_1 = population at t_0			
5				+ dN/dt (the amount of population			
6				change over 1 time period)			
7							

The cells can then be filled down. The first three steps have been filled here.

	A	B	C	D	E	F	G
1	r	t (period)	N	K	K-N/K	dN/dt	
2	0.15	0	2.00	100	0.98	0.29	
3		1	2.29		0.98	0.34	
4		2	2.63		0.97	0.38	
5		3	3.01				
6							

15. On a spreadsheet, replicate the data above (or download from **BIOZONE's Resource Hub**) and fill the cells down to about 60 time periods.

(a) Plot t vs N and describe the shape of the curve: _____

(b) Around which time period does the curve on the spreadsheet above begin to flatten out? _____

(c) Use the logistic equation and mathematical reasoning to explain the changes in population growth rate (dN/dt):

_____

_____

_____

©2018 **BIOZONE** International
ISBN: 978-1-927309-55-1
Photocopying Prohibited

7 Modeling Population Growth

ELABORATE: Computer simulations can be used to model population growth

Population growth can be simulated using spreadsheets or computer programs. This activity uses Populus, a Javascript program, which will run on Mac or Windows platforms. It models continuous and discrete population growth as well as the effects of competition. In this activity you will model continuous density-independent (exponential) and density-dependent (logistic) growth. Using Populus, you can also model discrete growth, which uses λ instead of r, where λ is the discrete-time per capita growth rate. Discrete models are used for organisms with a discrete breeding season (e.g. annual plants and insects that breed once a year) because population growth occurs in 'steps' only within a discrete time period (not continuously) and there is no population growth outside those times.

Populus is shareware. Download it free:
https://cbs.umn.edu/populus/overview

(you can also download via **BIOZONE's Resource Hub**)

The opening screen looks like this.
▶ **Model** allows you to choose which type of simulation you want to run.
▶ **Preferences** lets you to load saved files and save new ones.
▶ **Help** loads a comprehensive PDF file covering all aspects of the program.

If it fills the entire screen grab the lower corner and resize it with the mouse.

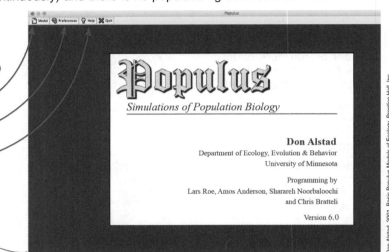

Don Alstad, 2001. Basic Populus Models of Ecology, Prentice Hall, Inc. ISBN-13 978-0130212894

Density independent growth
▶ Click on the Model in the menu bar
▶ Select Single-Species Dynamics
▶ Then choose **Density-Independent** Growth

Set the model type to continuous (as in continuous growth). This produces a single line in the output window. Discrete produces a series of points (as in discrete bursts of growth).

Set plot type to N vs t. This models population vs time.

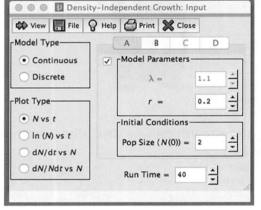

Up to four populations can be displayed on the one graph, using A, B, C, and D. Make sure the check box is ticked.

Set r to 0.2 and population size N to 2. Set run time 40. Click View to see the graph.

Questions 1-4 refer to density independent growth

1. What is the shape of the graph produced? _____

2. Describe what happens to the shape of the graph when:

   (a) r is increased to 0.4: _____

   _____

   (b) Population size is increased to 20: _____

   _____

   (c) Population size is increased to 20 but r is reduced to -0.2: _____

   _____

3. Set the parameters back to N = 2 and r = 0.2. Set the plot type to dN/dt vs N and view the plot. Describe the shape of the graph and explain what it means:

   _____

   _____

4. What is the value of r if the population doubles over one time period? _____

SAVE AND PRINT ALL YOUR SIMULATIONS AND ATTACH THEM TO THIS PAGE

LS2.A CE SC

©2018 **BIOZONE** International
ISBN: 978-1-927309-55-1
Photocopying Prohibited

Density dependent growth

▶ Click on the Model in the menu bar
▶ Then select Single-Species Dynamics
▶ Then select **Density-Dependent Growth**

▶ As before set the model type to continuous.
▶ Produce a plot for N = 5, K = 500, r = 0.2, and t to 50.

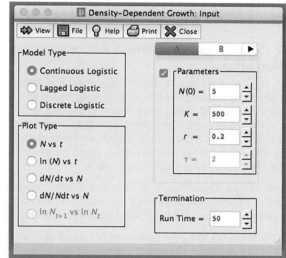

Questions 5-9 refer to density dependent growth

5. Describe what happens to the shape of the graph when:

   (a) r is increased to 0.4: _____

   _____

   (b) Population size is increased to 50: _____

   _____

   _____

   (c) Reset the parameters and plot a graph of dN/dt vs N. Describe the shape of this graph and explain what it means:

   _____

   _____

6. The standard logistic growth curve assumes the effect of the population size immediately affects the population growth rate. Now set the graph type to Lagged Logistic. This introduces a time lag between the population size and its effect on growth rate. Set the parameters to N = 5, K = 500, r = 0.2, and t to 50. Set the time lag T to 4 and view the graph. What is the effect of the time lag on population growth?

   _____

   _____

7. (a) Now set r to 0.5 and t to 150. Describe the shape of the graph: _____

   _____

   (b) What kind of organisms would show this type of growth and what characteristics do you think they would show?

   _____

   _____

8. (a) Keep T at 4 and set r to 0.2 view the graph. Describe the shape of the graph now: _____

   _____

   (b) What kind of organisms would show this type of growth and what characteristics do you think they would show?

   _____

   _____

9. Keeping r at 0.2, vary T between 1 and 10. How does increasing the lag affect how the population cycles around K?

   _____

   _____

SAVE AND PRINT ALL YOUR SIMULATIONS AND ATTACH THEM TO THIS PAGE

©2018 **BIOZONE** International
ISBN: 978-1-927309-55-1
Photocopying Prohibited

8 The Carrying Capacity of an Ecosystem

ENGAGE: What happened when wolves were introduced to Coronation Island?

Coronation Island is a small, 116 km² island off the Alaskan coast. A resident black tailed deer population had overgrazed the island and, as a result, very little forest understory remained intact and many common plant species were absent. The forest was quite open and park-like and not dense like a typical South Alaskan forest. Researchers noted that the deer on the island were smaller than those in other populations and that several died each year from malnutrition.

In 1960, the Alaska Department of Fish and Game released two breeding pairs of timber wolves onto the island. Their aim was to control the black-tailed deer (top right), which had been overgrazing the land.

Initially, wolves fed off the deer (lower right), bred successfully, and deer numbers fell. The island's vegetation began to return and by 1964 the vegetation was quite abundant. However, within a few years the deer population crashed. The wolves ran out of food (deer) and began eating each other, causing a drop in wolf numbers. Within 8 years, only one wolf remained on the island and the deer were once again abundant.

By 1983, wolves were absent and the deer numbers were high.

Black tailed deer

Gray wolf consuming deer

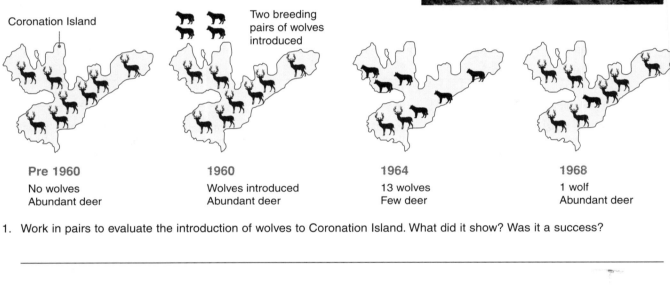

Coronation Island	Two breeding pairs of wolves introduced		
Pre 1960	**1960**	**1964**	**1968**
No wolves	Wolves introduced	13 wolves	1 wolf
Abundant deer	Abundant deer	Few deer	Abundant deer

1. Work in pairs to evaluate the introduction of wolves to Coronation Island. What did it show? Was it a success?

_____

_____

_____

_____

_____

2. If the experiment was carried out at a larger scale on a bigger island with more resources, do you think the outcome would have been different? What is your reasoning?

_____

_____

_____

_____

_____

_____

_____

    LS2.A CE SPQ EM SC

©2018 **BIOZONE** International
ISBN: 978-1-927309-55-1
Photocopying Prohibited

EXPLORE: What factors influence how many individuals an ecosystem can support?

The Coronation Island example on the previous page showed how biotic factors can influence population numbers. Abiotic factors are also important in regulating population size. Drought (lower than average rainfall) is one such factor and can persist for many years.

During the late 90s and early 2000s drought occurred in Arizona and New Mexico. Researchers studied the effect of the drought on the ecosystem, in particular, how populations were affected by changes to habitat and availability of food and water. One of the aspects they investigated was how drought affected the survival rate of pronghorn fawns in Arizona. Their results are shown in the table below right.

Pronghorn fawn

3. (a) Plot the rainfall and fawn survival data on the grid below:

(b) Describe the plot for rainfall over time: _____

_____

_____

(c) Describe the fawn survival rate over time: _____

_____

_____

(d) Is there a relationship between rainfall and fawn survival?

_____

_____

_____

(e) Can you think of why rainfall might be correlated with fluctuations in pronghorn survival? _____

_____

_____

_____

Year	Rainfall (cm)	Fawns surviving to December per 100 females
1995	11	12
1996	3	0
1997	4	0
1998	19	32
1999	6	0
2000	5	15
2001	15	78
2002	2	9

Data after Bright, J., and J. Hervert. (2005). Adult and fawn mortality of Sonoran pronghorn Wildlife Society Bulletin, 33(1):43-50 pp.

NEED HELP? See Activity 97

©2018 **BIOZONE** International
ISBN: 978-1-927309-55-1
Photocopying Prohibited

EXPLORE: Population density can contribute to the regulation of its size

Population size is regulated by factors that limit population growth, either by increasing the death rate and/or reducing the birth rate. The action of these factors may or may not be influenced by population density. Factors that act independently of population density and affect all individuals more or less equally are called **density independent factors**. Factors that have a proportionately greater effect at higher population densities are called **density dependent factors**. They become much less important when the population density is low.

Density dependent factors

The effect of these on population size is influenced by population density. They include:

▶ Competition
▶ Predation
▶ Disease

Density dependent factors tend to be biotic and are less important when population density is low.

They regulate population size by decreasing birth rates and increasing death rates.

Density independent factors

The effect of these on population size does not depend on population density. They include catastrophic events such as:

▶ Volcanic eruptions, fire
▶ Drought, flood, tsunamis
▶ Earthquakes

Density independent factors tend to be abiotic.

They regulate population size by increasing death rates.

Death Valley (California) is a long, narrow basin 86 meters below sea level and surrounded by steep, high mountain ranges. These features contribute to the climate of the valley.

In summer, Death Valley is one of the hottest and driest places on Earth. The mean maximum temperature throughout the year is 33°C but the mean summer temperature is as high as 47°C. The highest recorded temperature there was 56.7°C.

Rainfall is scarce, with an average annual rainfall of only 61 mm. However, damaging flash floods can occur in Death Valley after only moderate amounts of rainfall. This is because the steep slopes surrounding the valley channel water to low spots, and the hard-packed ground doesn't soak up the water quickly. Despite the extreme conditions in Death Valley, it supports over 400 different animal species and 1000 species of plants.

Devil's Golf Course, a salt pan, Death Valley

Brocken Inaglory CC3.0

4. The organisms in Death Valley are subject to both density dependent and density independent limiting factors. Work individually or in pairs to list likely density dependent and density independent factors at work in this ecosystem:

_____

_____

_____

_____

_____

5. Classify each of the following scenarios as examples of density dependent or density independent population regulation:

(a) A flash flood reduces the numbers of a chuckwalla (desert lizard) population: _____

(b) Moose population growth is limited due to lack of palatable vegetation: _____

(c) Roadrunner population growth is limited by the spread of a disease: _____

(d) A prolonged drought prevents wildflowers from setting seed: _____

(e) A wildfire destroys a population of redwood trees: _____

(f) A native fish declines after a sport fish species is introduced to the lake: _____

(g) A mussel population declines as a result of an explosion in its sea star predator: _____

©2018 **BIOZONE** International
ISBN: 978-1-927309-55-1
Photocopying Prohibited

EXPLAIN: Changes in carrying capacity influence population size

Now that we know abiotic and biotic factors influence population growth, we can develop and use a model to explain the process. Recall that the **carrying capacity** is the maximum number of organisms of a given species a particular environment can support indefinitely. An ecosystem's carrying capacity, and therefore the maximum population size it can sustain, is limited by its resources and is affected by both biotic and abiotic factors. Carrying capacity is determined by the most limiting factor, but this may not be the same over time (e.g. as a result of a change in food availability or a shift in climate). As a result, carrying capacity may also change over time.

6. In the space below draw a diagram (model) to map each step of the following scenario. Mark the carrying capacity as a line and adjust it as your model develops. Label each step (a-f) on your model:

   (a) A population moves into a new region of forest and rapidly increases in numbers due to abundant resources.

   (b) The population overshoots the carrying capacity.

   (c) Large numbers damage the environment and food becomes more limited, lowering the original carrying capacity.

   (d) The population becomes stable at the new carrying capacity.

   (e) The forest experiences a drought and the carrying capacity is reduced as a result.

   (f) The drought breaks and carrying capacity increases but is lower than before because of habitat damage incurred during the drought.

Population size

Time

7. Use your model to explain why population size is regulated around the environment's carrying capacity: _____

_____

_____

_____

_____

_____

_____

_____

_____

_____

9 Species Interactions Can Regulate Populations

ENGAGE: Species interactions may limit or maintain populations

▶ Within ecosystems, each species interacts with others in their community. In many of these interactions, at least one of the parties in the relationship is disadvantaged. Predators eat prey, parasites and pathogens exploit their hosts, and individuals compete for limited resources. These interactions, together with abiotic factors (such as temperature), limit population growth and prevent any one population from becoming too large.

▶ Not all species interactions have a negative impact on population growth. Some species participate in mutually beneficial relationships that can enable them to maintain population sizes much greater than could be achieved by either species alone, e.g. the mutualistic relationships of some flowering plants with their pollinating insects. Indeed, some species depend on their mutualistic relationships for survival.

Type of interaction between species				
Mutualism	**Exploitation**			**Competition**
	Predation	**Herbivory**	**Parasitism**	
All species benefit. Mutualism can involve more than two species. **Examples:** Flowering plants and their insect pollinators. The flowers are pollinated and the insect gains food. Ruminants and their rumen protozoa and bacteria. The microbes digest the cellulose in plant material and produce short-chain fatty acids, which the ruminant uses as an energy source.	Predator kills the prey outright and eats it. Predators may take a range of species as prey or they may prey exclusively on another species. **Examples:** Praying mantis consuming insect prey. Canada lynx eating snowshoe hare. The ochre sea star feeding on its primary prey, mussels. They also eat chitons, barnacles, and limpets.	Herbivore eats parts of a plant. Some plants have defenses to limit damage. In others, such as grasses, growth is stimulated by grazing activity. **Example:** Giraffes browsing acacia trees. Browsing stimulates the acacia to produce toxic alkaloids, which cause the giraffe to move to another plant. African herbivores grazing savanna grasslands.	The parasite lives in or on the host, taking all its nutrition from it. The host is harmed but usually not killed. Heavy parasite loads may kill, especially in young animals. **Examples:** Tapeworm in a pig's gut absorbs nutrients from the pig host. Some plants (e.g. mistletoes) are semi-parasitic. They photosynthesize but rob the host plant of nutrients and water.	Species, or individuals, compete for the same limited resources. Both parties are detrimentally affected. **Examples:** Neighboring plants of the same and different species compete for light and soil nutrients. Insectivorous birds compete for suitable food in a forest. Tree-nesting birds with similar nesting requirements will compete for nest sites.
Honeybee and flower	Mantid eats cricket	Giraffe browses acacia	Pork tapeworm	Forest plants
A Benefits ⇄ B Benefits				

1. In the spaces above, draw a simple model to show whether each species/individual in the interaction described is harmed or benefits. The first one has been completed for you.

2. Ticks are obligate blood feeders (meaning they must obtain blood to pass from one life stage to the next). Ticks attach to the outside of hosts where they suck blood and fluids and cause irritation.

   (a) Identify this type of interaction: _____

   (b) How would the tick population be affected if the host became rare? _____

   _____

©2018 **BIOZONE** International
ISBN: 978-1-927309-55-1
Photocopying Prohibited

EXPLORE: Some species interactions are necessary for survival

The mutualistic relationships between species provide benefits to both species so that both populations are able to survive, reproduce, and increase their numbers more successfully than either could do alone. Some mutualistic relationships go even further and the relationship is so close that both species depend on it for survival. These are called obligate relationships. All mutualisms involve one species providing the other with something they need. Some involve a trade in resources. In others, a resource (e.g. food) is traded for a service (e.g. protection).

Reef building corals depend for survival on a mutualism with photosynthetic algae in their tissues. The corals obtain 75-80% of their required energy from the algae and the algae obtain a habitat and utilize the coral's nitrogenous waste and carbon dioxide. Higher ocean temperatures can cause corals to eject their algae and, without the algae, they eventually starve and die.

Some ant species "farm" aphids by protecting the aphids from predation by ladybugs. In return, the ants harvest the honeydew produced by the aphids. The ants stroke the aphids with their antennae, stimulating them to release the honeydew, which supplements their diet. Caretaker ants may also collect and store aphid eggs, and move the young to host plants when they hatch.

3. Which of the above mutualistic relationships would you classify as obligate and why? _____

_____

4. How might a non-obligate mutualistic relationship enable a population to increase more than it could on its own?

_____

_____

EXPLORE: Does resource competition limit the size of populations?

How can we know if populations are limited by lack of a crucial resource such as food? In the Florida Everglades, several species of wading birds compete for the same food sources (mainly fish). In these birds, the young depend on the parents for feeding and the young hatch in sequence. Older nestlings have priority access to food and younger, later hatched young often starve if food is limited (or they may be thrown from the nest by older stronger siblings).

In 2013, a study of nestling survival was carried out on four species of wading birds (wood stork, snowy egret, little blue heron, and tricolored heron) in the Florida Everglades. It found that nestling survival was high across all species and both early and late hatched young received increasing amounts of food as they grew through the season. The students concluded that food that year was not limiting. In other years however, food shortages may result in only the first hatched young receiving enough food to survive.

Wood stork

Snowy egret

Florida Atlantic University Undergraduate Research Journal, Spring 2015

5. Researchers hypothesize that sequential (called asynchronous) hatching of young in birds is what is known as a bet-hedging strategy to make sure that at least some young will survive times of stress. Bet-hedging occurs when an organism suffers reduced **fitness** (a measure of individual survival and reproductive success) in their usual conditions in exchange for increased fitness during times of stress (e.g. when food supplies are low).

As a group, discuss the bet-hedging hypothesis with reference to the wading birds of the Florida Everglades. Suggest how bet-hedging might aid survival of the population as a whole when food is limited. How might asynchronous hatching reduce both individual fitness and population size when food is not limiting? How could competition for food affect the population outcome for any one species? Present a summary of your discussion, e.g. as an oral presentation, a short essay, a bullet point list, or a mind map.

©2018 **BIOZONE** International
ISBN: 978-1-927309-55-1
Photocopying Prohibited

10 Predation Can Control Some Populations

ENGAGE: How does the ratio of predators to prey vary in a predator/prey relationship?

The photo below shows a ladybug feeding on aphids. A single ladybug can eat as many as 5000 aphids a year.

1. List three examples of predator-prey relationships you have observed locally:

    (a) _____

    (b) _____

    (c) _____

2. For each of the examples you listed, how many predators do there appear to be compared to the numbers of their prey?

    _____

    _____

    _____

    _____

3. Why do you think there is this ratio of predators to prey? Do you think it relates only to the predator-prey examples you listed? Why or why not?

    _____

    _____

    _____

    _____

    _____

4. Describe any other local examples of relationships between organisms that may influence population numbers?

    _____

    _____

    _____

    _____

    _____

  LS2.A CE SPQ EM

©2018 **BIOZONE** International
ISBN: 978-1-927309-55-1
Photocopying Prohibited

EXPLORE: Case studies in predator-prey numbers

In some areas of Northeast India, a number of woolly aphid species colonize and feed off bamboo plants. The aphids can damage the bamboo so much that it is no longer able to be used by the local people for construction and textile production.

Giant ladybug beetles (*Anisolemnia dilatata*) feed exclusively off the woolly aphids of bamboo plants. There is some interest in using them as biological control agents to reduce woolly aphid numbers, and limit the damage woolly aphids do to bamboo plants.

The graph below shows the relationship between the giant ladybug beetle and the woolly aphid when grown in controlled laboratory conditions.

Bamboo plants are home to many insect species, including ladybugs and aphids.

Aphids feed off the bamboo sap, and the ladybugs are predators of the aphids (below).

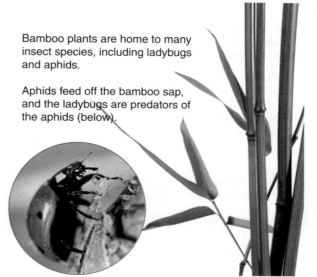

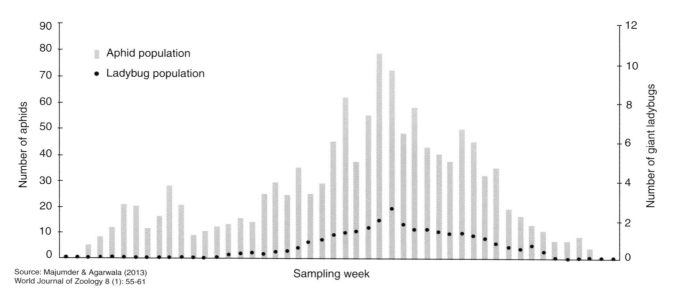

Source: Majumder & Agarwala (2013)
World Journal of Zoology 8 (1): 55-61

5. (a) On the graph above, mark the two points (using different colored pens) where the peak numbers of woolly aphids and giant ladybugs occurs.

(b) Do the peak numbers for both species occur at the same time? _____

(c) Why do you think this is? _____

_____

_____

6. (a) What is the response of the ladybug population when their prey (aphids) decline? _____

_____

_____

(b) What do you think would happen to prey (aphid) numbers if the predator (ladybug) numbers decreased?

_____

_____

(c) Can you think of any other factors that could affect the numbers of ladybugs and aphids? _____

_____

_____

_____

©2018 **BIOZONE** International
ISBN: 978-1-927309-55-1
Photocopying Prohibited

Ecologists often study specific populations to determine the extent of the interaction between a predator its prey species and how this interaction affects the wider ecosystem.

A census of a deer population on an island forest reserve Indicated a population of 2000 animals in 1960. In 1961, ten wolves (natural predators of deer) were brought to the island in an attempt to control deer numbers. The numbers of deer and wolves were monitored over the next nine years. The results of these population surveys are presented right.

7. (a) Plot a line graph for the results. Use one scale (on the left) for numbers of deer and another scale (on the right) for the number of wolves. Use different symbols or colors to distinguish the lines and include a key.

    (b) What does the plot show?_____

    _____

    _____

Island population surveys (1961-1969)		
Year	Wolf numbers	Deer numbers
1961	10	2000
1962	12	2300
1963	16	2500
1964	22	2360
1965	28	2244
1966	24	2094
1967	21	1968
1968	18	1916
1969	19	1952

(c) Suggest a possible explanation for the pattern in the data: _____

_____

_____

_____

(d) In the space below draw a diagram to predict what would happen if deer numbers increased again:

11 Organisms Compete for Limited Resources

ENGAGE: What is competition?

Sea anemones compete for space

Male deer compete for mates

Individuals compete for food

▶ No organism exists in isolation. Each organism interacts with other organisms and with the physical (abiotic) components of the environment. Competition is an important interaction in biological systems.

▶ **Competition** occurs when two or more organisms attempt to access to the same limited resource (e.g. food).

▶ In situations involving competition, the resources available for growth, reproduction, and survival for each competitor are reduced. Competition therefore has a negative effect on both competitors and limits population numbers.

1. In small groups, make a list of all the types of things organisms can compete for: _____

_____

_____

_____

_____

_____

EXPLORE: Competition occurs both within and between species

▶ Intraspecific competition: this is competition between members of the same species for the same resources.

▶ Interspecific competition: this is competition between members of different species for the same resources.

Intraspecific competition

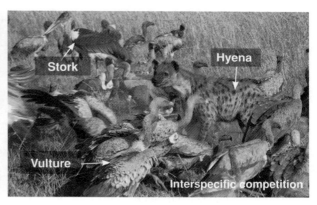

Stork

Hyena

Vulture

Interspecific competition

2. Using the examples above as inspiration:

(a) Describe a different example of intraspecific competition: _____

_____

_____

(b) Describe a different example of interspecific competition: _____

_____

_____

©2018 **BIOZONE** International
ISBN: 978-1-927309-55-1
Photocopying Prohibited

EM SPQ CE LS2.A

EXPLORE: How could intraspecific competition limit population size?

3. The photo right shows monarch (wanderer) caterpillars on a swan plant. They are of different ages but are competing for the same resources. Monarchs are one of very few consumers of the swan plant, which produces a sap that is toxic to most insects.

(a) What resources could the caterpillars be competing for? _____

_____

(b) How do you think intraspecific competition might affect population size over the summer breeding period?

_____

_____

_____

(c) As a group, discuss how you think competition might act to constrain the breeding period of monarchs to a particular time frame. What happens to eggs that are laid too early or too late? Summarize the conclusions of your discussion below:

_____

_____

_____

Scott McD1

EXPLAIN: Why is intraspecific competition more intense at higher densities?

Individuals of the same species have the same niche requirements (although they will have slightly different tolerances for variations in biotic and abiotic factors). This means they will all compete for exactly the same resources. In contrast, competing species usually have different requirements for at least some resources (e.g. different habitats or food preferences) or they may be able to exploit the same resources at different times. Both inter- and intraspecific competition can limit population size, but the effects are *usually* more marked for intraspecific competition, particularly at higher population densities. Exceptions include new competitive relationships established when a species is introduced to an area where it did not naturally occur and its niche overlaps with the native residents there.

4. The diagram right is a generalized model showing how population numbers are influenced by intraspecific competition for resources. In this model, the limited resource is food. Describe what is happening at each of points A and B to resource availability, population birth and death rates, and level of competition.

A: _____

_____

_____

_____

B: _____

_____

_____

_____

Population increases

B

A

Population decreases

5. In the space right, sketch a plot of population numbers over time to represent the model in question 4 as a simple mathematical model of how intraspecific competition limits population numbers:

6. Explain why the effect of intraspecific competition on population size is greater at higher population densities:

_____

_____

_____

©2018 **BIOZONE** International
ISBN: 978-1-927309-55-1
Photocopying Prohibited

EXPLAIN: Why does intraspecific competition influence population size?

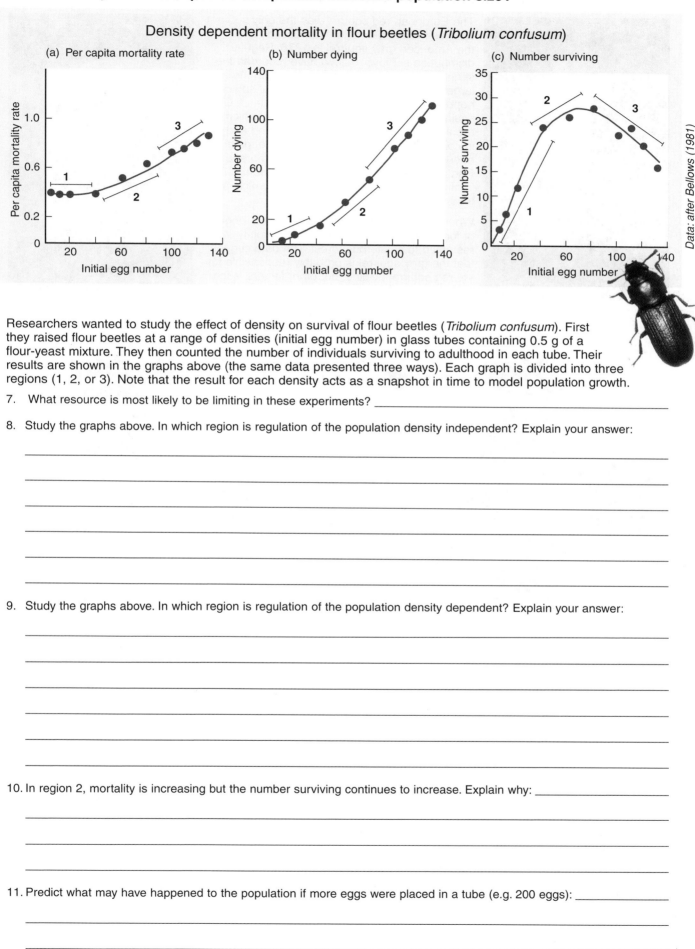

Density dependent mortality in flour beetles (*Tribolium confusum*)

(a) Per capita mortality rate

(b) Number dying

(c) Number surviving

Data: after Bellows (1981)

Researchers wanted to study the effect of density on survival of flour beetles (*Tribolium confusum*). First they raised flour beetles at a range of densities (initial egg number) in glass tubes containing 0.5 g of a flour-yeast mixture. They then counted the number of individuals surviving to adulthood in each tube. Their results are shown in the graphs above (the same data presented three ways). Each graph is divided into three regions (1, 2, or 3). Note that the result for each density acts as a snapshot in time to model population growth.

7. What resource is most likely to be limiting in these experiments? _____

8. Study the graphs above. In which region is regulation of the population density independent? Explain your answer:

_____

_____

_____

_____

_____

_____

9. Study the graphs above. In which region is regulation of the population density dependent? Explain your answer:

_____

_____

_____

_____

_____

_____

_____

10. In region 2, mortality is increasing but the number surviving continues to increase. Explain why: _____

_____

_____

_____

11. Predict what may have happened to the population if more eggs were placed in a tube (e.g. 200 eggs): _____

_____

_____

EXPLAIN: How has interspecific competition affected squirrel populations in the UK?

Red squirrel (*Sciurus vulgaris*)

The European red squirrel was the only squirrel species in Britain until the introduction of the American gray squirrel in 1876. Regular distribution surveys (below) have recorded the shrinkage in the range of the reds, with the larger, more aggressive gray squirrel displacing populations of reds over much of England. Gray squirrels can exploit tannin-rich foods, which are avoided by reds. In mixed woodland and in competition with grays, reds may not gain enough food to survive the winter and breed. Reds are also very susceptible to several viral diseases, including squirrelpox, which is transmitted by grays.

Whereas red squirrels once occupied a range of forest types, they are now almost solely restricted to coniferous forest. The data suggest that the gray squirrel is probably responsible for the red squirrel decline, but other factors, such as habitat loss, are also likely to be important.

Gray squirrel (*Sciurus carolinensis*)

Change in distributions of red and gray squirrels in the UK

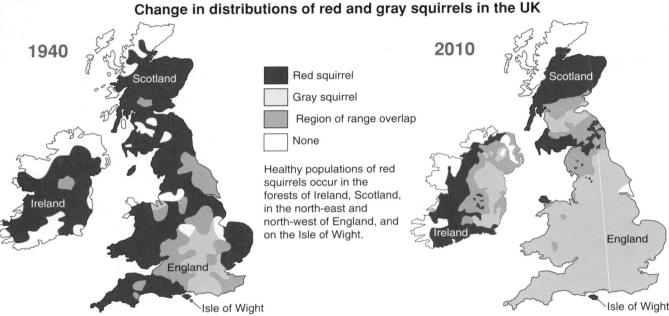

1940

Scotland

Ireland

England

Isle of Wight

Red squirrel

Gray squirrel

Region of range overlap

None

Healthy populations of red squirrels occur in the forests of Ireland, Scotland, in the north-east and north-west of England, and on the Isle of Wight.

2010

Scotland

Ireland

England

Isle of Wight

12. Compare the maps above and comment on how the distribution of red and gray squirrels has changed over time:

_____

_____

_____

_____

13. (a) What evidence is there that competition with gray squirrels is responsible for the decline in red squirrels in the UK:

_____

_____

_____

(b) Is the evidence conclusive? If not why not?_____

_____

_____

_____

©2018 **BIOZONE** International
ISBN: 978-1-927309-55-1
Photocopying Prohibited

ELABORATE: Use a model to show the effect of competition between two species

Populations of species living in the same area usually exploit different resources. This avoids competition that would affect the survival of the competing populations. When a new species (species 2) moves into an area and its resource use overlaps with an existing species (species 1), there will be resource competition. Species 2 may be better at acquiring the resource and so affect the population growth and viability of species 1. This effect of species 2 on species 1 can be quantified by the competition coefficient α. The greater the value of α, the greater the effect of species 2 on species 1. Similarly species 1 must in some way affect species 2 and this is termed β. By adding these terms into the logistic equation we can simulate what will happen when two competing species are placed in the same environment. This mathematical model is named the competitive Lotka-Volterra model after the mathematicians who developed it.

Here we are again using the Populus program.

From the menu bar select Model, then Multi-species Dynamics, then Lotka-Volterra Competition.

To start set the parameters for both species to the following:

▶ N = 10, r = 0.4, K = 100, α and β= 0.5.

▶ Set the Termination Condition to Run until steady state.

▶ View the graph as N vs t and note its shape.

Competition coefficients in the Lotka-Volterra model:

▶ α quantifies the per capita reduction in the population growth of species 1 caused by species 2

▶ β quantifies the per capita reduction in the population growth of species 2 caused by species 1

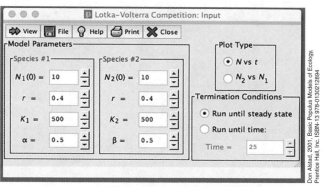

The Lotka-Volterra equations are a simple mathematical model of the population dynamics of competing species.

Don Alstad. 2001. Basic Populus Models of Ecology. Prentice Hall, Inc. ISBN-13 978-0130212894

14. (a) Describe the shape of the graph: _____

_____

(b) What is the effect of competition on the populations? Do they reach K? _____

_____

_____

_____

15. Now set α to 0.9 and view the graph again. Describe the effect of increasing α on the population growth of each species:

_____

_____

_____

16. Now set α back to 0.5 and set r for species 1 to 0.6. View the graph and describe the effect on the population growth each of species:

_____

_____

_____

17. Using this model we can simulate what would happen when an aggressively competing species is introduced (either deliberately or by accident) into an area where a native species utilizes the same resource but is not an aggressive competitor. Set the parameters for species 1 to N = 1000, r = 0.15, K = 1000 and α = 0.9. Set the parameters for species 2 to N = 4, r = 0.85, K = 1000, and β = 0.1.

(a) Before you run the simulation, predict what will happen: _____

_____

(b) Were you correct?_____

(c) What level of α allows the populations reach a similar size? _____

(d) What level of β allows the populations reach a similar size? β = _____

SAVE AND PRINT ALL YOUR SIMULATIONS AND ATTACH THEM TO THIS PAGE

©2018 **BIOZONE** International
ISBN: 978-1-927309-55-1
Photocopying Prohibited

12 Human Activity Alters Populations

ENGAGE: Human activity can affect the environment and its organisms

The human population is expanding rapidly. It is currently sitting at ~7.6 billion and is expected to increase to 8.6 billion by mid-2030 and continue to rise. Land must be developed in order to provide enough food, housing, and resources to sustain the human population's growth. The photo above shows land being cleared in the Porter Ranch area of Los Angeles to make way for residential buildings.

1. Working in pairs or a small group, list some ways that humans alter the environment: _____

   _____

   _____

   _____

   _____

   _____

   _____

2. In what way could the activities you listed above have a negative effect the populations of organisms living in the environment?

   _____

   _____

   _____

   _____

   _____

   _____

3. Can you think of any benefits of concentrating human populations to within confined urban developments?

   _____

   _____

   _____

   _____

   _____

LS2.C   CE   EM

©2018 **BIOZONE** International
ISBN: 978-1-927309-55-1
Photocopying Prohibited

EXPLORE: The changing face of California's grasslands

Much of California's natural environment has become reduced as a result of human activity. For example, native grasslands (below) once covered a quarter of the state of California, but since European settlement 8.9 million hectares of grassland has been reduced to less than 90,000 hectares.

Many factors have contributed to the loss of native grasslands, including an increase in agriculture on grasslands, the invasion of introduced plant species, and urban development.

Grasslands are among 21 of the most endangered ecosystems in the United States. Extensive loss of habitat (e.g. through roading development) reduces the resources available to resident native species and, as a result, California has the highest number of federally listed threatened and endangered species in America. Humans may also suffer through a loss of ecosystem services (gains acquired by humans from their surroundings ecosystems). Ecosystem services include nutrient cycling, carbon storage, provision of food and water, and purification of air and water.

CA EP&Cs II: The expansion and operation of human communities influence the geographic extent, composition, biological diversity, and viability of natural systems (II c)

Natural grasslands provide habitat to thousands of species.

▶ Destruction of natural grassland puts the wellbeing of native species at risk and threatens the biological diversity and long term sustainability of the ecosystem.

▶ The health and wellbeing of humans also depends on the ecosystem services provided by grasslands.

Natural grassland at Santa Susana Pass State Historic Park, Los Angeles County, California

Changes to California's grasslands at a glance

▶ 99% decrease in native grassland (8.9 million ha to 90,000 ha).

▶ 26% of native grassland was destroyed between 1945-1980.

▶ 90% decreases in North coastal bunchgrass.

▶ 99% decrease in needlegrass steppe.

▶ Introduced plant species increase by over 8000%.

▶ 99% of temporary pools are lost.

▶ 90% loss of native coastal prairie communities.

EXPLORE: How do roads affect populations?

The road and grass strips alongside it create a different habitat to the forest they intercept.

The environmental impact of roads

Increased mortality: Death rates (mortality) within animal populations increase when a road is built. Animals are killed or injured when they cross roads during regular movements, use road surfaces for basking, or scavenge carcasses and death rates are higher when animals are highly mobile or migratory.

Pollution: Roads and their traffic introduce light, noise, and chemical pollution to the surrounding environment. Chemical pollutants can be toxic. If absorbed by an organism, they may disrupt metabolism, delay development, or cause death. Changes to the noise and lighting environment associated with roading can disrupt communication between animals (e.g. when locating a mate) and confuse behavior that depends on normal light-dark cues.

Habitat changes: Roads can create barriers, breaking up natural habitats and preventing animals from carrying out their normal movements. They also restrict the area they have in which to find food, water and mates. As a result, population numbers may decline. Roads change the nature of the habitat (left), and often create easy routes for the invasion of weeds and pests into new areas. These changes potentially affect the survival of resident populations.

4. Discuss how human changes to habitat can affect the natural environment and populations reliant on them: _____

_____

_____

_____

_____

_____

©2018 **BIOZONE** International
ISBN: 978-1-927309-55-1
Photocopying Prohibited

EXPLAIN: How did expansion of LAX affect the biodiversity of California's coastal ecosystems?

Los Angeles International airport (LAX) is a huge airport, its four runways and buildings covering 1416 ha. Airport construction began in the 1920s and many expansions and upgrades have taken place since (below, right).

Much of the land LAX is built upon was once coastal prairie (below, left), a habitat rich in wildflowers, home to extensive wildlife, and a source of temporary pools, which provided habitat and water to some of the resident species. The last remaining coastal prairie in the area was destroyed in the 1960s to make way for further airport development.

Urbanization and industrialization of land destroys the natural habitat necessary to maintain high levels of biodiversity and promote ecological stability. It can result in a fragile ecosystem unable to provide its usual ecosystem services.

Natural habitat: coastal grasslands, California
Adbar CC3.0

Developed landscape: Los Angeles International Airport
Don Ramey Logan CC4.0

The ecologically rich **El Segundo sand dunes** once covered 1300 ha within the Los Angeles County, bordered on one side by the Pacific Ocean and on the other by the LA coastal prairie. During the 1970s, 90% of the dunes were destroyed in land developments, endangering the survival of many species, including the El Segundo blue butterfly. This species is only found in the dune system and once numbered 750,000 in three populations. Development of the dunes resulted in the loss of the El Segundo blue's only food source, native coastal buckwheat. The butterfly's numbers declined rapidly to fewer than 500 and it was placed on the Federal Endangered Species list in 1976. In 1986, the Los Angeles World Airports (LAWA) began the LAX dunes restoration project in an attempt to save the butterfly. The removal of invasive plants and replanting of coastal buckwheat, along with intensive management, has seen the population increase to several thousand, although it still remains on the endangered species list.

The El Segundo blue butterfly is a subspecies of the square-spotted blue, pictured here on coastal buckwheat
Steve Berardi CC2.0

5. (a) Explain why the destruction of the El Segundo sand dunes almost caused the extinction of the El Segundo blue?

_____

_____

_____

   (b) Why is a species with a narrow niche (such as the El Segundo blue butterfly) at a greater risk of extinction through habitat loss than a species with a broader niche?

_____

_____

_____

_____

6. Working in small groups, use the information on this page and links provided on BIOZONE's Resource Hub to analyze the efforts of LAWA in trying to save the El Segundo blue butterfly. Have they been successful? What challenges have they faced and what challenges may set their efforts back? Summarize your findings.

_____

_____

_____

©2018 **BIOZONE** International
ISBN: 978-1-927309-55-1
Photocopying Prohibited

13 Producers, Consumers, and Food Webs

ENGAGE: *Tyrannosaurus rex* was at the top of the food chain

Tyrannosaurus rex was an apex (top) predator of the late Cretaceous period (ending 65 million years ago), living throughout what is now North America. At 12.3 meters long and weighing 8.4 tonnes it was an active predator and one of the largest land predators to have ever existed. *T. rex* obtained its food by hunting herbivorous dinosaurs including ceratopsians and sauropods, and sometimes members of its own species. The herbivorous dinosaurs dominated the landscape and obtained food by eating a wide variety of plant-based materials such as ferns, horsetails, club-mosses, conifers, cycads, and ginkgoes. Over the time dinosaurs existed (from the Triassic period 252 million years ago to the end of the Cretaceous period 65 million years ago) there were 66 known carnivorous dinosaurs and 185 known herbivorous dinosaurs in North America.

1. (a) Based on the information above, sketch a simple diagram to show the feeding relationships between the plants, herbivorous dinosaurs and the carnivorous *T. rex* described above.

(b) Life on Earth needs energy to live. Where do you think the plants in the example above obtained their energy from?

_____

2. What do you think would happen to the Cretaceous ecosystem described above if the number of plants fell significantly?

_____

_____

_____

_____

©2018 **BIOZONE** International
ISBN: 978-1-927309-55-1
Photocopying Prohibited

EM LS2.B

EXPLORE: Some organisms produce their own food, others do not

Organisms can be classified into two broad groups by how they obtain their food. Those that make their own food are **producers**. Those that cannot make their own food are **consumers**. Consumers may eat animals (carnivores), plants (herbivores), or both plants and animals (omnivores). Decomposers and detritivores are also consumers.

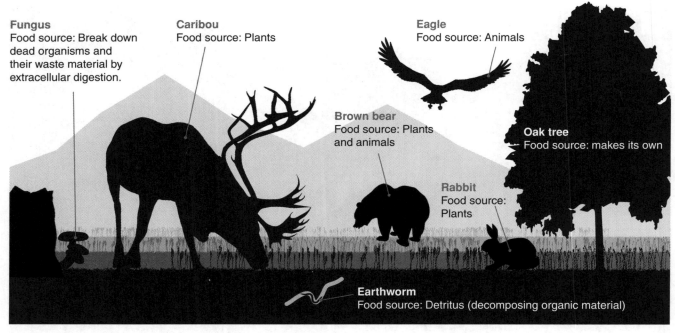

Fungus
Food source: Break down dead organisms and their waste material by extracellular digestion.

Caribou
Food source: Plants

Eagle
Food source: Animals

Brown bear
Food source: Plants and animals

Oak tree
Food source: makes its own

Rabbit
Food source: Plants

Earthworm
Food source: Detritus (decomposing organic material)

3. Study the diagram above and identify the organisms that are:

(a) Consumers: _____

(b) Producers: _____

Producers make their own food	Consumers cannot make their own food

Producers are also called **autotrophs**, which means self feeding. Plants, algae, and some bacteria are producers. Most producers are **photo**autotrophs and use the energy in sunlight and the carbon in carbon dioxide to make their food through a process called **photosynthesis**. Photosynthesis transforms **sunlight energy** into chemical energy. The chemical energy is stored in glucose molecules and released when the glucose is metabolized in an energy yielding process called respiration. Some producers are **chemo**autotrophs and use the chemical energy in inorganic molecules (e.g. H_2S) to make their food.

Consumers (**heterotrophs**) are organisms that cannot make their own food and must obtain their energy (and carbon) by consuming other organisms (by eating or through extracellular digestion). Animals, fungi, and some bacteria are consumers. Consumers rely on producers for survival, even if they do not consume them directly. Herbivores, such as the rabbit in the diagram, gain their energy by eating plants. Although higher level consumers, such as the eagle, may feed off herbivores, they rely indirectly on plants to sustain them. Without plants, the rabbit would not survive and the eagle could not eat it.

4. Explain why all consumers (even carnivores) are reliant on producers for survival: _____

_____

_____

©2018 **BIOZONE** International
ISBN: 978-1-927309-55-1
Photocopying Prohibited

EXPLORE: Energy and matter move through the ecosystem via food chains

Food chains

▶ Organisms in ecosystems interact through their feeding (or trophic) relationships. These interactions can be shown in a **food chain**, which is a simple model to illustrate how energy and matter, in the form of food, pass from one organism to the next. Each organism in the chain is a food source for the next.

Trophic levels

▶ The levels of a food chain are called **trophic** (feeding) **levels**. An organism is assigned to a trophic level based on its position in the food chain. Organisms may occupy different trophic levels in different food chains or during different stages of their life.

▶ Arrows link the organisms in a food chain. The direction of the arrow shows the flow of energy and matter through the trophic levels. At each link, energy is lost (as heat) from the system. This loss of energy limits how many links can exist. Most food chains begin with producers, which use the energy in sunlight to make their own food. Sunlight is therefore the ultimate source of energy for life on Earth. Producers are eaten by primary consumers (herbivores). Secondary (and higher level) consumers eat other consumers, as shown below.

Great white heron spears a fish to eat

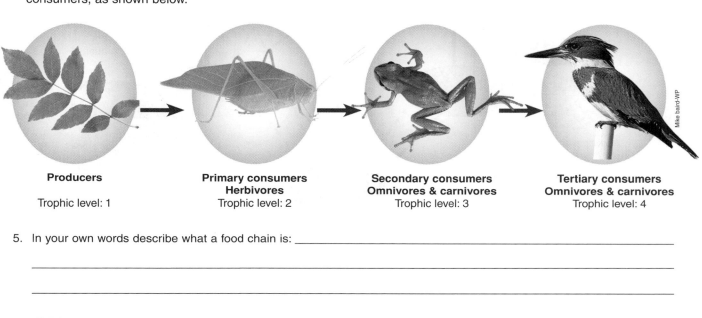

Producers	**Primary consumers** **Herbivores**	**Secondary consumers** **Omnivores & carnivores**	**Tertiary consumers** **Omnivores & carnivores**
Trophic level: 1	Trophic level: 2	Trophic level: 3	Trophic level: 4

5. In your own words describe what a food chain is: _____

_____

_____

_____

6. (a) A simple food chain for a cropland ecosystem is pictured below. Label the organisms with their trophic level and trophic status (e.g. primary consumer).

Corn   Mouse   Corn snake   Hawk

Trophic level: _____ _____ _____ _____

Trophic status: _____ _____ _____ _____

(b) What is the ultimate energy source for both the food chains pictured on this page? _____

(c) Why are there are rarely more than five or six links in a food chain? _____

_____

_____

©2018 **BIOZONE** International
ISBN: 978-1-927309-55-1
Photocopying Prohibited

EXPLAIN: How are food chains interconnected to make food webs?

If we show all the connections between all the food chains in an ecosystem, we can create a web of trophic interactions called a food web. A **food web** is a model to illustrate the feeding relationships between the organisms in a community.

▶ Food webs are simplified representations of real ecosystems and frequently do not (or cannot) show all the interactions occurring in a real system. The flow of energy through the trophic linkages is in one direction as opposed to the cyclic flow of matter. The biomass (mass of biological material) in the system represents stored energy and decreases from the base of each food chain to the top because energy is lost to the environment with each transfer. A simplified food web for a lake community is shown below.

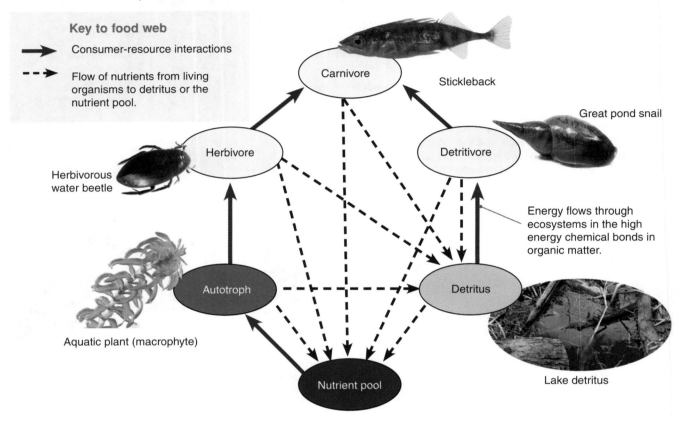

7. Explain why an ecosystem with only a few different types of organisms would have a less complex food web than an ecosystem with many different types of organisms?

_____

_____

_____

8. (a) A herbicide spill kills all the aquatic plants in the food web above. Predict the likely effect of this on the food web:

_____

_____

(b) What would be the likely effect on the food web of removing the great pond snail? _____

_____

_____

(c) Why is it difficult to accurately predict the effect on the lake ecosystem of removing one trophic element (producer or consumer) from this simplified food web?

_____

_____

_____

_____

©2018 **BIOZONE** International
ISBN: 978-1-927309-55-1
Photocopying Prohibited

ELABORATE: Use your knowledge of food webs to construct a model food web for a lake

The organisms below are typical of those found in many lakes. For simplicity, only a small number of organisms are represented. Real lake communities may have a great many species interacting through their trophic relationships. Your task is to assemble the organisms below into a food web in a way that illustrates their feeding relationships.

Autotrophic protists (algae)
Chlamydomonas (above left), and some diatoms (above right) are photosynthetic.

Macrophytes
Aquatic green plants are photosynthetic.

Protozan (e.g. *Paramecium*)
Diet: Mainly bacteria and microscopic green algae such as *Chlamydomonas*.

Asplanchna (zooplankton)
A large, carnivorous rotifer.
Diet: Protozoa and young zooplankton (e.g. *Daphnia*).

Daphnia (zooplankton)
Small freshwater crustacean.
Diet: Planktonic algae.

Common carp
Diet: Mainly feeds on bottom living insect larvae and snails, but will also eat some plant material (not algae).

Three-spined stickleback
Common in freshwater ponds and lakes. **Diet:** Small invertebrates such as *Daphnia* and insect larvae.

Diving beetle (adults and larvae)
Diet: Aquatic insect larvae and adult insects. They will also scavenge from detritus. Adults will also take fish fry.

Herbivorous water beetle
Diet: Adults feed on macrophytes. Young beetle larvae are carnivorous, feeding primarily on pond snails.

Dragonfly larva
Large aquatic insect larvae.
Diet: Small invertebrates including *Hydra*, *Daphnia*, insect larvae, and leeches.

Great pond snail
Diet: Omnivorous. Main diet is macrophytes but will eat decaying plant and animal material also.

Leech
Fluid feeding predators.
Diet: Small invertebrates, including rotifers, small pond snails, and worms.

Pike
Diet: Smaller fish and amphibians. They are also opportunistic predators of rodents and small birds.

Mosquito larva
Diet: Planktonic algae.

Hydra
A small, carnivorous cnidarian.
Diet: small *Daphnia* and insect larvae.

Detritus
Decaying organic matter (includes bacterial decomposers).

©2018 **BIOZONE** International
ISBN: 978-1-927309-55-1
Photocopying Prohibited

9. From the information provided for the lake food web components on the previous page, construct twelve different **food chains** to show the feeding relationships between the organisms. Some food chains may be shorter than others and most species will appear in more than one food chain. An example has been completed for you.

   Example 1: Macrophyte ⟶ Herbivorous water beetle ⟶ Carp ⟶ Pike

   (a) _____

   (b) _____

   (c) _____

   (d) _____

   (e) _____

   (f) _____

   (g) _____

   (h) _____

   (i) _____

   (j) _____

   (k) _____

   (l) _____

10. Use the food chains you created above to help you to draw up a food web for this community in the box below. Use the information supplied on the previous page to draw arrows showing the flow of energy between species (only energy **from** (not to) the detritus is required).

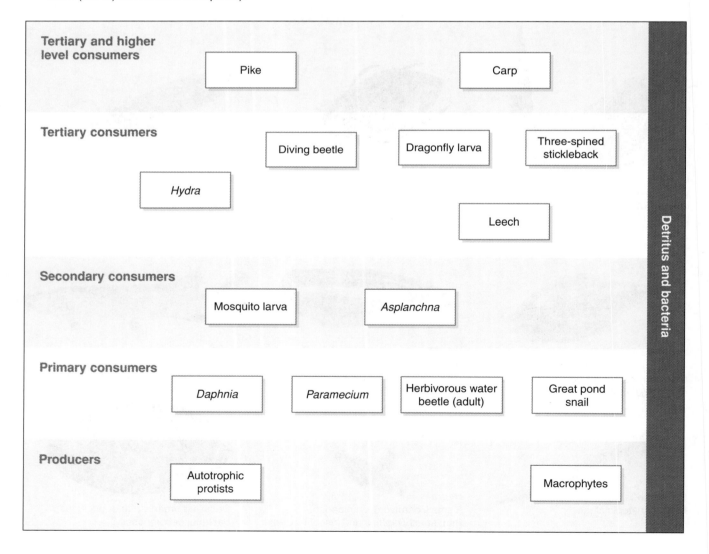

©2018 **BIOZONE** International
ISBN: 978-1-927309-55-1
Photocopying Prohibited

14 Energy in Ecosystems

ENGAGE: Energy cannot be created nor destroyed

The first law of thermodynamics states that energy can neither be created nor destroyed; energy can only be transferred or changed from one form to another. For example, at first glance turning on a light seems to produce energy; however, it is simply a conversion of energy as electrical energy is converted into light and heat energy. No new energy has been created.

The photo on the right shows how energy can change within a system. The mass of snow at the top of the peak contains potential (stored) energy. When the mass of snow becomes sufficiently large and unstable it rolls down the mountain as an avalanche. When this occurs the potential energy is converted into kinetic energy. Although the two energy forms are different, the amount of energy in each is the same (e.g. the potential energy in one tonne of static snow contains the same amount of energy as one tonne of falling snow).

1. Working in pairs, list other examples of energy conversions you encounter on a regular basis: _____

   _____

   _____

   _____

   _____

   _____

2. The average American consumes 900 kg of food each year, but does not gain 900 kg in weight. What do you think happens to the food (and ultimately the energy) that is consumed? Capture your thoughts and ideas in the space below:

3. Consider a simple food chain where a cow eats grass and the cow is consumed by a human in a hamburger. How do you think the concept of **energy transfer** could be applied to this simple food chain?

   _____

   _____

   _____

4. What do you think the relationship might be between the energy availability in an ecosystem and population size?

   _____

   _____

   _____

©2018 **BIOZONE** International
ISBN: 978-1-927309-55-1
Photocopying Prohibited

EM LS2.B

EXPLORE: Energy inputs and outputs

Recall from the previous activity that the Sun is the ultimate energy source for life on Earth. Through the process of photosynthesis, plants (and other photoautotrophs) convert light energy into chemical energy stored within the bonds of glucose. Glucose provides the chemical energy necessary to support life's processes.

▶ The **gross primary production** (GPP) of any ecosystem will depend on the capacity of the producers to capture light energy and fix carbon in organic compounds.

▶ The **net primary production** (NPP) is then determined by how much of the GPP goes into plant biomass, after the respiratory needs of the producers are met. This will be the amount available to the next trophic level.

▶ The primary production of an ecosystem is distinct from its productivity, which is the amount of production per unit time (a rate). However because values for production (accumulated biomass) are usually given for a certain period of time in order to be meaningful, the two terms are often used interchangeably.

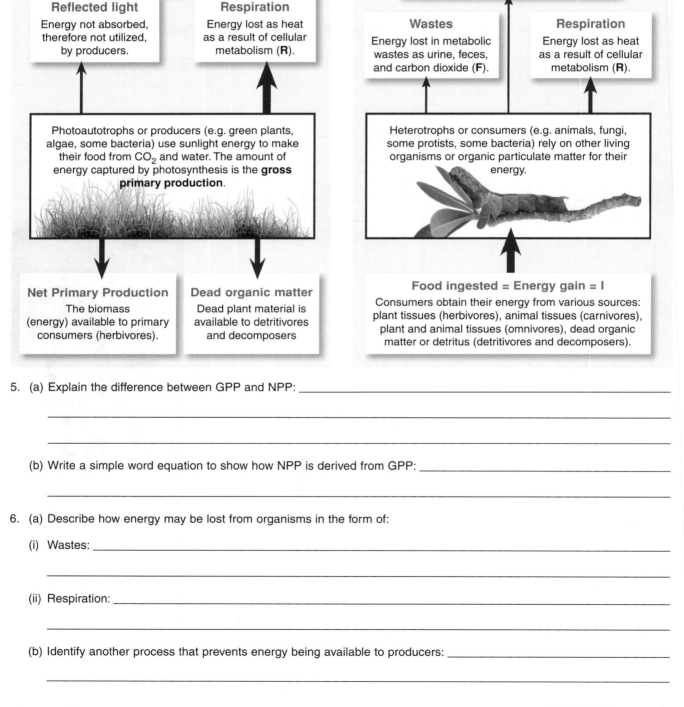

Reflected light
Energy not absorbed, therefore not utilized, by producers.

Respiration
Energy lost as heat as a result of cellular metabolism (**R**).

Photoautotrophs or producers (e.g. green plants, algae, some bacteria) use sunlight energy to make their food from CO_2 and water. The amount of energy captured by photosynthesis is the **gross primary production**.

Net Primary Production
The biomass (energy) available to primary consumers (herbivores).

Dead organic matter
Dead plant material is available to detritivores and decomposers

Net production
Biomass (energy) available to next trophic level (**N**). For herbivores, this amount will be the **net secondary production**.

Wastes
Energy lost in metabolic wastes as urine, feces, and carbon dioxide (**F**).

Respiration
Energy lost as heat as a result of cellular metabolism (**R**).

Heterotrophs or consumers (e.g. animals, fungi, some protists, some bacteria) rely on other living organisms or organic particulate matter for their energy.

Food ingested = Energy gain = I
Consumers obtain their energy from various sources: plant tissues (herbivores), animal tissues (carnivores), plant and animal tissues (omnivores), dead organic matter or detritus (detritivores and decomposers).

5. (a) Explain the difference between GPP and NPP: _____

_____

_____

(b) Write a simple word equation to show how NPP is derived from GPP: _____

_____

6. (a) Describe how energy may be lost from organisms in the form of:

(i) Wastes: _____

_____

(ii) Respiration: _____

_____

(b) Identify another process that prevents energy being available to producers: _____

_____

©2018 **BIOZONE** International
ISBN: 978-1-927309-55-1
Photocopying Prohibited

EXPLAIN: The ten percent rule

▶ As energy flows through an ecosystem from one trophic level to the next only 10% (on average) of the energy is transferred to the next trophic level. This is because each time energy is transferred (as food) between trophic levels some energy is given out as heat during cellular respiration. This means the amount of energy available to the next trophic level is less than at the previous level.

▶ Potentially, we can account for the transfer of energy from its input (as solar radiation) to its release as heat from organisms, because energy is conserved.

▶ The percentage of energy transferred from one trophic level to the next is called the **trophic efficiency**. It varies between 5% and 20% and measures the efficiency of energy transfer. An average figure of 10% trophic efficiency is often used. This is called the **ten percent rule**, illustrated below.

Calculating available energy

The energy available to each trophic level will equal the amount entering that trophic level, minus total losses from that level (energy lost as heat + energy lost to detritus).

Heat energy is lost from the ecosystem to the atmosphere. Other losses become part of the detritus and may be utilized by other organisms in the ecosystem.

Energy absorbed from the previous trophic level

100

Energy lost as heat **65** ← | **Trophic level** | → **15** Energy lost to detritus

20

Energy passed on to the next trophic level

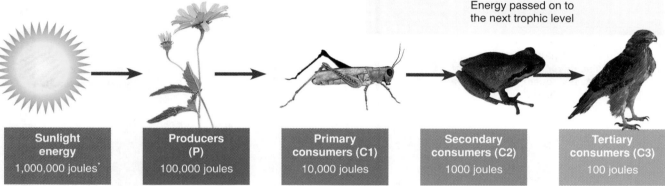

Sunlight energy	Producers (P)	Primary consumers (C1)	Secondary consumers (C2)	Tertiary consumers (C3)
1,000,000 joules*	100,000 joules	10,000 joules	1000 joules	100 joules

*Note: joules are units of energy

7. (a) What percentage of the original energy in this food chain is present at the tertiary consumer level? _____

_____

(b) On average, how much energy is transferred between trophic levels? _____

(c) Suggest how the relative numbers of organisms at each trophic level may change as a result of the energy available:

_____

_____

_____

(d) What effect do you think the ten percent rule has on regulating population size?_____

_____

_____

8. Explain why there is less energy available at each successive trophic level? _____

_____

_____

_____

_____

9. Why must ecosystems receive a continuous supply of energy from the Sun? _____

_____

_____

©2018 **BIOZONE** International
ISBN: 978-1-927309-55-1
Photocopying Prohibited

Quantifying energy transfer and trophic efficiency in an ecosystem

10. Identify the process occurring at each of points 1-4 on the diagram.

1: _____

2: _____

3: _____

4: _____

11. Calculate the amount of energy transferred to each trophic level. Write your answers in the spaces labeled (a)-(d).

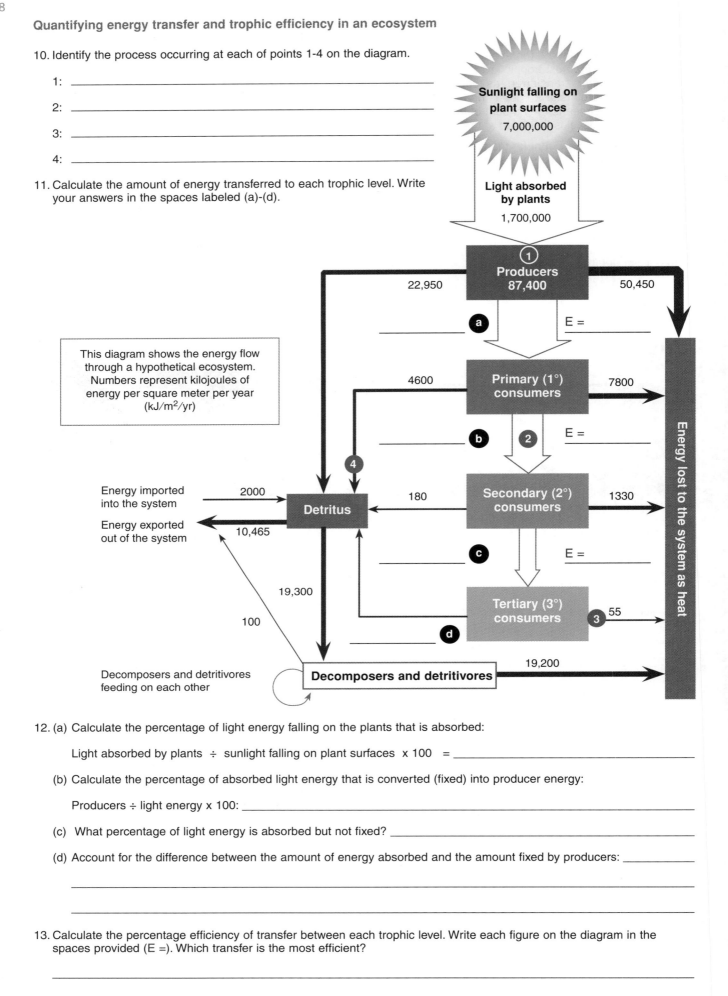

This diagram shows the energy flow through a hypothetical ecosystem. Numbers represent kilojoules of energy per square meter per year (kJ/m²/yr)

Sunlight falling on plant surfaces 7,000,000

Light absorbed by plants 1,700,000

① **Producers** 87,400

22,950   50,450

a   E =

Primary (1°) consumers

4600   7800

b   ②   E =

Secondary (2°) consumers

180   1330

c   E =

Tertiary (3°) consumers

③ 55

d

Energy imported into the system   2000

Detritus

Energy exported out of the system   10,465

19,300

100

Decomposers and detritivores feeding on each other

Decomposers and detritivores   19,200

Energy lost to the system as heat

12. (a) Calculate the percentage of light energy falling on the plants that is absorbed:

   Light absorbed by plants ÷ sunlight falling on plant surfaces x 100 = _____

(b) Calculate the percentage of absorbed light energy that is converted (fixed) into producer energy:

   Producers ÷ light energy x 100: _____

(c) What percentage of light energy is absorbed but not fixed? _____

(d) Account for the difference between the amount of energy absorbed and the amount fixed by producers: _____

   _____

   _____

13. Calculate the percentage efficiency of transfer between each trophic level. Write each figure on the diagram in the spaces provided (E =). Which transfer is the most efficient?

   _____

©2018 **BIOZONE** International
ISBN: 978-1-927309-55-1
Photocopying Prohibited

EXPLAIN: How are ecological pyramids used to represent relationships between trophic levels?

The energy, biomass, or numbers of organisms at each trophic level in any ecosystem can be represented by an **ecological pyramid**. The first trophic level is placed at the bottom of the pyramid and subsequent trophic levels are stacked on top in their 'feeding sequence'. Ecological pyramids provide a convenient model to illustrate the relationship between different trophic levels in an ecosystem.

▶ Pyramid of numbers shows the numbers of individual organisms at each trophic level.

▶ Pyramid of biomass measures the mass of the biological material at each trophic level.

▶ Pyramid of energy shows the energy contained within each trophic level. Pyramids of energy and biomass are usually quite similar in appearance.

▶ This generalized ecological pyramid (right) shows a conventional pyramid shape, with a large number of producers at the base, and decreasing numbers of consumers at each successive trophic level.

▶ Ecological pyramids for this plankton-based ecosystem have a similar appearance regardless of whether we construct them using energy, or biomass, or numbers of organisms.

▶ Units refer to biomass or energy. The images provide a visual representation of the organisms present.

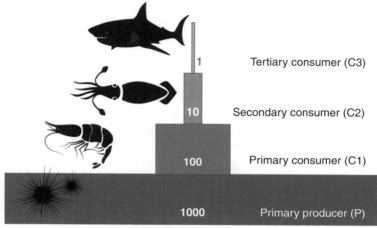

Not to scale

14. (a) What major group is missing from the pyramid above? _____

(b) Explain the significance of this group in a food web: _____

_____

_____

EXPLAIN: Which pyramid provides the most useful information?

There are benefits and disadvantages to each type of pyramid.

▶ A pyramid of numbers provides information about the number of organisms at each level, but it does not account for their size, which can vary greatly. For example, one large producer might support many small consumers.

▶ A pyramid of biomass is often more useful because it takes into account the amount of biological material (biomass) at each level. The number of organisms at each level is multiplied by their mass to produce biomass.

▶ While number and biomass pyramids provide information about an ecosystem's structure, a pyramid of energy provides information about function (how much energy is fixed, lost, and available to the next trophic level).

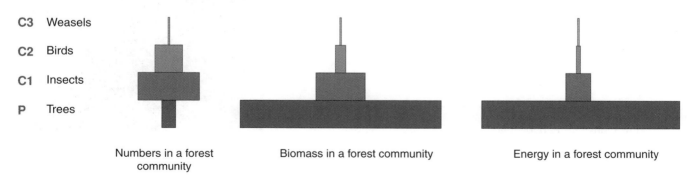

C3 Weasels
C2 Birds
C1 Insects
P Trees

Numbers in a forest community   Biomass in a forest community   Energy in a forest community

15. What is the advantage of using a biomass or energy pyramid rather than a pyramid of numbers to express the relationship between different trophic levels?

_____

_____

_____

©2018 **BIOZONE** International
ISBN: 978-1-927309-55-1
Photocopying Prohibited

A pyramid of numbers can sometimes be inverted

In some ecosystems (e.g. a forest ecosystem, right) a few, large producers can support all the organisms at the higher trophic levels. This is due to the large size of the producers; large trees can support many individual consumer organisms. A pyramid of energy for this system would be a conventional pyramid shape because the energy of the producers is enough to support the consumer levels.

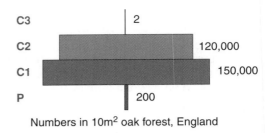

C3	2
C2	120,000
C1	150,000
P	200

Numbers in 10m² oak forest, England

16. The forest community above has relatively few producers. How can it support a large number of consumers?

_____

_____

_____

17. The table (below) shows the number of organisms at each trophic level of a grassland community.

   (a) Draw the pyramid of numbers for this data in the space below:

| Numbers in a grassland community ||
Trophic level	Number of organisms
Producer	1,500,000
Primary consumer	200,000
Secondary consumer	90,000
Tertiary consumer	1

   (b) Do you think the pyramid of energy would be similar or different? Explain your answer: _____

_____

_____

18. Would a pyramid of energy ever have an inverted (upturned) shape? Explain your reasoning: _____

_____

_____

_____

ELABORATE: Sustaining the mountain lion

The cougar (*Puma concolor*) is also called the mountain lion. Its primary prey are deer, and a single mountain lion will consume 510 kg of deer in a year. In turn, a deer eats 1655 kg of vegetation annually. The amount of vegetation present varies with location but 2.6 km² of land produces 344 kg of suitable forage each year.

Cougar (*Puma concolor*)

19. (a) If an average deer weighs 68 kg, how many deer are needed to feed a mountain lion in a year?

_____

   (b) How many km² of habitat are needed to support one deer?

_____

   (c) How many km² of habitat are needed to support the number of deer required to sustain one cougar for a year?

_____

   (d) Predict what would happen at the consumer trophic levels if drought reduced the carrying capacity of the area:

_____

_____

©2018 **BIOZONE** International
ISBN: 978-1-927309-55-1
Photocopying Prohibited

15 | Nutrient Cycles

ENGAGE: Composting food waste models a simple nutrient cycle

Compost heap

Soil

Vegetable garden and hen

All things on Earth are made of up of chemical elements, such as carbon, nitrogen, and phosphorus. These elements are transferred through the Earth's systems through a series of biological and chemical processes. In composting, food scraps, lawn clippings, and other green waste are broken down (decomposed). Decomposition produces a rich organic material called humus, and releases minerals in forms (e.g. ammonium) that can then be taken by the roots of plants. Application of humus to soil returns chemical elements to the biosphere as the plants use them to grow.

1. Identify the biotic and abiotic components in the example above: _____

_____

_____

2. How could the chemical elements from the producers be transferred to the consumer level?

_____

_____

3. How could the chemical elements in the consumer be transferred back to the abiotic system? _____

_____

_____

_____

EXPLORE: Earth's biogeochemical cycles

▶ Elements such as carbon, hydrogen, nitrogen, and oxygen cycle through the Earth's systems. A series of biological, geological and chemical processes cycles elements through the biotic and abiotic components of a system. This process is generally called a nutrient cycle or a biogeochemical cycle.

▶ Matter is conserved throughout all these transformations, but it may pass from one system to another and it may be transformed into different chemical forms as it cycles.

▶ It is important to remember that energy is required to drive the cycling of matter from one system to another.

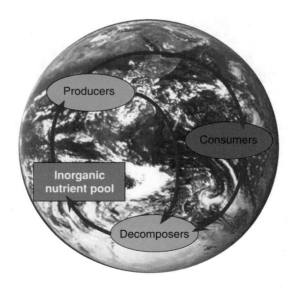

4. The diagram above shows a simplified model of the cycling of chemical elements on Earth. Add arrows to indicate the transfer of energy in this system (including inputs and outputs).

©2018 **BIOZONE** International
ISBN: 978-1-927309-55-1
Photocopying Prohibited

EM | LS2.B

A generalized biogeochemical cycle

Chemical matter can be stored in different parts of the cycle for varying lengths of time, e.g. a carbon atom will stay in the ocean, on average, more than 500 years.

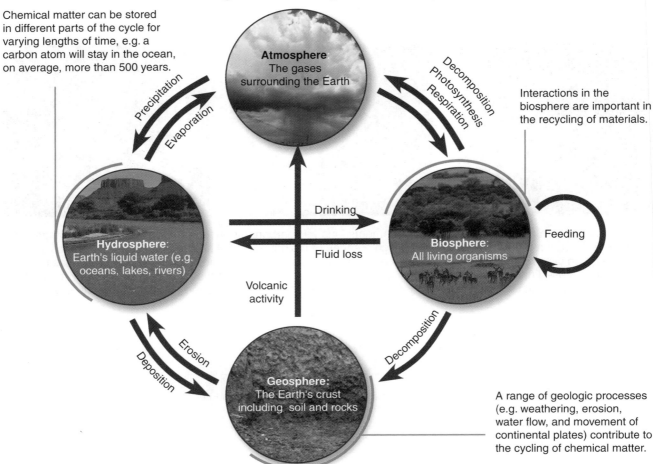

Atmosphere
The gases surrounding the Earth

Precipitation

Evaporation

Decomposition
Photosynthesis
Respiration

Interactions in the biosphere are important in the recycling of materials.

Drinking

Fluid loss

Feeding

Hydrosphere:
Earth's liquid water (e.g. oceans, lakes, rivers)

Biosphere:
All living organisms

Volcanic activity

Erosion

Deposition

Decomposition

Geosphere:
The Earth's crust including soil and rocks

A range of geologic processes (e.g. weathering, erosion, water flow, and movement of continental plates) contribute to the cycling of chemical matter.

Carbon, hydrogen, oxygen, nitrogen, calcium, phosphorus, and sulfur are essential to sustaining life. Together with trace elements such as zinc and iron, they form the major molecules of living things and are needed for metabolic processes. The elements are constantly recycled through the biotic and abiotic components of the Earth's systems.

Coal

Water droplets on a leaf

Application of nitrogen fertilizer

Carbon combines with many other elements and so is able to form a wide variety of molecules. Carbon is found in all the major macromolecules of life (e.g. lipids, carbohydrates, and proteins). It is also the basis of the fossil fuels used to fuel motor vehicles, heat homes, and power many industrial processes.

Water is essential to all life on Earth and contains hydrogen and oxygen as H_2O. Not only is water important in cell chemistry, but it has a fundamental role in the distribution of other elements around the Earth. Major macromolecules (large multi-unit molecules) also contain significant amounts of oxygen and hydrogen.

Nitrogen is found in the genetic material of life (DNA and RNA) and is a major component of amino acids and proteins. In the world around us, nitrogen is the major gas in our atmosphere (78%) and (as nitrate) is an essential plant nutrient. It used widely by humans as fertilizer to help boost crop production.

5. Why is it important that matter is cycled through an ecosystem? _____

_____

_____

_____

_____

_____

©2018 **BIOZONE** International
ISBN: 978-1-927309-55-1
Photocopying Prohibited

EXPLAIN: Biogeochemical cycles interact

Although each element has its own unique cycle, no cycle occurs in isolation because there is interaction between them. For example, the water (hydrologic) cycle has a crucial role in the movement of elements between the spheres and also provides a large reservoir for carbon.

The carbon cycle (below) can be divided into interconnected sub-cycles; the biological carbon cycle and the geochemical carbon cycle. In this example, our focus is on the biological component of the cycle (red arrows).

▶ Carbon enters the biological carbon cycle when autotrophs convert atmospheric carbon dioxide (CO_2) into glucose via **photosynthesis**. Carbon is released back into the atmosphere as CO_2 when living organisms carry out **cellular respiration**. Thus, photosynthesis and cellular respiration drive the biological cycling of carbon.

▶ Carbon is passed along the food chain when the carbon in plant material is eaten by primary consumers (herbivores) and when they are subsequently consumed by higher order consumers.

A simplified carbon cycle

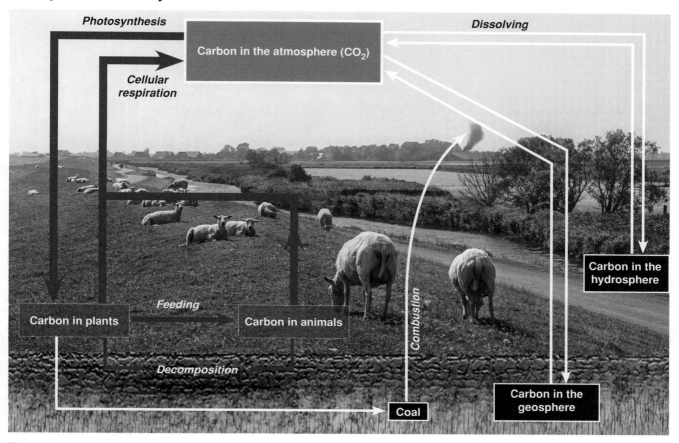

The oxygen cycle

The oxygen cycle describes the movement of oxygen between the biotic and abiotic components of ecosystems. It is closely linked to the carbon cycle because most producers use CO_2 in photosynthesis and produce oxygen as a waste product. The oxygen is used in cellular respiration and CO_2 is produced as a waste product.

6. Use the information above to construct a simple diagram to show transformations in the oxygen cycle. What do you notice about the cycling of carbon and oxygen?

©2018 **BIOZONE** International
ISBN: 978-1-927309-55-1
Photocopying Prohibited

EXPLAIN: Elements change form as they cycle

As elements cycle they change their chemical form and these transformations affect their availability to living things. Nitrogen transformations affect its ability to be utilized by living organisms. Before nitrogen (N) can be incorporated into DNA and proteins it must be transformed into a form that is available to plants (and therefore to consumers). Bacteria play a very important role in this, transferring nitrogen between the abiotic and biotic systems.

▶ The Earth's atmosphere is about 78% nitrogen gas, which is very stable and unreactive and so is unavailable to most living organisms. Atmospheric nitrogen can be fixed (captured) and transformed into usable forms (e.g. nitrate) via biological or non-biological pathways. Nitrogen can then enter food chains via plant uptake of nitrate nitrogen.

▶ Oxidation of N_2 by lightning forms nitrogen oxides. These dissolve in rain, forming nitrates, which enter soil. In biological pathways, **nitrogen fixing bacteria** in the soil, water, and in the roots of some plants convert nitrogen gas to ammonia. The ammonia undergoes oxidation by soil bacteria (**nitrification**) eventually to nitrate. The reverse process, **denitrification**, is carried out by anaerobic bacteria and returns nitrogen to the atmosphere.

The nitrogen cycle

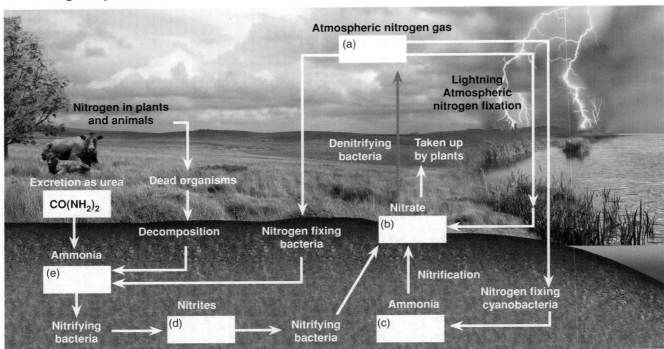

7. (a) In the boxes (a-e) above write the appropriate chemical form of nitrogen (urea has been completed for you):

   (b) Identify the form of nitrogen that can up taken up by plants: _____

8. Use the model above to explain the importance of bacteria in the cycling of nitrogen: _____

   _____

   _____

   _____

9. What role do plants play in the availability of nitrogen to animals? _____

   _____

   _____

10. (a) What biological process produces ammonia from atmospheric nitrogen: _____

    (b) What biological process produces nitrate in the soil: _____

    (c) What biological process releases nitrogen to the atmosphere? _____

11. Nitrogen fertilizers often contain nitrogen as urea. It is applied to land to boost the productivity of crops. Outline what must happen to the urea before the nitrogen in it can be used by the plants:

    _____

    _____

©2018 **BIOZONE** International
ISBN: 978-1-927309-55-1
Photocopying Prohibited

16 Humans Intervene in Nutrient Cycles

ENGAGE: Human activities can disrupt nutrient cycles

Carbon dioxide is released into the atmosphere when fossil fuels are burned.

Human activity influences many biogeochemical cycles, including the carbon, nitrogen, phosphorus, and sulfur cycles. The disturbances may cause significant alterations to cycles by altering reservoirs and nutrient fluxes.

For example, many industrial processes are driven by the combustion (burning) of fossil fuels. This returns carbon dioxide to the atmosphere at a rate much faster than would otherwise occur and depletes natural carbon sinks, changing the normal fluxes of the carbon cycle.

1. Can you identify other examples of how the carbon cycle is disrupted by human activity?

   _____

   _____

   _____

   _____

EXPLORE: Human activity affects the nitrogen cycle

The largest human interventions in the nitrogen cycle occur through farming and effluent (liquid waste or sewage) discharges. Other interventions include burning, which releases nitrogen oxides into the atmosphere, and irrigation and land clearance, which leach nitrate ions from the soil.

Farmers apply organic nitrogen fertilizers to the land in the form of green crops and manures, replacing the nitrogen lost through harvesting. Until the 1950s, atmospheric nitrogen could not be made available to plants except through microbial nitrogen fixation. During World War II, Fritz Haber developed the Haber process, combining atmospheric nitrogen and hydrogen to form gaseous ammonia. The ammonia is converted into ammonium salts and sold as inorganic fertilizer.

The Haber process made inorganic nitrogen fertilizers readily available at relatively low cost, increasing crop yields and revolutionizing farming. However, it comes at a large energetic cost (around 1-2% of the world's annual energy supply is used in the Haber process).

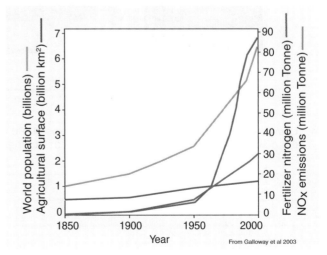

Humans intervene in the nitrogen cycle through application of fertilizers (top) and removal of biomass by harvesting (bottom).

2. Study the graph (right).

   (a) Describe the trend in nitrogen fertilizer use since 1850:

   _____

   _____

   (b) Emissions of nitrogen in various forms (termed NO_x) have increased over time. What do you think could be contributing to this:

   _____

   _____

   _____

   _____

From Galloway et al 2003

©2018 **BIOZONE** International
ISBN: 978-1-927309-55-1
Photocopying Prohibited

EM · LS2.C

EXPLAIN: What are the consequences of altering nutrient cycles?

Application of nitrogen fertilizers at rates that exceed uptake by plants leads to accumulation of nitrogen in the soil. This nitrogen can then leach into groundwater and run off into surface waters. This extra nitrogen load is one of the causes of accelerated enrichment (eutrophication) of lakes and coastal waters. An increase in algal production also increases decomposer activity, depleting oxygen and leading the deaths of fish and other aquatic organisms. Many aquatic microorganisms also produce toxins, which may accumulate in the water, fish, and shellfish. The rate at which nitrates are added has increased faster than the rate at which nitrates are returned to the atmosphere as N_2 gas. This has led to the widespread accumulation of nitrogen.

CA EP&Cs III: Human practices can alter the cycles and processes that operate within natural systems (III c)

Heavy use of nitrogen fertilizers disrupts the natural cycling and balance of the nitrogen cycle.

▶ Excess soil nitrogen can kill plants with lower nitrogen requirements. This can encourage growth of non-native plants at the expense of native species. Biodiversity and community composition can be altered.

▶ Eutrophication lowers oxygen levels in water and can kill many species, reducing biodiversity. At the same time, eutrophic conditions allow undesirable species, tolerant of low oxygen, to increase in numbers. Humans may also be harmed if the high nitrogen levels enter the drinking water system.

3. (a) How does excess nitrogen enter waterways? _____

_____

(b) How do increased levels of nitrogen in water affect the organisms and the biodiversity of the affected environment:

_____

_____

_____

_____

Global changes in nitrogen inputs and outputs between 1860 and 1995 in million tonne

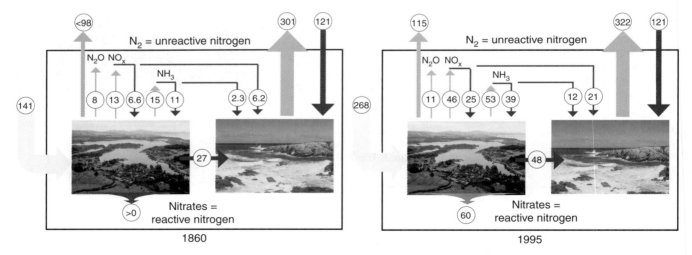

4. Using the quantitative models of nitrogen fluxes (1860 and 1995) above, calculate the increase in nitrogen deposition in the oceans from 1860 to 1995 and compare this to the increase in release of nitrogen from the oceans:

_____

_____

_____

_____

©2018 **BIOZONE** International
ISBN: 978-1-927309-55-1
Photocopying Prohibited

EXPLORE: Human activity releases mercury into the environment

Mercury (Hg) is an extremely toxic heavy metal. It occurs naturally in the Earth's crust and is released into the environment by volcanic eruptions. Human activities add mercury to the environment (right).

CA EP&Cs IV: The by-products of human activity are not readily prevented from entering natural systems and may be beneficial, neutral, or detrimental in their effect (IV b)

Anthropogenic activities contribute significant levels of mercury into the environment. It has many detrimental effects.

▶ Mercury compounds are toxic and dangerous at low levels.

▶ Mercury passes along food chains, becoming more concentrated at higher trophic levels. This process is called biomagnification.

▶ Mercury in California's environment is largely a legacy of past gold mining.

▶ The Environmental Protection Agency (EPA) now has regulations to reduce mercury release into the environment.

Global mercury emissions from human activity

Source	Mercury (tonne/year)	Contribution (%)
Coal and oil combustion	810	
Non-ferrous metal production	310	
Pig iron and steel production	43	
Cement production	236	
Caustic soda production	163	
Mercury production	50	
Artisanal gold mining	400	
Waste disposal	187	
Coal bed fires	32	
Vinyl chloride monomer production	24	
Other	65	
Total	2320	100

Data source: Pirrone *et al* (2009) Mercury Fate and Transport in the Global Atmosphere (p1-47)

5. (a) Complete the table above by calculating the percentage contribution of each anthropogenic mercury source:

(b) Identify the three highest anthropogenic sources of mercury: _____

_____

6. Research how much mercury is released by natural sources and compare it to levels released by anthropogenic sources:

_____

_____

The California gold rush: A lasting legacy of mercury

The Californian gold rush in the mid 1800s introduced large amounts of mercury into the environment. This was because mercury was used to extract gold during the mining process. Between 10-30% of the mercury used in the process was lost and ended up in the waterways and therefore the food chain. Many sites have undergone remediation (repair) to remove mercury, but other mining sites still leach mercury into the environment. For example, mercury used in mining areas of the Sierra Nevada continues to enter water bodies today, including the Sacramento Delta and the San Francisco Bay.

A sluice (water channel) is often used to recover gold and other minerals from stream bed mining operations.

7. Access the EPA's National Summary of Impaired Waters and TMDL Information site (see **BIOZONE's Resource Hub** for the link). Select California from the list and then select cause of impairment (mercury). Write down the sizes of the assessed waters with mercury impairment and record them below. As individuals or as a class or group, choose one type of waterway, e.g. rivers, and compare the extent of mercury impairment to that in other states for which there are equivalent data. Plot your results by hand or using a spreadsheet and attach the graph to this page. How does California compare to other states? Are the results different if you choose another type of waterway? Record your comments under your graph.

_____

_____

©2018 **BIOZONE** International
ISBN: 978-1-927309-55-1
Photocopying Prohibited

EXPLAIN: How does mercury accumulate in a food chain?

▸ Mercury can end up in lakes, rivers, and the ocean via surface run off or when atmospheric mercury settles in water.

▸ Mercury is very persistent in the environment and is excreted only very slowly. This allows it to build up in the tissues of organisms (a process called bioaccumulation).

▸ Mercury is converted into highly toxic **methylmercury** through biogeochemical interactions. Methylmercury is absorbed six times more easily than inorganic mercury, and is easily taken up by cells.

▸ Methylmercury passes through the food chain becoming more concentrated at each trophic level. This is called **biomagnification**. When a predatory fish consumes a smaller fish it acquires the mercury in its tissues. In this way mercury levels increase up the trophic levels of a food chain.

▸ The consumption of contaminated fish and shellfish is the most common source of methylmercury for humans.

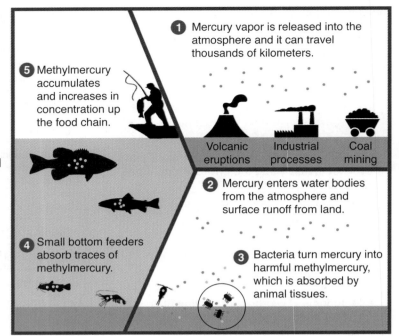

1 Mercury vapor is released into the atmosphere and it can travel thousands of kilometers.

5 Methylmercury accumulates and increases in concentration up the food chain.

Volcanic eruptions Industrial processes Coal mining

2 Mercury enters water bodies from the atmosphere and surface runoff from land.

4 Small bottom feeders absorb traces of methylmercury.

3 Bacteria turn mercury into harmful methylmercury, which is absorbed by animal tissues.

8. (a) Outline the process by which mercury enters and passes through the food chain: _____

_____

_____

_____

_____

(b) Explain why organisms at the top of the food chain have higher levels of mercury than those at lower trophic levels:

_____

_____

_____

_____

The table on the right provides food chain information and mercury concentrations in the tissues of four organisms in a marine food chain in the Persian Gulf.

9. (a) Fill in the right hand column of the table by calculating the percentage increase in mercury between each of the consumers:

(b) Is this what you would expect?

_____

_____

_____

Mercury concentrations in consumers in a simple marine food chain

Organism	Feeding habitat	Mercury concentration (μg/g)	Percentage increase per trophic level
Spinycheek grouper: benthic (bottom dwelling) fish	Eats mainly benthic invertebrates, detritus, and plants	0.82	–
Spottail needlefish: pelagic (open sea) fish	Eats mainly benthic fish, invertebrates, plants, and crustaceans	1.64	
Blue crab (estuarine and open water)	Fish predator, consumes mainly pelagic fish, but also detritus and plants	2.22	
Eurasian teal (sea bird)	Consumes mainly crab, fish, shrimp, and marine invertebrates	11.5	

Data source: Hosseini *et.al* (2014) J Marine Sci Res Dev 4:2

(c) Explain the results: _____

_____

_____

©2018 **BIOZONE** International
ISBN: 978-1-927309-55-1
Photocopying Prohibited

17 | Group Behavior Improves Survival

ENGAGE: Moving as a cohesive group can assist survival of individuals and populations

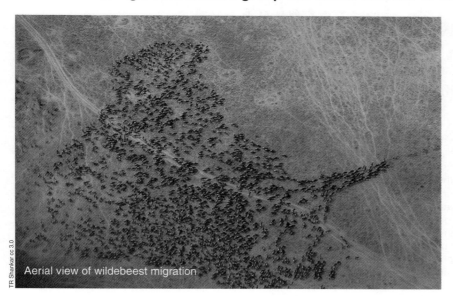

Aerial view of wildebeest migration

TR Shankar cc 3.0

Some animals are solitary, interacting only to reproduce, but many live in groups, which may be non-social (as in fish schools) or highly organized social groupings, in which individuals within the groups interact to varying degrees.

Grouping together in flocks, schools, and herds provides advantages in terms of protection from physical factors and predation. Highly social groupings may provide additional survival advantages such as division of labor within the group and more efficient gathering and use of resources.

The annual migration of wildebeest in the Serengeti, Africa, provides a startling example of the coordinated movement of a herd with the common goal of seeking fresh grazing and water.

1. (a) Give an example of an animal species you have observed that is generally found in a cohesive group: _____

_____

(b) Does the group move together? Is there any kind of organization visible? _____

_____

_____

(c) What survival advantages do you think might arise from the group behavior? _____

_____

_____

EXPLORE: Cohesive groups gain survival benefits

Schooling, herding, or flocking animals gain benefits from being part of a group, including increased protection from predators and reduced energy expenditure during movement. In groups, each individual behaves in a way that helps its own survival regardless of the others in the group. Apparently complex behavior within the group results from the application of a few simple rules: 1) move towards the group or others in the group, 2) avoid collision with external objects or others in the group, and 3) align your movement with the movement of the others in the group. If every individual moves according to these rules, the school, flock, or herd will stay as a dynamic cohesive unit, changing according to the movement of others and the cues from the environment.

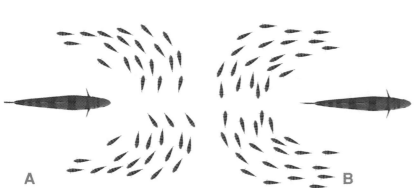

A

B

Schooling dynamics: In schooling fish or flocking birds, every individual behaves according to a set of rules. In A, each fish moves away from the predator while remaining close to each other. The school splits to avoid the predator (A), before moving close together again behind the predator (B).

In a **flash expansion**, each individual moves directly away from the predator. Collisions have never been observed, suggesting each fish is able to sense the direction of movement of the fish next to it.

©2018 **BIOZONE** International
ISBN: 978-1-927309-55-1
Photocopying Prohibited

 EM SSM LS2.D

School of fish

A large flock of auklets (a small seabird)

Schooling in fish

Prey fish may school for defensive reasons. Schooling fish also benefit by better hydrodynamics within the school so less energy is expended in swimming. Schooling help avoid predators because:

▶ The movement of the school causes confusion.
▶ There is a reduced probability of individual capture.
▶ There are more prey than can be eaten (predator satiation).
▶ Predator detection is more efficient (the many eyes principle).

Herding in mammals

Herding is common in hoofed mammals, especially those on grasslands such as the African savanna. A herd provides protection because while one animal is feeding, another will have its head up looking for predators. In this way, each individual benefits from a continual supply of lookouts (the many eyes principle). During an attack, individuals move closer to the center of the herd, as those on the outside are more frequently captured. The herd moves as one group, driven by individual needs.

Flocking in birds

Flocks follows similar rules to schools. In flight, each bird maintains a constant distance from others and keeps flying in the average direction of the group. Flocks can be very large, with thousands of birds flying together as a loosely organized unit, e.g. starlings flocking in the evening or queleas flocking over feeding or watering sites. While most flocks are non-social, flamingos form social groups with many hundreds of birds. The large colonies help the birds avoid predators, maximize food intake, and use suitable nesting sites more efficiently.

2. Briefly outline how animals in a school or flock maintain a cohesive group: _____

_____

_____

_____

EXPLORE: Moving as a group

Migration is the long distance movement of individuals from one place to another and usually involves a two way journey between two locations. Migration usually occurs on a seasonal basis and for a specific purpose, e.g. feeding, breeding, or over-wintering. Migration requires navigation over long distances to reach the correct destination.

▶ Migration is an energy expensive and high risk behavior, so its advantages must outweigh the disadvantages. The destination must provide enough of a required resource, such as food, water, or suitable breeding sites, to enhance survival of individuals and their offspring.

▶ Although some animals migrate individually, most migrate in large groups (right). Migrating as a group helps navigation through application of the "**many-wrongs principle**" in which combining many inaccurate navigational compasses produces a more accurate single compass. Thus, if an animal navigates by itself with a slightly inaccurate internal compass, or inaccurately interprets environmental cues, it may arrive in the wrong location. In a group, each member can adjust its heading according to the movement of the others, thus an average direction is produced and each member is more likely to arrive in the correct place.

▶ Graph right: Increasing group size (up to ~10) decreases the time taken to reach a navigational target when the group moves as a social unit. Groups larger than ~10 gain no further efficiencies and very large groups (>60) may be slightly worse off. Non-social groups take longer with increasing size (relative to the theoretical 'no collision avoidance' line) because of the need to avoid collisions with others in the group.

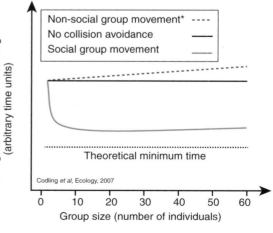

Non-social group movement* - - - -
No collision avoidance ———
Social group movement ———

Average time taken to reach target (arbitrary time units)

Theoretical minimum time

Codling *et al*, Ecology, 2007

0   10   20   30   40   50   60
Group size (number of individuals)

* Individuals do not interact except to avoid collision.

©2018 **BIOZONE** International
ISBN: 978-1-927309-55-1
Photocopying Prohibited

Group navigation: Modeling the many-wrongs principle

The effect of the many-wrongs principle on the accuracy of group navigation can be modeled and tested using human subjects. In the experiment described below, researchers designed a series of experiments to test the influence of group size and directional uncertainly on navigation.

Method

Target landmarks (1-16) were arranged around the perimeter of a circle with a 10 m diameter (right). Before each trial, a destination point was randomly chosen by the experimenter.

Within a trial, participants were each given a card with a consecutive sequence of landmarks printed on them (e.g. 15, 16, 1, 2 or 2, 3, 4, 5). Each participant's card always contained the target number. For example if the target was 2, all the participants' cards had sequences containing the number 2.

The participants all started in the center of the circle (spots labeled A-J) and were instructed to begin walking towards any number printed on their card. However, they were not allowed to communicate with each other and they had to stay together as a group. This was achieved by having each person stay within an arm's length of another person.

Researchers varied the number of people in the groups (1, 2, 3, 6, or 10) and varied the number of numbers in the sequence to reflect navigational uncertainty (in °). Longer sequences reflected a higher level of navigational uncertainty than smaller sequences.

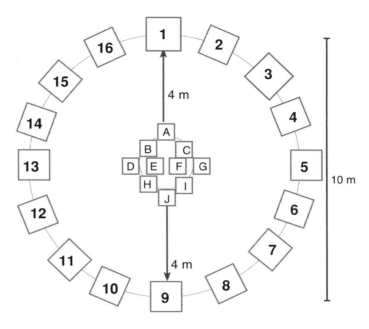

Results

The results showing median navigational accuracy for the largest directional uncertainty tested (112.5°) are presented in the graph (right). A measure of spread in the data is included.

Do it yourselves!

You can replicate this experiment for yourselves as a class. Go to **BIOZONE's resource hub** for complete instructions. What did you find? Did you reach similar conclusions?

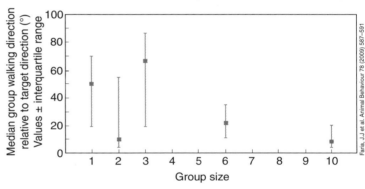

Navigational accuracy at 112.5° directional uncertainly

Faria, JJ et al. Animal Behaviour 78 (2009) 587–591

3. Why do you think the participants were not allowed to communicate with each other? _____

_____

_____

4. (a) What could have happened if the participants were not required to stay in contact with each other? _____

_____

(b) How might this have affected the time taken for the group to reach the target? _____

_____

_____

5. (a) Is there a significant difference between any of the groups (remember to account for spread in the data)?

_____

(b) Do the results support the many-wrongs principle? Explain: _____

_____

_____

_____

©2018 **BIOZONE** International
ISBN: 978-1-927309-55-1
Photocopying Prohibited

EXPLAIN: Why do migrating birds fly in V formation?

Migrating flocks generally fly in a V formation (photo, right). There are two complementary explanations for why birds fly in formation.

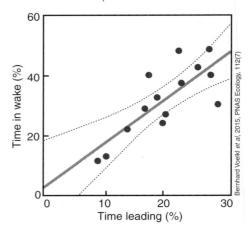
Migrating geese

- ▶ The first reason is to help navigation, coordination, and communication between individuals. The V formation places the birds in close proximity, allowing them to hear and see others.

- ▶ The second reason is that the V formation provides the best aerodynamics for all in the flock except the leader and those trailing at the tips of the V. Each bird gains lift from the movement of the air caused by the bird ahead of it and this saves energy. Moreover, birds regularly take turns at the front where they receive no energy savings.

The wingtip vortex provides lift for the trailing bird.

How do we know?

- ▶ **Evidence #1**: Research on great white pelicans has shown that the V formation helps the birds conserve energy. As the wing moves down, air rushes from underneath the wing to above it, causing an upward moving vortex behind the wing. This provides lift to the bird flying behind and to one side, requiring it to use less effort to maintain lift. Energy savings come from increased gliding.

- ▶ **Evidence #2**: Research on the critically endangered Northern bald ibis showed that the flying behavior is cooperative and the amount of time a bird is leading a formation is correlated with the time it can spend flying behind another bird. Overall, individuals spent an average of 32% of their time behind another bird, and a similar amount of time leading a formation (right).

Graph: y-axis "Time in wake (%)" from 0 to 60; x-axis "Time leading (%)" from 0 to 30. Source: Bernhard Voelkl et al, 2015, PNAS Ecology, 112(7)

6. (a) What are the advantages of migration to a population? _____

_____

_____

(b) Can you think of any risks involved in an animal undertaking a migratory journey? _____

_____

_____

7. How does grouping together increase navigational efficiency and how does this benefit the individual and the species?

_____

_____

_____

_____

8. (a) Using the information above, explain how flying in V formation helps an individual bird during migration: _____

_____

_____

_____

(b) How would this help survival of the population as whole? _____

_____

_____

(c) What evidence is there that migrating birds cooperate during their migration and how would this help them?

_____

_____

©2018 **BIOZONE** International
ISBN: 978-1-927309-55-1
Photocopying Prohibited

18 Individuals in Groups Often Cooperate

ENGAGE: Orca improve the chances of a successful hunt when they work in groups

Drone technology has captured footage showing how orca (*Orcinus orca*) use cooperative hunting behavior to successfully catch prey. Off the coast of Norway, a pod of five orca were filmed swimming calmly with synchronized surfacing behavior. One of the pod suddenly showed drastic behavioral changes, turning sharply and diving. The behavior was quickly adopted by other members of the pod. During this time they stayed in a relatively restricted area. The drone footage revealed that the pod were circling a seal on the ocean's surface. Their circling prevented it escaping. When the seal dived to attempt an escape, the orca also dived. They remained submerged for long periods of time, surfacing one by one to breathe. This ensured that the seal was being closely monitored at all times. Eventually the oldest female surfaced with the dead prey. One by one the orca took turns in eating the seal, taking a few bites and then dropping the carcass for the next orca to feed.

1. The orca example above illustrates how working together cooperatively can improve the chances of a successful outcome. Can you think of any other examples where cooperative behavior helps animals achieve the desired outcome:

_____

_____

_____

_____

_____

_____

2. (a) Describe how the cooperative behavior of the orca contributed to their successful hunt: _____

_____

_____

_____

_____

(b) Once captured, the pod shared the prey amongst themselves. What do you think the benefit of this is:

i) to the individual: _____

_____

_____

ii) to the survival of the population: _____

_____

_____

3. What do you think the benefits of group hunting are over hunting individually: _____

_____

_____

©2018 **BIOZONE** International
ISBN: 978-1-927309-55-1
Photocopying Prohibited

EM SSM LS2.D

EXPLORE: Examples of cooperative behaviors within species

▶ **Cooperative behavior** involves behavior where two or more individuals work together to achieve a common goal such as defending the group, obtaining food, or rearing young. Examples include hunting as a team (e.g. wolf packs, chimpanzee hunts), responding to the actions of others with the same goal (e.g. migrating mammals), or acting to benefit others (e.g. mobbing in small birds). Cooperation occurs most often between members of the same species.

▶ Altruism is an extreme form of cooperative behavior in which one individual disadvantages itself for the benefit of another. Altruism is often seen in highly social animal groups. Most often the individual who is disadvantaged receives benefit in some non-material form (e.g. increased probability of passing genes onto the next generation).

Coordinated behavior is used by many social animals for the purpose of both attack (group hunting) and defense. Cooperation improves the likelihood of a successful outcome, e.g. a successful kill.

Animals may move en masse in a coordinated way and with a common goal, as in the mass migrations of large herbivores. Risks to the individual are reduced by the group behavior.

Kin selection is altruistic behavior towards relatives. In meerkats, individuals from earlier litters remain in the colony to care for new pups instead of breeding themselves. They help more often when more closely related.

4. How do cooperative interactions enhance the survival of both individuals and the group they are part of? _____

_____

_____

_____

_____

5. Why do you think animals are more likely to help each other if they are related? _____

_____

_____

_____

EXPLORE: Evidence of cooperation between species

▶ Many small birds species will cooperate to attack a larger predatory species, such as a hawk, and drive it off. This behavior is called mobbing. It is accompanied by mobbing calls, which can communicate the presence of a predator to other vulnerable species, which benefit from and will become involved in the mobbing.

▶ One example is the black-capped chickadee (right), a species that often forms mixed flocks with other species. When its mobbing calls in response to a screech owl were played back, at least ten other species of small bird were attracted to the area and displayed various degrees of mobbing behavior. The interspecific communication helps to coordinate the community anti-predator mobbing behavior.

6. What evidence is there that unrelated species can act cooperatively? Why would they do this? _____

_____

_____

_____

_____

©2018 **BIOZONE** International
ISBN: 978-1-927309-55-1
Photocopying Prohibited

EXPLAIN: Why does survival increase when animals live in groups?

Living in a group can improve the survival of the members, e.g. improving foraging success or decreasing the chances of predation. Animals such as meerkats, ground squirrels, and prairie dogs decrease the chances of predation by using sentries, which produce alarm calls to alert others when a predator approaches.

Gunnison's prairie dogs

Gunnison's prairie dogs live in large communities called towns in the grasslands of western North America. The towns are divided into territories which may include up to 20 individuals. During their foraging, above-ground individuals may produce alarm calls (right) if a predator approaches, at which nearby prairie dogs will take cover.

However, whether or not an alarm call is given depends on the relatedness of the individuals receiving the call to the individual giving it. Gunnison's prairie dogs put themselves at risk when giving an alarm call by attracting the attention of the predator. Apparently altruistic (self-sacrificing) behavior involving close relatives is called **kin selection**.

A Gunnison's prairie dog sentinel screams a warning

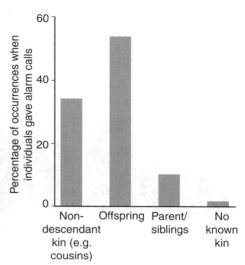

7. Use the prairie dog example to explain how living in a group improves individual and population survival: _____

_____

_____

_____

_____

_____

White fronted bee-eaters

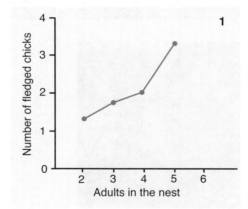

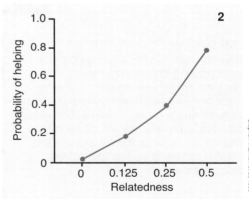

Stephen T. Emlen et al. 1995

White fronted bee-eaters (left) live in family groups which include a breeding pair and non-breeding pairs. All adults help provide for the chicks. Graph 1 shows the relationship between the number of adults in the nest and the number of chicks fledged. Graph 2 shows how relatedness affects the amount of help the pairs give the chicks.

8. The level of help between group members often depends on relatedness. Using the example above, explain how relatedness to the helper affects the level of help given:

_____

_____

_____

_____

©2018 **BIOZONE** International
ISBN: 978-1-927309-55-1
Photocopying Prohibited

EXPLAIN: How does cooperative defense aid survival?

Group defense is a key strategy for survival in social or herding mammals. Forming groups during an attack by a predator decreases the chances of being singled out, while increasing the chances of a successful defense.

Group defense in musk oxen

In the Siberian steppes, which are extensive grasslands, musk oxen must find novel ways to protect themselves from predators. There is often no natural cover, so they must make their own barrier in the form of a defensive circle. When wolves (their most common predator) attack, they shield the young inside the circle. Lone animals have little chance of surviving an attack as wolves hunt in packs.

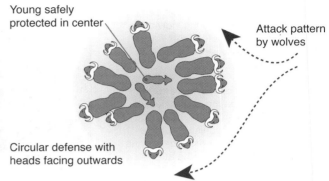

Young safely protected in center

Attack pattern by wolves

Circular defense with heads facing outwards

Japanese honeybees

Japanese honeybees are often attacked by the aggressive Asian giant hornet (right), which also steals the bee colony's honey. When a hornet scout enters the honeybee hive the honeybees mob it with more than 100 bees, forming a bee-ball (far right). The center of the ball can reach 50°C, literally baking the scout to death.

I. Kenpei cc 3.0

Takahashi cc: 2.1

Red colobus monkey defense

Red colobus monkeys are a common target during chimpanzee hunts. They counter these attacks by fleeing (especially females with young), hiding, or mounting a group defense. The group defense is usually the job of the males and the more defenders there are the greater the likelihood of the defense being successful.

Afamari

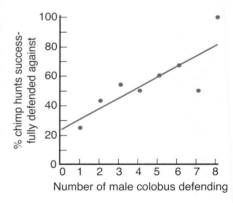

9. Use the evidence presented here to explain the advantages of cooperating as a group to defend against predators:

_____

_____

_____

_____

_____

_____

©2018 **BIOZONE** International
ISBN: 978-1-927309-55-1
Photocopying Prohibited

10. (a) Explain the relationship between number of male colobus monkeys mounting a defensive attack and its success:

_____

_____

(b) How many colobus males are needed to effectively guarantee a successful defense against chimpanzees?

_____

EXPLAIN: How does cooperative food gathering aid survival?

▶ Cooperating to gather food can be much more efficient that finding it alone. It increases the chances of finding food or capturing prey.

▶ Cooperative hunting will evolve in a species if the following circumstances apply:
   • If there is a sustained benefit to the hunting participants
   • If the benefit for a single hunter is less than that of the benefit of hunting in a group
   • Cooperation within the group is guaranteed

Foraging leafcutter ants

Bottlenose dolphins

Mountain caracara

Cooperative food gathering in ants often involves division of labor. Leafcutter ant societies are based on an ant-fungus mutualism. Foragers cut parts of leaves and transport the fragments to the nest where other worker castes use them to cultivate a fungus. Workers that tend the fungus gardens have smaller heads than the foragers. Other castes defend the foragers during their work. The very largest workers defend the nest itself.

Role specialization also occurs in bottlenose dolphins. In groups of three to six, a designated 'driver' dolphin herds fish in circles towards the others, which are tightly grouped to form a barrier. The driver slaps the water with its flukes, causing the fish to leap out of the water where they can be easily caught. Similarly, orca will cooperate to create a wave-wash large enough to knock their seal prey off icebergs where they are resting.

The mountain caracara in Peru (above) forages in groups of three or four, looking for prey hidden around rocks. Working together, the birds are able to overturn rocks far bigger than any individual could move. If a bird finds a rock that is worth turning over, it produces a high pitched call to attract the others. In most cases, only one bird (usually the initial caller) benefits from overturning the rock. However, the other birds may benefit when other rocks are overturned later (reciprocal altruism).

Army ants foraging

Like leafcutter ants, army ants (below) have several distinct worker castes. The smaller castes collect small prey, and larger porter ants collect larger prey. The largest workers defend the nest. Through group cooperation, the tiny ants are able to subdue prey much larger than themselves, even managing to kill and devour animals such as lizards and small mammals. This would not be possible if they hunted as individuals.

▶ There are two species of army ant that have quite different raiding patterns (below): *Eciton hamatum* whose columns go in many directions and *Eciton burchelli*, which is a swarm-raider, forming a broad front. Both species store food at various points along the way.

Eciton worker castes

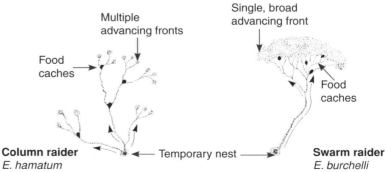

Multiple advancing fronts

Single, broad advancing front

Food caches

Food caches

Column raider
E. hamatum

Temporary nest

Swarm raider
E. burchelli

©2018 **BIOZONE** International
ISBN: 978-1-927309-55-1
Photocopying Prohibited

11. What is cooperative food gathering? _____

_____

_____

12. What conditions favor cooperative food gathering? _____

_____

_____

13. Using examples, describe how cooperative food gathering provides an advantage to survival or reproduction: _____

_____

_____

_____

_____

_____

_____

_____

_____

_____

ELABORATE: Chimpanzees benefit from cooperative hunting

Cooperative hunting in chimpanzees

Chimpanzees benefit from cooperative hunting. Although they may hunt alone, they also form hunting groups of up to six members or more. Chimpanzee hunts differ from the cooperative hunting of most other animals in that each chimpanzee in the hunt has a specific role in the hunt, such as a blocker or ambusher. Studies of chimpanzee hunting show that different groups employ different hunting strategies.

The hunt information in table 1 was gathered from chimpanzees in the Tai National Park in Ivory Coast.

Number of hunters	Number of hunts	Hunting success (%)	Meat per hunt (kg)	Net benefit per hunter (kJ)
1	30	13	1.23	4015
2	34	29	0.82	1250
3	39	49	3.12	3804
4	25	72	5.47	5166
5	12	75	4.65	3471
6	12	42	3.17	1851
>6	10	90	9.27	5020

Christophe Boesch 1994

The hunt information in table 2 was gathered from chimpanzees in the Gombe Stream National Park in Tanzania.

Number of hunters	Number of hunts	Hunting success (%)	Meat per hunt (kg)	Net benefit per hunter (kJ)
1	30	50	1.23	4245
2	13	61	1.85	3201
3	9	78	1.61	1837
4	7	100	2.86	2494
5	1	100	3.00	2189
6	2	50	2.00	861

Christophe Boesch 1994

©2018 **BIOZONE** International
ISBN: 978-1-927309-55-1
Photocopying Prohibited

14. Use the information in the table to discuss the differences between the two groups of chimpanzee in the extent of cooperation and how it relates to hunting success. You should plot graphs to help illustrate reasons for differences:

_____

_____

_____

_____

_____

_____

_____

_____

_____

_____

Sharing and bonding in chimpanzees

In Tai chimpanzees, hunting is a chance to form social bonds. Study the information below showing the number of chimpanzees taking part in a hunt and eating afterwards and the mean (average) number of bystanders during the hunt and eating afterwards.

15. Explain what the information is showing and discuss the reasons why this might occur:

Number of hunters	Mean number of hunters eating	Mean number of bystanders	Mean number of bystanders eating
1	0.7	3.5	3.0
2	1.6	3.6	2.6
3	2.5	3.6	3.0
4	2.5	2.7	2.1
5	3.5	2.7	2.3
6	4.7	2.4	2.2

Christophe Boesch 1994

_____

_____

_____

_____

_____

_____

19 An Endless Swarm Revisited

The anchoring phenomenon for this chapter was the swarming behavior of locusts. Now that you have completed this chapter you should be able to apply the knowledge you have gained to answer the following questions.

1. (a) Describe how the emergence of a large number of locusts would affect a pyramid of numbers at that time within the grassland ecosystem they inhabit:

_____

_____

_____

   (b) What effect might this increase in locust numbers have on food web stability in the ecosystem? Explain: _____

_____

_____

_____

_____

2. (a) How does the movement of a locust swarm benefit the locust population? _____

_____

_____

_____

   (b) Explain, using the concept of carrying capacity, why the incredibly large numbers of locusts contained within a swarm can not be sustained indefinitely by the environment:

_____

_____

_____

_____

3. (a) Swarming increases the level of intraspecific competition in the locust population. Describe the resources that the insects are competing for:

_____

   (b) Do you think a locust swarm increases the level of interspecific competition in an ecosystem? Explain your answer:

_____

_____

©2018 **BIOZONE** International
ISBN: 978-1-927309-55-1
Photocopying Prohibited

20 Summative Assessment

Analyzing a model predator-prey system

Mathematical models predict that predator and prey populations will form stable cycles of population increase and decrease. Early ecologists set out to verify these population cycles in small model ecosystems. Two researchers, Gause and Huffaker, each worked on this question. Their results gave great insight into the nature of predator-prey interactions and the factors that control population size.

Gause's experiments

▶ Gause's experiments examined the interactions of two protists, *Paramecium* and its predator *Didinium* in simple test tube 'microcosms'. When *Didinium* was added to a culture of *Paramecium*, it quickly ate all the *Paramecium* and then died out. When sediment was placed in the microcosm, *Paramecium* could hide, *Didinium* died out and the *Paramecium* population recovered.

Huffaker's experiments

▶ Huffaker built on Gause's findings and attempted to design artificial systems that would better model a real system. He worked on two mite species, the six spotted mite and its predator. Oranges provided both the habitat and the food for the prey.

▶ In a simple system, such as a small number of oranges grouped together, predators quickly ate all the prey and then died out.

▶ Huffaker then created a more complex system with arrays of 120 oranges (below). The amount of available food on each orange was controlled by sealing off parts of each orange with wax. Patchiness in the environment was created using balls (representing unsuitable habitat). Sticks aided dispersal of prey mites and petroleum jelly was used to form barriers to predatory mite dispersal. In this system, the predator and prey coexisted for three full cycles (> year). In the diagram below, the arrays depict the distribution and density of the populations at the arrowed points. The circles represent oranges or balls and the dots the predatory mites.

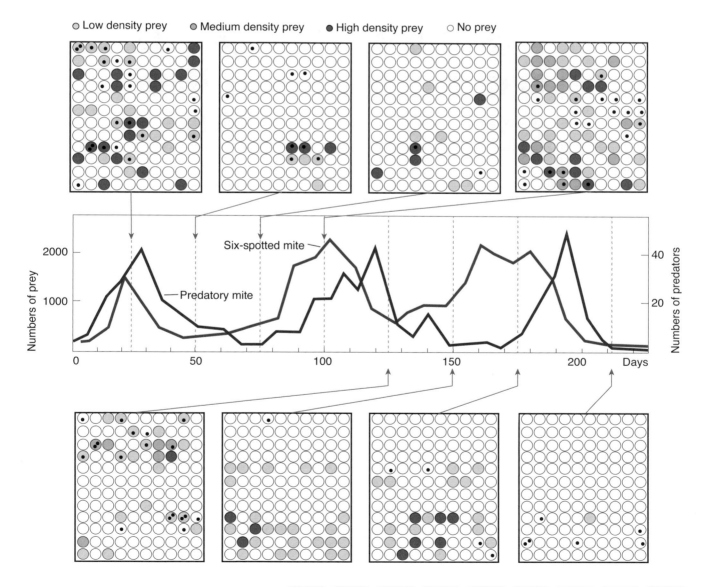

○ Low density prey ◉ Medium density prey ● High density prey ○ No prey

EM   CE   LS2.D   LS2.C   LS2.B   LS2.A

72

1. What did Gause's simple microcosm experiments tell us about the role of predation in limiting prey populations?

_____

_____

2. (a) Mark the three population cycles completed in Huffaker's experiment on the plot on the previous page.

   (b) In a different color, mark the lag in the predator population response to change in prey numbers.

   (c) What does the lag represent? _____

   _____

3. (a) What was the purpose of patchiness in the environment Huffaker created? _____

   _____

   _____

   (b) How well do you think Huffaker's model system approximated a real ecosystem? _____

   _____

   _____

   (c) Study the two artificial environments below, set up as per Huffaker's mite experiment. Which of the two set ups do you think would produce a larger number of predator prey cycles. Justify your answer.

Rubber balls

Oranges

Bridges

A

B

   _____

   _____

   _____

   _____

   _____

4. Using Huffaker's experimental evidence, discuss the factors that could affect diversity and long term stability of predator and prey populations in real ecosystems:

   _____

   _____

   _____

   _____

   _____

   _____

   _____

   _____

©2018 **BIOZONE** International
ISBN: 978-1-927309-55-1
Photocopying Prohibited

The gross primary production of any ecosystem will be determined by the efficiency with which solar energy is captured by photosynthesis. The efficiency of subsequent energy transfers will determine the amount of energy available to consumers. These energy transfers can be quantified using measurements of dry mass.

In questions 5-12, you will calculate and analyze energy and biomass transfers in real and experimental systems.

Corn field

Mature pasture

5. The energy budgets of two agricultural systems (4000 m² area) were measured over a growing season of 100 days. The results are tabulated right.

(a) For each system, calculate the percentage efficiency of energy utilization (how much incident solar radiation is captured by photosynthesis):

Corn: _____

Mature pasture: _____

(b) For each system, calculate the percentage losses to respiration:

Corn: _____

Mature pasture: _____

(c) For each system, calculate the percentage efficiency of NPP:

Corn: _____

Mature pasture: _____

	Corn field	Mature pasture
	kJ x 10⁶	kJ x 10⁶
Incident solar radiation	8548	1971
Plant utilization		
Net primary production (NPP)	105.8	20.7
Respiration (R)	32.2	3.7
Gross primary production (GPP)	138.0	24.4

(d) Which system has the greatest efficiency of energy transfer to biomass? _____

Estimating NPP in *Brassica rapa*

Brassica rapa (right) is a fast growing brassica species, which can complete its life cycle in as little as 40 days if growth conditions are favorable. A class of students wished to estimate the gross and net primary productivity of a crop of these plants using wet and dry mass measurements made at three intervals over 21 days.

▸ Seven groups of three students each grew 60 *B. rapa* plants in trays under controlled conditions. On day 7, each group removed a random selection of 10 plants with roots intact. The 10 plants were washed, blotted dry, and weighed collectively to determine **wet mass**.

▸ The 10 plants were placed in a ceramic drying bowl and placed in a drying oven at 200°C for 24 hours, then weighed (to determine **dry mass**).

▸ The procedure was repeated on day 14 and again on day 21 with a further 10 plants (randomly selected).

▸ The results for group 1 are presented in Table 1 below. You will complete the calculation columns.

Table 1: Group 1's results for growth of 10 *B. rapa* plants over 21 days

Age in days	Wet mass of 10 plants (g)	Dry mass of 10 plants (g)	Percent biomass	Energy in 10 plants (kJ)	Energy per plant (kJ)	NPP (kJ/plant/day)
7	19.6	4.2				
14	38.4	9.3				
21	55.2	15.5				

6. Calculate percent biomass using the equation: % biomass = dry mass ÷ wet mass x 100. Enter the values in Table 1.

7. Each gram of dry biomass is equivalent to 18.2 kJ of energy. Calculate the amount of energy per 10 plants and per plant for plants at 7, 14, and 21 days. Enter the values in Table 1.

8. Calculate the Net Primary Productivity per plant, i.e. the amount of energy stored as biomass per day (kJ/plant/day). Enter the values in Table 1. We are using per plant in this exercise as we do not have a unit area of harvest.

©2018 **BIOZONE** International
ISBN: 978-1-927309-55-1
Photocopying Prohibited

9. The other 6 groups of students completed the same procedure and, at the end of the 21 days, the groups compared their results for NPP. The results are presented in Table 2, right.

   Transfer group 1's NPP results from Table 1 to complete the table of results and calculate the mean NPP for *B. rapa*.

Table 2: Class results for NPP of *B. rapa* over 21 days

Time in days (d)	Group NPP (kJ/plant/day)							Mean NPP
	1	2	3	4	5	6	7	
7		1.05	1.05	1.13	1.09	1.13	1.09	
14		1.17	1.21	1.25	1.21	1.25	1.17	
21		1.30	1.34	1.30	1.34	1.38	1.34	

10. (a) What is happening to the NPP over time?

   _____

   (b) Explain why this is happening: _____

   _____

   _____

11. What would you need to know to determine the gross primary productivity of *B. rapa*?_____

   _____

12. In a second set of experiments, students determined the net secondary production and respiratory losses using 12 day old cabbage white larvae feeding on Brussels sprouts. Of the NPP from the Brussels sprouts that is eaten by the larvae, some is used in cellular respiration, some is available to secondary consumers (this is net secondary production) and some is lost as waste (frass). The students accurately measured and recorded the wet mass of 10 larvae and ~ 30 g Brussels sprouts and placed the larvae and Brussels sprouts into an aerated container. After three days, the plant material, larvae, and frass were separated and the wet mass and the dry mass (after drying) determined for each. Students were given the energy values (in kJ) per gram of plant, animal, or waste material. Assume the proportion biomass of Brussels sprouts and larvae on day 1 is the same as the calculated value on day 3. Results are given below.

   (a) Complete the calculations in the highlighted cells of Table 3 below.

   (b) Net secondary production per larva (S) is:

   _____

   (c) Write the equation to calculate the percentage efficiency of energy transfer from producers to consumers (use the notation provided) and calculate the value here:

   _____

   (d) Is this what you would expect? _____

   Explain: _____

   _____

   _____

   _____

   (e) Write the equation to calculate respiratory losses per larva (use the notation provided):

   _____

   (f) Calculate the respiratory losses per larva:

   _____

   (g) What percentage of the energy consumed by each larva is lost in respiration?

   _____

   (h) Why must we assume the proportion biomass of Brussels sprouts and larvae on day 1 is the same as the calculated value on day 3?

   _____

Table 3: Results for secondary production experiment

	Day 1	Day 3	
Brussels sprouts			
Wet mass of sprouts	30 g	11 g	g consumed =
Dry mass of sprouts	_	2.2 g	
Plant proportion biomass (dry/wet)			
Plant energy consumed (wet mass x proportion biomass x 18.2 kJ)			kJ consumed per 10 larvae =
Plant energy consumed ÷ no. of larvae			kJ consumed per larva (E) =
Larvae			
Wet mass of 10 larvae	0.3 g	1.8 g	g gained =
Wet mass per larva			g per larva =
Dry mass of 10 larvae	_	0.27 g	
Larva proportion biomass (dry/wet)			
Energy production per larva (wet mass x proportion biomass x 23.0 kJ)			kJ gained per larva (S) =

Frass	**Day 3**
Dry mass frass from 10 larvae	0.5 g
Frass energy (waste) = frass dry mass x 19.87 kJ	
Dry mass frass from 1 larva (F)	

©2018 **BIOZONE** International
ISBN: 978-1-927309-55-1
Photocopying Prohibited

Sentinel behavior in meerkats

Meerkats are highly social carnivores that live in mobs consisting of a dominant (alpha) breeding pair and up to 40 subordinate helpers of both sexes who do not normally breed but are usually related to the alpha pair. They are known for their sentinel behavior, watching for predators and giving alarm calls when they appear.

The graphs right show the likelihood of female or male meerkats standing sentinel when pups are either in the burrow or outside in the sentinel's group. The scale represents a statistical measure from a large number of observations. Error bars are ± standard error.

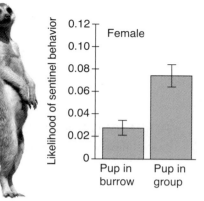
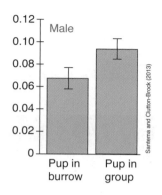

13. Discuss the evidence that meerkat sentinel behavior is altruistic in its nature: _____

_____
_____
_____
_____
_____
_____
_____

14. The graph below shows the effect of pigeon flock size on the success of hawk attacks:

Hawk predation on pigeons

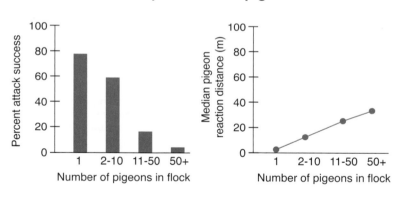

(a) What is the effect of flock size on attack success by hawks? _____

_____

(b) What is the effect of pigeon flock size on the ability to detect hawks? _____

_____

(c) Explain why living in a flock increases the probability of survival of each individual and the population as a whole. Include references to predation, feeding, and navigation.

_____
_____
_____
_____
_____
_____

©2018 **BIOZONE** International
ISBN: 978-1-927309-55-1
Photocopying Prohibited

Instructional Segment 2

History of the Earth's Atmosphere: Photosynthesis and Respiration

Activity number

Anchoring Phenomenon

One of these worlds is not like the others: The stability of artificial ecosystems relative to the Earth.

21 33

How do living things acquire energy and matter for life?

☐ 1 Living organisms use cellular respiration as the process by which organic molecules are broken down to release energy and form molecules of ATP. ATP is an energy transfer molecule and its hydrolysis provides the energy to do cellular work. Use a model to show how ATP provides energy to carry out life's functions.

22

☐ 2 Recall the differences between producers (autotrophs) and consumers (heterotrophs) and explain the relationship between them.

13

☐ 3 Photosynthesis is the process that captures light energy and converts it into stored chemical energy. Use a model to show how photosynthesis transforms light energy into stored chemical energy. Show inputs and outputs of matter and the transfer and transformation of energy by producers.

23 25 34

How do organisms store energy?

☐ 4 Life on Earth is carbon based. Carbon can form covalent bonds with four other atoms and can form both single and double bonds. The same raw materials can be reconfigured to produce molecules, such as glucose and starch, with different chemical potential energy. Carbon-based molecules are therefore able to store and release energy during these changes.

23

☐ 5 Construct and revise an explanation based on evidence for how organisms build a wide range of organic molecules (e.g. amino acids, nucleic acids, fats) from the carbon, hydrogen, and oxygen in glucose.

23

How do organisms use the raw materials they ingest?

☐ 6 Using the models you have constructed so far, explain the ways in which heterotrophs use the raw materials they ingest. Describe what happens to the heat energy released from the chemical reactions of metabolism and explain why ecosystems rely on a constant influx of energy from the Sun?

24 34

How are photosynthesis and cellular respiration connected?

☐ 7 Build a model of glucose to show how it can be split up to form CO_2, O_2, and H_2O. Use your model to illustrate the chemical process of cellular respiration, showing that bonds in oxygen and food molecules are broken and bonds in new compounds are formed, resulting in a net transfer of energy.

25

☐ 8 Compare aerobic and anaerobic pathways for ATP production in different organisms. Construct and revise explanations based on evidence (e.g. the results of simple experiments of yeast fermentation) for the flow of energy and cycling of matter in aerobic and anaerobic conditions.

24 34

☐ 9 **ENGINEERING CONNECTION: WASTEWATER TREATMENT FACILITIES**
Obtain information about the stages involved in the treatment of human waste in wastewater treatment facilities. Make a model of the waste water treatment process and use it to investigate the effect of aeration to the rate of waste water treatment. Use your findings to construct an explanation about the effect of changing oxygen concentration of the rate of biological processes in a waste water treatment plant.

26

☐ 10 Photosynthesis and cellular respiration provide most of the energy needed to carry out essential life processes. Construct a physical model of these processes to illustrate the connections between them, emphasizing the inputs and outputs of energy and matter.

25

☐ 11 Develop a model to illustrate the role of photosynthesis and cellular respiration in the cycling of carbon among the biosphere, atmosphere, hydrosphere, and geosphere. Include reference to the chemical, physical, geological, and biological processes by which carbon is exchanged among these spheres.

29 30 34

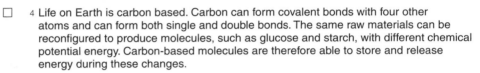

Dartmouth College Electron Microscope Facility

How has the cycling of energy and matter changed over Earth's history?

☐ 12 How have scientists determined the age of the Earth and what was it like when it was first formed? At what point in Earth's history did the cycles of energy and matter we see now begin? Apply scientific reasoning and evidence from ancient Earth materials, meteorites, and other planetary surfaces to construct an account of Earth's formation and early history.

27 34

☐ 13 Explain how the evolution of photosynthetic organisms marked the beginning of the biosphere's interaction with the global carbon cycle. Use evidence as well as your own model of photosynthesis (see #3) to construct an argument about the coevolution of life and other Earth systems. Your argument should emphasize the dynamic causes, effects, and feedback between the biosphere and Earth's other systems and use suitable illustrative and supporting examples, e.g. microbial life and the formation of soil, leading to the evolution of land plants.

28

☐ 14 Explain how scientists can track the movement of carbon through the Earth's spheres. Develop a quantitative model to show how the cycling of carbon among the ocean, atmosphere, soil, and biosphere (including humans) provides the foundation for living organisms. Your model should include reservoirs of carbon accumulation, processes by which carbon is exchanged within and between reservoirs, and the relative importance of these processes and reservoirs.

29 30 34

☐ 15 Explore various representations of the carbon cycle to build conceptual models of the processes involved in exchanges between reservoirs of carbon. Use simple physical models to investigate interactions between the atmosphere and hydrosphere and construct explanations about changes occurring in ocean chemistry as a result of differences between the amount of carbon stored and released by the oceans.

29 30 34

☐ 16 Explain how the lowering of ocean pH (what is widely termed 'ocean acidification') can cause major damage to the ocean's coral reefs and planktonic organisms (those at the base of ocean food chains). Discuss the implications of these changes to ocean biodiversity and productivity.

31

☐ 17 Analyze computational models to explore the relationships between different Earth systems and quantify the impact of human activities on these systems [see IS6].

31 & IS6

☐ 18 Explain how carbon has come to accumulate in the largest reservoir, the geosphere, describing the role of heat and pressure during the burial of dead organic matter in the formation of fossil fuels (coal, oil, and natural gas). Use published work to generate a model (drawing) summarizing the stages of oil formation.

32

☐ 19 Discuss the advantages and disadvantages associated with the use of fossil fuels, including environmental issues associated with their extraction and use and the impact of carbon extraction on the natural carbon cycle.

32

21 One of These Worlds is Not Like the Others

ANCHORING PHENOMENON: The stability of artificial ecosystems relative to the Earth

A blown-glass aquarium marketed as a miniaturized self-sustaining closed ecological system. Each sealed sphere contains air, water, three shrimps, algae and a branch of soft coral (dead). Shrimps may survive three or more years, but usually for less time than they do in the wild.

The Earth. A planet with an active core supporting plate tectonics and a magnetic field, an atmosphere of 78% nitrogen, 21% oxygen, and 1% carbon dioxide and other gases, billions of tonnes of water, and millions of species of living organisms. The Earth has existed for around 4.5 billion years and supported life for around 3.8 billion years.

Biosphere 2. A 1.27 hectare self contained facility containing a rainforest, ocean, desert, mangrove swamp, grassland, and agricultural area. It initially ran for two years, supporting numerous plants and animals, including eight humans. There were many problems with oxygen and carbon dioxide levels, plants and animals dying, and high populations of pest species.

1. The three photos show three sealed or isolated systems at various scales. Only one of these appears to function indefinitely, supporting the life on it without need of replenishing the system.

   (a) Discuss as a class how the shrimp/algae system illustrates the interaction of photosynthesis and respiration, as well as illustrating how a simple ecosystem works. Decide if this system is sustainable. Summarize your ideas below:

   _____

   _____

   _____

   (b) Comment on the effect of scaling up the shrimp/algae system to the size of Biosphere 2. How successful was this? Could it ever be totally successful (e.g. for future Moon or Mars colonization)? Summarize your ideas below:

   _____

   _____

   (c) Discuss how the scale of the Earth and its many interacting systems form an indefinitely self-sustaining system whereas the other two systems are only partially successful in this respect. Summarize your ideas below:

   _____

   _____

   _____

SPQ   SSM

©2018 **BIOZONE** International
ISBN: 978-1-927309-55-1
Photocopying Prohibited

22 Energy in the Cell

ENGAGE: During exercise the cells of the human body use energy

1. In groups or as a class, discuss the different ways the body uses energy. Do some activities use more or less energy than others? Summarize your ideas below:

_____

_____

_____

2. In groups or as a class, discuss where the body uses its energy. What sources of energy deliver energy quickly to where it is needed? Summarize your ideas below:

_____

_____

_____

3. In groups or as a class, discuss how you think the cells of the body utilize energy sources (food) to produce usable energy for the cell and the body as a whole:

_____

_____

_____

4. The photo shows an athlete after a race, refuelling with an energy gel. The information describes the nutritional profile of the 50 g gel she is holding.

(a) How much energy is in the gel per **100 grams**? _____

(b) What does the gel appear to be mostly made of?

_____

(c) Use the following information to answer the questions below:
Energy in fat = 37 kJ/g, protein = 17 kJ/g, carbohydrate = 17 kJ/g.

Where does most of the energy in the gel appear to come from? Use calculations in your answer:

_____

(d) What does this tell you about the most readily available energy sources for the body?

_____

_____

Energy per gel:	436 kJ
Protein:	<1 g
Fat - Total:	0 g
Saturated:	0 g
Carbohydrate - Total:	25.6 g
Sugars:	4.2 g

©2018 **BIOZONE** International
ISBN: 978-1-927309-55-1
Photocopying Prohibited

EM   LS1.C

EXPLORE: Energy for metabolism

▸ All organisms require energy to be able to perform the metabolic processes of life.

▸ This energy is obtained by **cellular respiration**, a set of metabolic reactions that ultimately convert biochemical energy from 'food' into the energy-carrying molecule **adenosine triphosphate (ATP)**.

▸ The steps of cellular respiration take place in the cell cytoplasm and in cellular organelles called **mitochondria**.

▸ ATP is considered to be a universal energy carrier, transporting chemical energy within the cell for use in metabolic processes such as biosynthesis, cell division, cell signaling, thermoregulation, cell movement, and active transport of substances across membranes.

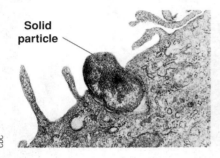

Energy is needed to transport molecules and substances across cellular membranes against their concentration gradient, e.g. engulfing solid particles, (above).

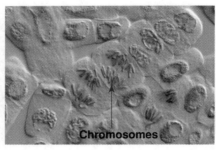

Cell division (mitosis) (above), requires energy to proceed. ATP provides energy for the formation of the mitotic spindle and the separation of chromosomes.

The maintenance of body temperature requires energy. Both heating and cooling the body require energy by shivering and secretion of sweat.

The mitochondrion

A mitochondrion is bounded by a double membrane. The inner and outer membranes are separated by an inter-membrane space, separating the regions of the mitochondrion into compartments in which the different reactions of cellular respiration occur. The inner membrane is much more folded than the outer membrane and has many enzymes attached to it.

The number of mitochondria per cell varies:

Cell type	No. of mitochondria	% volume of cell
Skin	200	-
Liver	1000-2000	20
Heart	5000+	35+
Muscle	1700	8

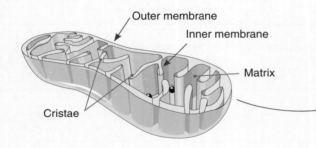

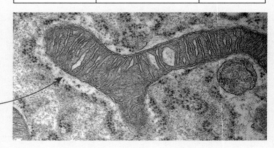

5. (a) Name the process that produces useful energy in the cell: _____

   (b) Where does this take place? _____

   (c) What molecule transports energy in the cell? _____

6. Discuss how the number of mitochondria and the total mitochondrial volume varies in different cell types. Why is this? What can the number and volume of mitochondria tell you about the cell? As a class or group find out about other cell types not listed above. Summarize your findings and discussion below:

   _____

   _____

   _____

   _____

   _____

   _____

©2018 **BIOZONE** International
ISBN: 978-1-927309-55-1
Photocopying Prohibited

EXPLAIN: How ATP provides energy

▶ The ATP (adenosine triphosphate) molecule (right) is made of three key components:

- A purine base (**adenine**)
- A pentose sugar (**ribose**)
- Three **phosphate groups**.

▶ ATP acts as a store of energy within the cell. The bonds between the phosphate groups contain electrons in a high energy state, which store a large amount of energy that is released during a chemical reaction. The removal of one phosphate group from ATP results in the formation of adenosine diphosphate (ADP).

▶ The energy in the ATP molecule is transferred to a target molecule (e.g. a protein) by a hydrolysis reaction. Water is split as part of the reaction to remove the terminal phosphate on ATP, forming ADP and an inorganic phosphate molecule (Pi). These are recycled to ATP during cellular respiration.

▶ When the Pi molecule combines with a target molecule, energy is released. Much of this energy is wasted as heat. The rest of the energy is transferred to the target molecule, allowing it to do work, e.g. joining with another molecule.

ATP molecule

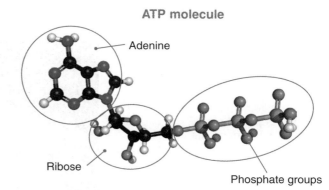

- Adenine
- Ribose
- Phosphate groups

Adenine + ribose = adenosine

ADP molecule

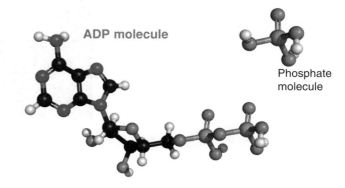

Phosphate molecule

7. ATP is often likened to a rechargeable battery. In the space below draw a diagram to model this analogy, showing how ATP transfers energy and is replenished:

8. When a phosphate group is added to a target molecule the energy of the target molecule is raised. This may allow it to join with another molecule. Beginning with the diagram on the left, complete the model to show how molecule A joins with molecule B:

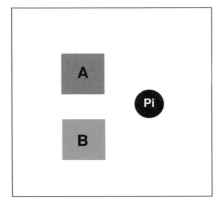

©2018 **BIOZONE** International
ISBN: 978-1-927309-55-1
Photocopying Prohibited

23 Photosynthesis

ENGAGE: Man in a box

In 2012, researchers carried out a larger version of Joseph Priestly's famous 1771 'mouse in a bell jar experiment'. 274 plants were placed in a sealed container with oxygen-depleted air (12.4% oxygen). A healthy 47 year man entered the container for 48 hours. The container was kept in constant light.

At the end of the experiment the oxygen content had risen to 18.1%.

Martin D. Thompson A. Stewart I, Gilbert E, Hope K, Kawai G, Griffiths A - Extrem Physiol Med CC 2.0

1. As a class or group, research the original bell jar experiment by Joseph Priestly. What was he trying to show? What did he conclude?

_____

_____

_____

_____

2. Compare the scale of Priestly's original 1771 experiment with the experiment shown above. Do you think the ratio of plant to animal biomass is similar or different? How many plants were needed for the man compared to the mouse?

_____

_____

_____

_____

_____

_____

3. Why is this experiment an important example of Earth's ecosystems? _____

_____

_____

_____

_____

_____

4. Identify the gases involved in the experiment. Why did the percentage of oxygen in the air increase over 48 hours?

_____

_____

_____

_____

   LS1.C EM

©2018 **BIOZONE** International
ISBN: 978-1-927309-55-1
Photocopying Prohibited

EXPLORE: Photosynthesis and carbon dioxide

Bromothymol blue is an indicator that is blue in basic (alkaline) solutions, blue/green in neutral water and yellow/green in acidic solutions. Adding carbon dioxide to water containing bromothymol blue causes the solution to turn yellow/green as the carbon dioxide dissolves into the water. The solution will eventually turn blue/green again as carbon dioxide escapes from solution and returns to the air or is removed from solution.

A student set up the following experiment:

Carbon dioxide was bubbled through 100 mL of water in a beaker containing bromothymol blue until the solution turned from blue/green to yellow/green. The solution was then divided equally into four test tubes.

▶ Into two test tubes were placed a sprig of *Cabomba*. The test tubes were then sealed.

▶ One of the test tubes containing the *Cabomba* was covered with aluminum foil to block out the light.

▶ One of the test tubes with no *Cabomba* in it was also covered with aluminum foil.

▶ The four test tubes where then left in the light for one hour where the *Cabomba* could photosynthesize.

5. In the space right draw the set up for this experiment.

6. Predict the outcome of the experiment: What color change will occur in each of the four test tubes below? The first is done for you:

(a) Test tube with *Cabomba* not covered: Turned blue/green   (c) Test tube with no *Cabomba* not covered: _____

(b) Test tube with *Cabomba* covered: _____   (d) Test tube with no *Cabomba* covered: _____

7. What caused the change in color in (a) above? _____

_____

_____

8. Explain your answers for (b-d) above: _____

_____

_____

_____

9. From this experiment what can you say about photosynthesis and carbon dioxide? _____

_____

_____

©2018 **BIOZONE** International
ISBN: 978-1-927309-55-1
Photocopying Prohibited

EXPLORE: Photosynthesis and light

Some students who carried out the investigation on the previous page noticed that the *Cabomba* produced bubbles when in the light, but not in the dark. Clearly light was causing *Cabomba* to produce a gas during photosynthesis.

The students decided to test the gas produced and to investigate if the intensity of the light affected the rate of gas production.

They set up the following experiment:

▶ 0.8-1.0 g of *Cabomba* stem was weighed. The stem was cut and inverted to ensure a free flow of the gas produced.

▶ The stem was placed into a beaker filled with a 20°C solution of 0.2 mol/L sodium bicarbonate ($NaHCO_3$) to supply CO_2. An inverted funnel and a test tube filled with the $NaHCO_3$ solution collected the gas produced.

▶ The beaker was placed at distances (20, 25, 30, 35, 40, 45 cm) from a 60W light source and the light intensity measured with a lux (lx) meter at each interval. One beaker was not exposed to the light source (5 lx).

▶ Before recording data, the stem was left to acclimatize to the new light level for 5 minutes. Bubbles were counted for a period of three minutes at each distance.

Experimental set up

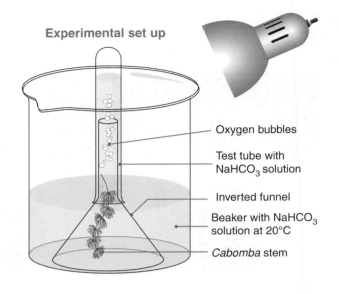

- Oxygen bubbles
- Test tube with $NaHCO_3$ solution
- Inverted funnel
- Beaker with $NaHCO_3$ solution at 20°C
- *Cabomba* stem

Results

Light intensity in lx (distance)	Bubbles counted in three minutes	Bubbles per minute
5	0	
13 (45 cm)	6	
30 (40 cm)	9	
60 (35 cm)	12	
95 (30 cm)	18	
150 (25 cm)	33	
190 (20 cm)	35	

The students tested one sample of the gas with a flaming splint. The splint burned brightly for a few seconds. A second sample tested with a glowing splint caused the splint to reignite.

10. From the results what gas was produced by the *Cabomba* plant? _____

11. Complete the table (above) by calculating the rate of gas production (bubbles of gas per minute):

12. Use the data to draw a graph on the grid above of the bubbles produced per minute vs light intensity

13. From this experiment what can you say about photosynthesis, light, and the gas produced? _____

_____

_____

14. Although the light source was placed set distances from the *Cabomba* stem, light intensity in lux was recorded at each distance rather than distance *per se*. Explain why this would be more accurate:

_____

_____

_____

_____

©2018 **BIOZONE** International
ISBN: 978-1-927309-55-1
Photocopying Prohibited

EXPLORE: The product of photosynthesis

Students investigating photosynthesis came across an experiment testing starch in leaves. The experiment is described below:

▶ A plant was placed in darkness for 48 hours. Several leaves were then covered with aluminum foil to block the light and the plant was placed back into the light for 24 hours.

▶ An uncovered and covered leaf were then randomly selected. Each leaf was boiled in water for two minutes.

▶ Each leaf was then placed into test tubes containing ethanol at its boiling point for ten minutes to remove the chlorophyll (green plant pigment) from the leaf. This produced a white leaf. The leaves were then placed back into hot water for 20 seconds to soften them.

▶ Iodine solution was then placed on the leaves using an eyedropper. The covered leaf remained the same color as the iodine solution. The uncovered leaf turned blue/black.

Plant with foil covered leaves

Color of iodine solution	
Starch	**No starch**
blue/black	brown

15. What does the result of the test suggest about the product of photosynthesis? _____

_____

_____

16. Why was the plant placed into darkness for 48 hours before covering some of the leaves and leaving the plant in light?

_____

_____

Glucose has the chemical formula $C_6H_{12}O_6$. Starch is a large molecule (polymer) made up of repeating glucose molecules. Plants use starch to store glucose when glucose is being made faster than it is being used. Starch is stored within chloroplasts and in storage organelles within the cytoplasm.

Starch granules in the chloroplast of a plant cell.

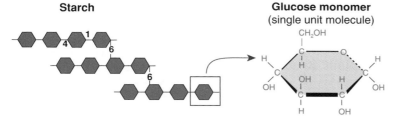

Starch

Glucose monomer (single unit molecule)

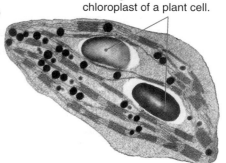

17. Use the information above to review your answer to 15. _____

_____

_____

EXPLORE: Water and photosynthesis

Are the CO_2 used in photosynthesis and O_2 produced by photosynthesis related? It is easy to think CO_2 is converted to O_2 by removal of carbon. However, in the 1930s, observations of other kinds of photosynthetic organisms led to the hypothesis that O_2 was generated from splitting water. Experiments carried out using isotopes of oxygen tested this hypothesis. The results are shown on the right:

18. Where do the results show the oxygen comes from?

_____

_____

19. Where does the carbon in CO_2 go during photosynthesis?

_____

_____

Experiment 1

Water labeled with oxygen 18

$H_2^{18}O, CO_2$

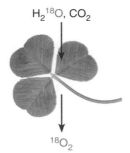

$^{18}O_2$

The oxygen released is labeled

Experiment 2

Carbon dioxide labeled with oxygen 18

$H_2O, C^{18}O_2$

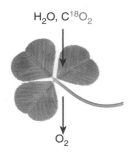

O_2

The oxygen released is not labeled

©2018 **BIOZONE** International
ISBN: 978-1-927309-55-1
Photocopying Prohibited

EXPLAIN: Photosynthesis

From your earlier explorations you will have noted that photosynthesis involves light, carbon dioxide (CO_2), oxygen (O_2), water (H_2O), and glucose ($C_6H_{12}O_6$).

20. Write a word equation for the process of photosynthesis:

_____

21. Using the information from the previous pages give reasons or evidence for the placement of reactants and products in your equation (i.e. what is the evidence that reactants are reactants and products are products?).

_____

_____

_____

_____

_____

_____

_____

_____

EXPLAIN: Chloroplasts

Photosynthesis takes place in disk-shaped organelles called **chloroplasts** (4-6 µm in diameter). The inner structure of chloroplasts is characterized by a system of membrane-bound compartments called **thylakoids** arranged into stacks called **grana** linked together by **stroma lamellae**.

Pigments on these membranes called **chlorophylls** capture light energy by absorbing light of specific wavelengths. Chlorophylls reflect green light, giving leaves their green color.

Chloroplasts are usually aligned with their broad surface parallel to the cell wall to maximize the surface area for light absorption.

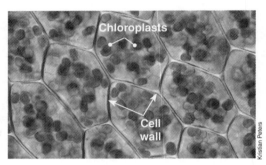

A mesophyll leaf cell contains 50-100 chloroplasts.

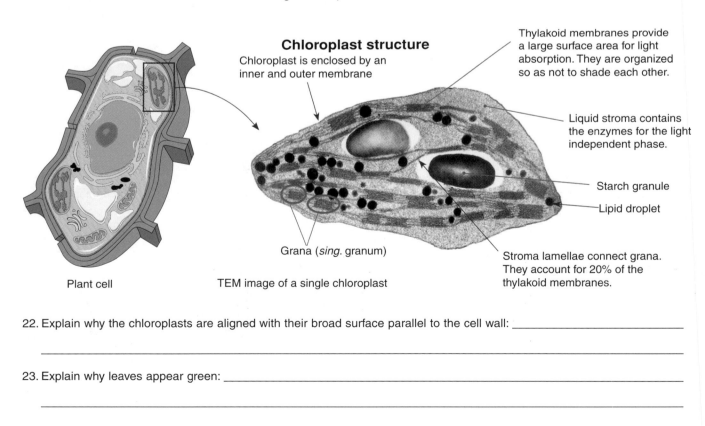

Chloroplast structure

Chloroplast is enclosed by an inner and outer membrane

Thylakoid membranes provide a large surface area for light absorption. They are organized so as not to shade each other.

Liquid stroma contains the enzymes for the light independent phase.

Starch granule

Lipid droplet

Grana (*sing.* granum)

Stroma lamellae connect grana. They account for 20% of the thylakoid membranes.

Plant cell

TEM image of a single chloroplast

22. Explain why the chloroplasts are aligned with their broad surface parallel to the cell wall: _____

_____

23. Explain why leaves appear green: _____

_____

©2018 **BIOZONE** International
ISBN: 978-1-927309-55-1
Photocopying Prohibited

EXPLAIN: Photosynthesis is actually two sets of reactions

▶ Photosynthesis has two phases, the light dependent phase and the light independent phase.

▶ In the reactions of the **light dependent phase**, light energy is converted to chemical energy (ATP and NADPH). This phase occurs in the thylakoid membranes of the chloroplasts.

▶ In the reactions of the **light independent phase**, the chemical energy is used to synthesize carbohydrate. This phase occurs in the stroma of chloroplasts.

Light dependent phase (LDP):

In the first phase of photosynthesis, chlorophyll captures light energy, which is used to split water, producing O_2 gas (waste), electrons and H^+ ions, which are transferred to the molecule NADPH. ATP is also produced. The light dependent phase occurs in the thylakoid membranes of the grana.

Light independent phase (LIP):

The second phase of photosynthesis occurs in the stroma and uses the NADPH and the ATP to drive a series of enzyme-controlled reactions (the **Calvin cycle**) that fix carbon dioxide to produce triose phosphate. This phase does not need light to proceed.

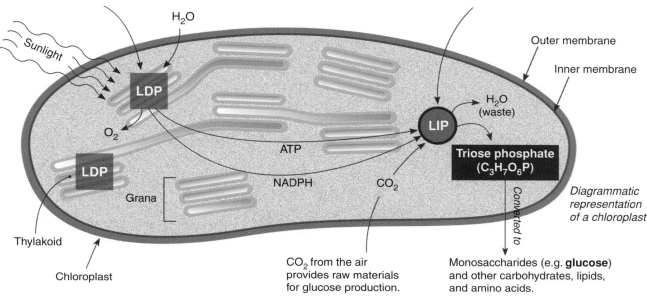

Diagrammatic representation of a chloroplast

CO_2 from the air provides raw materials for glucose production.

Monosaccharides (e.g. **glucose**) and other carbohydrates, lipids, and amino acids.

24. Explain how the light dependent and light independent reactions are linked: _____

_____

_____

_____

25. Using all the information from this activity complete the model of photosynthesis below by filling in the boxes:

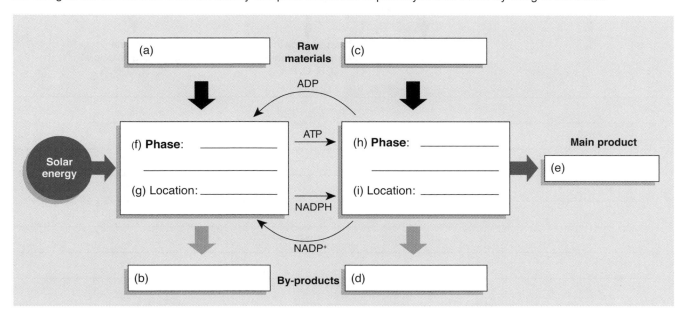

©2018 **BIOZONE** International
ISBN: 978-1-927309-55-1
Photocopying Prohibited

26. The chemical equation for photosynthesis is $6CO_2 + 6H_2O \rightarrow C_6H_{12}O_6 + 6O_2$. Explain how this equation is actually a summary of the process occurring in the light dependent and light independent reactions. At what point is energy added to the equation.

_____

_____

_____

_____

_____

_____

_____

_____

ELABORATE: The fate of glucose

27. The model below shows the possible fates of glucose resulting from photosynthesis. Note that glucose by itself can form different polymers (large multi-unit molecules) depending on the type of bond(s) linking the glucose monomers (single units) together. The elements present in each type of molecule are identified in red. Use the model to justify why photosynthesis could arguably be called the most important process on the planet:

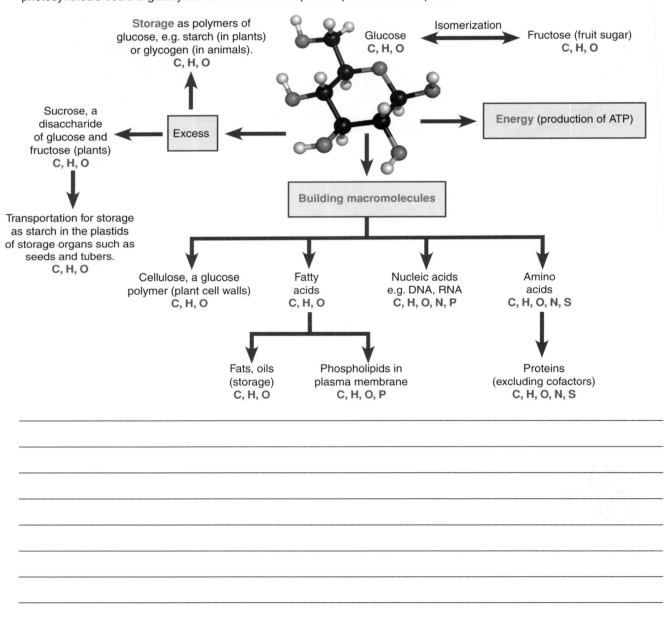

_____

_____

_____

_____

_____

_____

_____

_____

©2018 **BIOZONE** International
ISBN: 978-1-927309-55-1
Photocopying Prohibited

24 Cellular Respiration

ENGAGE: Free-diving

"Constant weight without fins" is a discipline of the free-diving sport "constant weight apnea" in which a diver descends and ascends as far as possible without the use of fins, weights, or a line. It is recognized as one of the most challenging disciplines because of the physical effort needed to ascend from the dive. The current record holder, as of July 21 2016, is New Zealander William Trubridge, who reached 102 meters. At that depth the pressure on his body is more than 11 times atmospheric pressure. His total dive time was 4 mins, 14 seconds.

Free-diving requires you to hold your breath, which can be a difficult process for anything longer than a minute. Although oxygen (O_2) is a gas needed for respiration, it is not the gas that the body measures to tell you when to breathe. Instead the body measures the concentration of carbon dioxide (CO_2) in the blood. When holding your breath, CO_2 builds up in the blood and produces a signal that creates the urge to breathe. Dive training involves teaching the body to relax (so that energy is not wasted and CO_2 is produced slowly) and to tolerate higher and higher levels of CO_2 (so that the breath can be held longer). The current record for breath holding (static apnea) is 11 minutes, 54 seconds, set in 2014.

Although O_2 is not the gas that regulates breathing, it is the gas needed to maintain consciousness. Blackout occurs when the pressure of O_2 in the lungs and the blood is too low and a person suddenly loses consciousness. When ascending from depth, this is called deep-water blackout or ascent blackout and may occur very close to or at the surface. As the body ascends, the pressure on the lungs reduces, causing a reduction in the pressure of oxygen in the body. At 4 atmospheres (4 bar, or about 30 m depth) 2% volume of oxygen in the lungs produces pO_2 of 80 mbar. But at the surface, that 2% oxygen drops to a pO_2 of just 20 mbar, which is too low for the body to maintain consciousness. A diver may seem fine until just below the surface, and then suddenly sink away.

Oxygen is needed for cellular respiration. It is the final acceptor of electrons taken from glucose during its break down. Without O_2, very little energy can be gained from glucose because the highest energy-yielding steps of the process are absent. CO_2 is released during respiration.

Igor Liberti CC 4.0

William Trubridge ascends from Dean's Blue Hole in the Bahamas

1. How long can you hold your breath? (don't turn purple!) _____

2. What makes you want to breathe? _____
   _____
   _____

3. Suggest why humans can't or don't need to hold their breath for very long: _____
   _____
   _____
   _____

4. Why is oxygen needed for respiration, and why would training to relax extend your ability to hold your breath?
   _____
   _____
   _____
   _____

©2018 **BIOZONE** International
ISBN: 978-1-927309-55-1
Photocopying Prohibited

EM   LS2.B   LS1.C

EXPLORE: Oxygen and respiration

▶ Respiration is the process by which cells convert energy in glucose to usable energy, which is stored in the molecule ATP. The process uses oxygen and produces carbon dioxide and water.

▶ The experiment below shows how the amount of oxygen used during respiration can be measured:

Measuring respiration with a simple respirometer

▶ Soda lime or potassium hydroxide is placed into the chamber to absorb any CO_2 produced during respiration. The respirometer measures O_2 consumption.

▶ Respiring organisms (e.g. germinating seeds) are placed into the chamber and the screw clip is closed. The position of the colored bubble is measured (time zero reading).

▶ The colored bubble in the capillary tube moves in response to the change in O_2 consumption. Measuring the movement of the bubble (e.g. with a ruler or taped graph paper) allows the change in volume of gas to be estimated.

▶ Changes in temperature or atmospheric pressure may change the readings and give a false measure of respiration when using a simple respirometer.

▶ **Differential respirometers** (not shown) use two connected chambers (a control chamber with no organisms and a test chamber). Changes in temperature or atmospheric pressure act equally on both chambers. Observed changes are therefore only due to respiration.

A respirometer measures the amount of oxygen consumed and the amount of carbon dioxide produced during cellular respiration. The diagram below shows a **simple respirometer**.

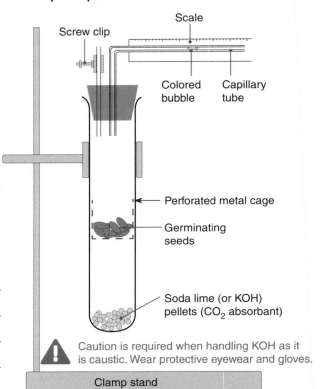

Caution is required when handling KOH as it is caustic. Wear protective eyewear and gloves.

5. Why does the bubble in the capillary tube move?

_____

_____

_____

_____

_____

6. A student used a simple respirometer (like the one above) to measure respiration in maggots. Their results are presented in the table (right). The maggots were left to acclimatize for 10 minutes before the experiment was started.

(a) Calculate the rate of respiration and record this in the table. The first two calculations have been done for you.

NEED HELP?
See Activity 91

(b) Plot the rate of respiration on the grid, below right.

(c) Describe the results in your plot: _____

_____

_____

_____

_____

Time (minutes)	Distance bubble moved (mm)	Rate (mm/min)
0	0	–
5	25	5
10	65	
15	95	
20	130	
25	160	

(d) Why was there an acclimatization period before the experiment began?

_____

_____

7. Why would it have been better to use a differential respirometer? _____

_____

_____

_____

©2018 **BIOZONE** International
ISBN: 978-1-927309-55-1
Photocopying Prohibited

EXPLORE: Cellular respiration

▶ Cellular respiration can be divided into four major steps, each with its own set of chemical reactions. Every step, except the link reaction, produces ATP. The four steps are: glycolysis, the link reaction, the Krebs cycle, and the electron transport chain (ETC). The last three are carried out in the mitochondrion, pictured below.

▶ The ETC uses the energy in electrons to move H^+ ions from the matrix to the intermembrane space (darker pink area below). The H^+ ions flow back into the matrix via the enzyme ATP synthase, with uses this flow to produce ATP.

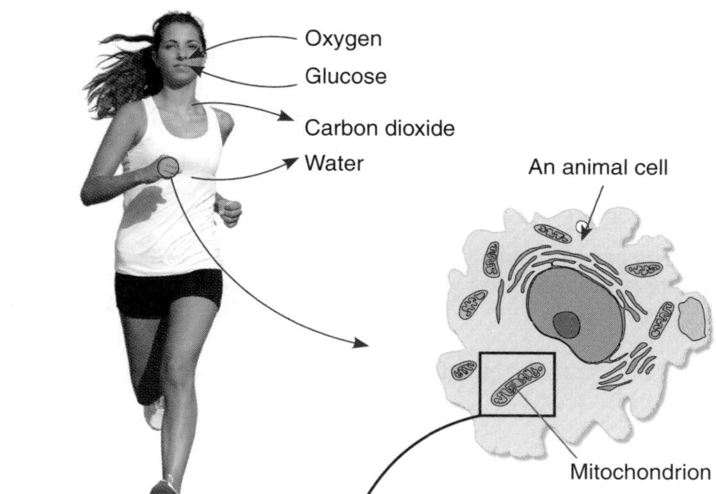

Oxygen
Glucose
Carbon dioxide
Water

An animal cell

Mitochondrion

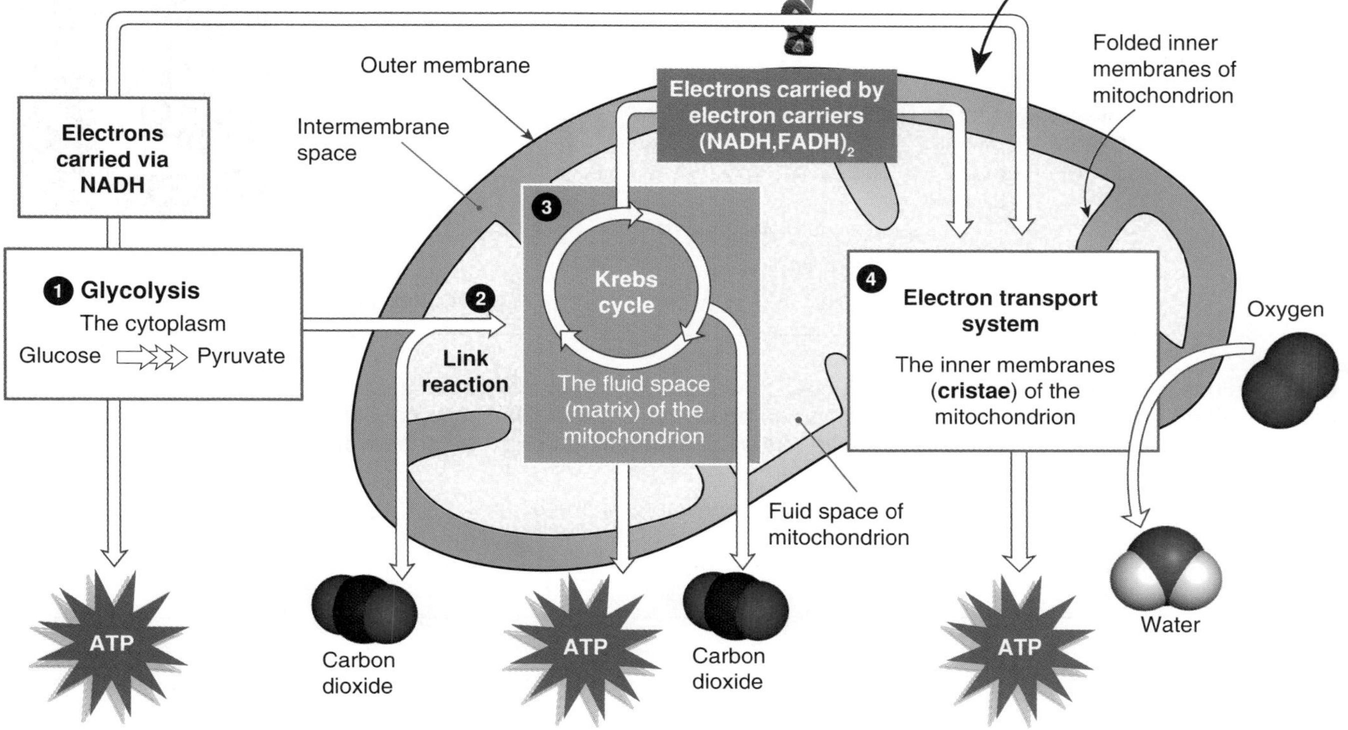

Folded inner membranes of mitochondrion

Outer membrane

Intermembrane space

Electrons carried by electron carriers (NADH,FADH)$_2$

Electrons carried via NADH

❶ Glycolysis
The cytoplasm
Glucose ⟹ Pyruvate

❷ Link reaction

❸ Krebs cycle
The fluid space (matrix) of the mitochondrion

❹ Electron transport system
The inner membranes (**cristae**) of the mitochondrion

Oxygen

ATP

Carbon dioxide

ATP

Carbon dioxide

Fuid space of mitochondrion

ATP

Water

8. (a) Write the word equation for aerobic cellular respiration: _____

(b) Write the chemical equation for aerobic cellular respiration: _____

9. Which of the four steps in cellular respiration yield ATP? _____

10. Which of the four steps of cellular respiration occur in the mitochondria? _____

11. (a) What are the main reactants (inputs) for cellular respiration? _____

(b) What are the main products (outputs) for cellular respiration? _____

12. How are electrons carried from glycolysis and the Krebs cycle to the electron transport chain? _____

13. How do animals obtain glucose? _____

©2018 **BIOZONE** International
ISBN: 978-1-927309-55-1
Photocopying Prohibited

EXPLAIN: How does cellular respiration provide energy?

▸ A molecule's energy is contained in the electrons within the molecule's chemical bonds. During a chemical reaction, energy (e.g. heat) can break the bonds of the reactants.

▸ When the reactants form products, the new bonds within the product will contain electrons with less energy, making the bonds more stable. The difference in energy is usually lost as heat. However, some of the energy can be captured to do work.

▸ Glucose contains 16 kJ of energy per gram (2870 kJ/mol). The step-wise breakdown of glucose through a series of chemical reactions yields ATP. A maximum of 32 ATP molecules can be produced from 1 glucose molecule.

A model for ATP production and energy transfer from glucose

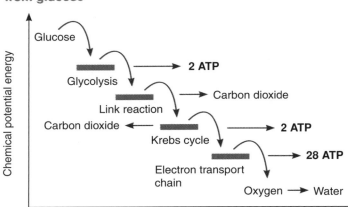

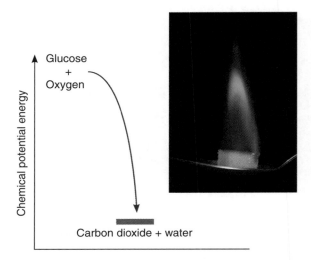

Energy released during the conversion of glucose to water and carbon dioxide can be captured by coupling it to reactions that produce ATP is successive steps (left above). In a direct combustion of glucose to water and carbon dioxide the same amount of energy is released, but all of it as heat and light (right above).

14. Explain how the energy in glucose is converted to useful energy in the body: _____

_____

_____

_____

_____

_____

15. One mole of glucose contains 2870 kJ of energy. The hydrolysis of one mole of ATP releases 30.7 kJ of energy. Calculate the percentage of energy that is transformed to useful energy in the body. Show your working.

_____

_____

_____

_____

_____

16. In what way is cellular respiration like the combustion of glucose, and in what way is it different? _____

_____

_____

_____

_____

_____

©2018 **BIOZONE** International
ISBN: 978-1-927309-55-1
Photocopying Prohibited

ELABORATE: Aerobic and anaerobic pathways for ATP production

In most energy-yielding pathways the initial source of chemical energy is glucose. The first step, glycolysis, is an almost universal pathway. The paths differ in what happens after glucose has been converted to pyruvate.

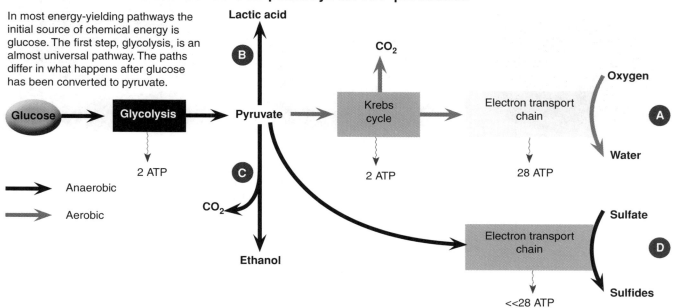

Legend:
→ Anaerobic
→ Aerobic

Lactic acid

B

CO₂

Krebs cycle

Oxygen

Electron transport chain

A

Water

Glucose → Glycolysis → Pyruvate

2 ATP

C

CO₂

Ethanol

2 ATP

28 ATP

Sulfate

Electron transport chain

D

Sulfides

<<28 ATP

A Aerobic respiration

Aerobic respiration produces the energy (as ATP) needed for metabolism. The rate of aerobic respiration is limited by the amount of oxygen available. In animals and plants, most of the time the oxygen supply is sufficient to maintain aerobic metabolism. Aerobic respiration produces a high yield of ATP per molecule of glucose (path A).

General equation

$$C_6H_{12}O_6 + 6O_2 \rightarrow 6CO_2 + 6H_2O$$

ATP yield: high (30-32 ATP)

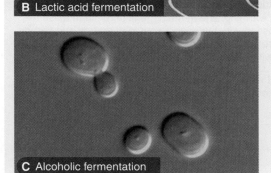

B Lactic acid fermentation

During maximum physical activity, when oxygen is limited, anaerobic metabolism provides ATP for working muscle. In mammalian muscle, metabolism of a respiratory intermediate produces lactate, which provides fuel for working muscle and produces a low yield of ATP. This process is called **lactic acid fermentation** (path B).

General equation

$$C_6H_{12}O_6 \rightarrow CH_3COCOOH + NADH$$

$$\leftrightarrow CH_3CHOHCOO^- + H+ + NAD^-$$

ATP yield: low (2 ATP)

The process of brewing utilizes the anaerobic metabolism of yeasts. Brewer's yeasts preferentially use anaerobic metabolism in the presence of excess sugars. This process, called **alcoholic fermentation**, produces ethanol and CO_2 from the respiratory intermediate pyruvate. It is carried out in vats that prevent entry of O_2 (path C).

General equation

$$C_6H_{12}O_6 \rightarrow CH_3CHO + CO_2$$

$$\rightarrow CH_3CH_2OH$$

ATP yield: low (2 ATP)

C Alcoholic fermentation

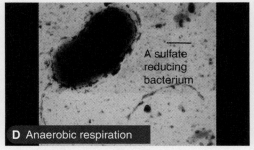

A sulfate reducing bacterium

D Anaerobic respiration

Many bacteria and archaea (ancient prokaryotes) are anaerobic, using molecules other than oxygen (e.g. nitrate or sulfate) as a terminal electron acceptor of their electron transport chain. These electron acceptors are not as efficient as oxygen (they have a lower reduction potential than oxygen so less energy is released per oxidized molecule) so the ATP yield from anaerobic respiration is generally low relative to aerobic metabolism (path D).

General equation (for a sulfur bacterium)

$$SO_4^{2-} \rightarrow \text{adenosine 5'}$$
$$\text{phosphosulfate} \rightarrow SO_3^{2-} \rightarrow H_2S$$

ATP yield: lower than aerobic respiration

©2018 **BIOZONE** International
ISBN: 978-1-927309-55-1
Photocopying Prohibited

Alcohol content 0%

Alcohol content 5%

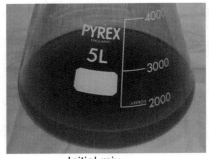

Foam

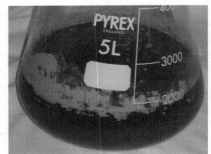

| Initial mix | after 4 days | After 7 days |

17. Making home brew beer is a simple process of adding a large amount of glucose (in brewing this is as D-glucose, called dextrose) to malt extract (containing the flavors and more complex sugars) and water, then adding a small amount of brewer's yeast. Look carefully at the photos above and explain what has happened at each stage based on your knowledge of cellular respiration and fermentation.

_____

_____

_____

_____

_____

ELABORATE: Anaerobic respiration in methane-producing bacteria

▸ Anaerobic systems use a molecule other than oxygen as the terminal electron acceptor.

▸ There are many different anaerobic pathways that living organisms use to produce usable energy.

▸ **Methanogenesis** (below) is a form of anaerobic respiration found in methane-producing bacteria in the stomach of ruminants (e.g. sheep and cows). The terminal electron acceptor in methanogenesis is not oxygen, but carbon and the pathways are complex involving many intermediates. The intermediate substances of methanogenesis (volatile fatty acids such as acetic acid) are an important energy source for the ruminant.

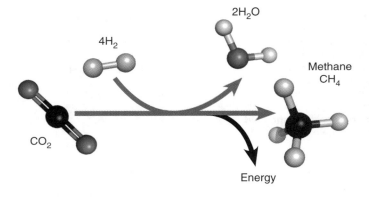

$2H_2O$

$4H_2$

Methane CH_4

CO_2

Energy

CSIRO cc3.0

Methane is a potent greenhouse gas and contributes to global warming. The methanogens in ruminants use the cellulose in vegetation as a source of energy and carbon. The methane generated is breathed out. These sheep are fitted with devices to measure the exhaled methane.

18. It has be found the chemical 3-nitrooxypropanol is able to inhibit the final step in methane production in methane producing bacteria. When fed to a ruminant, methane production is reduced and the ruminant gains weight.

(a) What is the effect of 3-nitrooxypropanol on methanogenesis?_____

_____

(b) Suggest why this causes a weight gain in the ruminant: _____

_____

_____

(c) Suggest how this might affect the way carbon cycles between the atmosphere and biosphere:

_____

_____

_____

©2018 **BIOZONE** International
ISBN: 978-1-927309-55-1
Photocopying Prohibited

25 Modeling Photosynthesis and Cellular Respiration

▶ The diagram below shows a stylized plant cell and the connections between photosynthesis and cellular respiration.

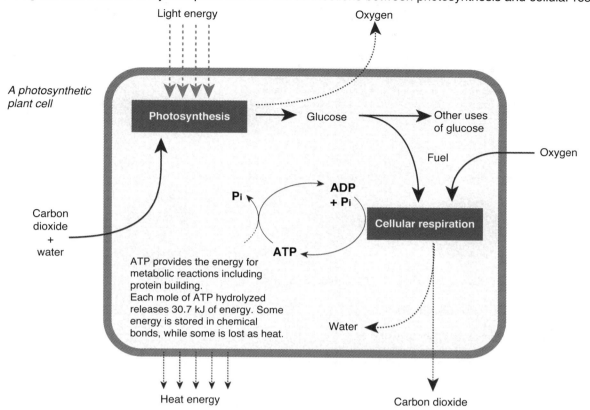

1. Circle the steps that would be missing if this was an animal and not a plant cell.

2. In what way are the processes pictured (photosynthesis and cellular respiration) opposites? _____

_____

_____

3. What is the ultimate source of energy in this system? _____

4. Complete the schematic diagram of the transfer of energy and the production of macromolecules below using the following word list: *water, ADP, protein, carbon dioxide, amino acid, glucose, ATP.*

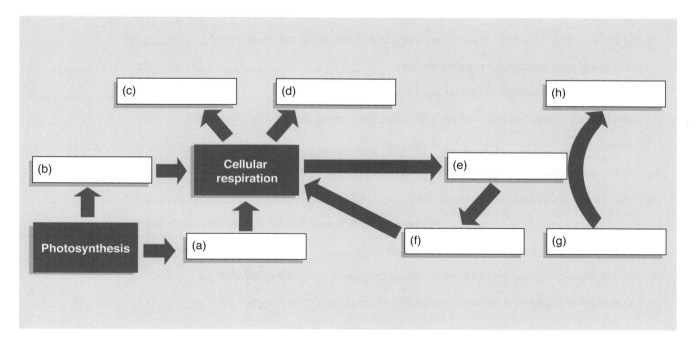

©2018 **BIOZONE** International
ISBN: 978-1-927309-55-1
Photocopying Prohibited

▸ During photosynthesis and cellular respiration, molecules are broken down and recombined to form new molecules.

▸ In this activity you will model the inputs and outputs of each of these processes using the atoms (carbon, hydrogen, and oxygen) on the next page. We have placed the atoms in boxes to make it easier to cut them out.

▸ At the end of this activity you will be able to see how the reactants (starting molecules) are recombined to form the final products.

Note: You can either work by yourself or team up with a partner. If you have beads or molecular models you could use these instead of the shapes on the next page.

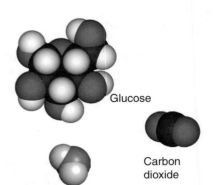

Glucose

Carbon dioxide

Water

5. Cut out the atoms and shapes on the following page. They are color coded as follows:

Carbon   **Hydrogen**   **Oxygen**

6. Rewrite the equation for **photosynthesis** here: _____

(a) State the starting reactants in photosynthesis: _____

(b) State the total number of atoms of each type needed to make the starting reactants:

Carbon: _____ Hydrogen: _____ Oxygen: _____

(c) Use the atoms you have cut out to make the starting reactants in photosynthesis.

(d) State the end products of photosynthesis: _____

(e) State the total number atoms of each type needed to make the end products of photosynthesis:

Carbon: _____ Hydrogen: _____ Oxygen: _____

(f) Use the atoms you have cut out to make the end products of photosynthesis.

(g) What do you notice about the number of C, H, and O atoms on each side of the photosynthesis equation? _____

_____

_____

(h) Name the energy source for this process and add it to the model you have made: _____

7. Rewrite the equation for **cellular respiration** here: _____

(a) State the starting reactants in cellular respiration: _____

(b) State the total number of atoms of each type needed to make the starting reactants:

Carbon: _____ Hydrogen: _____ Oxygen: _____

(c) Use the atoms you have cut out to make the starting reactants in cellular respiration.

(d) State the end products of cellular respiration: _____

(e) State the total number of atoms of each type needed to make the end products of cellular respiration:

Carbon: _____ Hydrogen: _____ Oxygen: _____

(f) Use the atoms you have cut out to make the end products of cellular respiration.

(g) Name the end products of cellular respiration that are utilized in photosynthesis: _____

©2018 **BIOZONE** International
ISBN: 978-1-927309-55-1
Photocopying Prohibited

| Sunlight energy | ⟶ | + | + | ATP |

©2018 **BIOZONE** International
ISBN: 978-1-927309-55-1
Photocopying Prohibited

This page is left blank deliberately

©2018 **BIOZONE** International
ISBN: 978-1-927309-55-1
Photocopying Prohibited

26 | Putting Biological Processes To Work

Treating sewage

▶ Sewage treatment is an important process, especially in large cities. Sewage contains organic waste, debris, and contaminants that must be removed before the water can be returned to the environment. Failure to treat sewage properly can lead to pollution of waterways and spread of disease.

▶ Sewage treatment plants typically treat sewage in three steps. Primary treatment involves removing debris in a pre-settling basin. Secondary treatment uses biological activity to remove contaminants from the liquid waste. Tertiary treatment usually involves a filtration and often chemical treatment of the water to remove any remaining contaminants. The diagram below shows a typical schematic for a treatment plant.

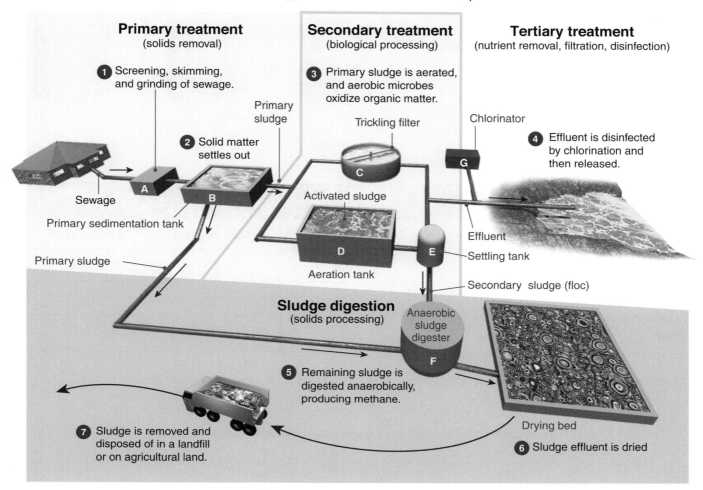

Primary treatment (solids removal)

1. Screening, skimming, and grinding of sewage.
2. Solid matter settles out

Primary sludge

Sewage

Primary sedimentation tank

A

B

Primary sludge

Secondary treatment (biological processing)

3. Primary sludge is aerated, and aerobic microbes oxidize organic matter.

Trickling filter

C

Activated sludge

D

Aeration tank

E

Tertiary treatment (nutrient removal, filtration, disinfection)

Chlorinator

G

4. Effluent is disinfected by chlorination and then released.

Effluent

Settling tank

Secondary sludge (floc)

Sludge digestion (solids processing)

Anaerobic sludge digester

F

5. Remaining sludge is digested anaerobically, producing methane.

7. Sludge is removed and disposed of in a landfill or on agricultural land.

Drying bed

6. Sludge effluent is dried

Modeling sewage treatment

Some students prepared a model of sewage in the following way. Two tea bags were cut open and the contents placed into 500 mL of hot water then left to cool to room temperature. Into this mixture was added 30 mL of cooking oil, 10 g of sucrose, and 10 g of coarse sand.

1. What do the following represent in the sewage model?

(a) Tea leaves: _____

(b) Tannins from tea leaves: _____

(c) Oil: _____

(d) Sugar: _____

(e) Sand: _____

An important part of treating sewage and waste water is the removal of solids and immiscible liquids (liquids that don't mix with water). These can be removed by physical methods (e.g. settling or scraping).

2. Which of the contents in the model sewage can be removed by physical methods: _____

_____

©2018 BIOZONE International
ISBN: 978-1-927309-55-1
Photocopying Prohibited

EM   ETS1.B   LS2.B

The second important part of treating sewage or waste water is removing dissolved nutrients. These are usually removed by microbes. This can be done aerobically in aeration tanks, or anaerobically in tanks that exclude oxygen.

3. Which of the contents in the model sewage can be removed by microbes? _____

_____

Some students decided to test ways to improve the removal of dissolved nutrients from sewage. *Brettanomyces* yeast, which respires aerobically, was used to represent the microbes in the sewage treatment process. Glucose was used to represent the dissolved nutrients. To test the effectiveness of their treatment they tested for the presence of glucose using glucose test paper.

The students tested various starting concentrations of yeast and increasing or decreasing the aeration in each mixture:

Each solution started with a glucose concentration of 100 g/L and was left for 48 hours in a water bath at 32°C before testing.

The results are shown below:

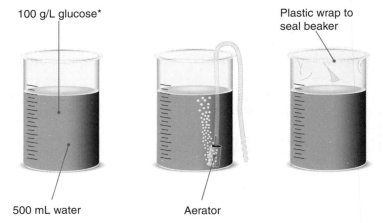

	Final glucose concentration (g/L)			
	2 mL of yeast	4 mL of yeast	6 mL of yeast	8 mL of yeast
No aeration (open)	20	15	10	6
Aerated	3	2	1.2	0.5
Sealed	34	30	20	15

4. (a) Which treatment was most effective at increasing the removal of glucose (dissolved nutrient)?

_____

   (b) Which letter(s) in the diagram on the previous page represents this step in sewage treatment?

_____

   (c) How could the most effective result above be replicated in a treatment plant? Explain your reasoning:

_____

_____

_____

_____

_____

5. The students decided to repeat the experiment using *Saccharomyces* yeast, which preferentially uses anaerobic metabolism in the presence of non-limiting glucose. Predict the outcome of the experiment and suggest which part of the sewage treatment process this may represent and its implications in it.

_____

_____

_____

_____

_____

©2018 **BIOZONE** International
ISBN: 978-1-927309-55-1
Photocopying Prohibited

27 How Old is the Earth?

ENGAGE: Lyell, Etna, some shellfish, and the age of the Earth

Mt Etna with Monti Rossi in middle ground.

The age of the Earth is currently accepted as around 4.5 billion years old. This was not always the way. Until the Renaissance period, starting around 500 years ago, the Earth was accepted as being quite young (anywhere between 7500 years old to 6000 years old). However discoveries over the last 500 years, and especially in the last 200 years, have produced evidence that the Earth and all the other planets and the Sun are very old.

In the summer of 1828, Charles Lyell travelled through France and Italy studying volcanoes. When Lyell reached Mount Etna in Sicily he observed numerous phenomena that could only be explained if the Earth was much older than people thought. Firstly, he noticed that Mt Etna was built up from many small cones formed by past eruptions, but only one of these cones, Monti Rossi (46 m high), had actually newly appeared in the last two hundred or so years. Given that Mt. Etna was more than 3000 m high this meant Mt Etna must have been tens if not hundreds of thousands of years old.

Lyell also discovered that shellfish sold in the fish markets of nearby towns were identical to fossil shellfish he found in limestone beds at the base of Mt. Etna. Importantly, Mt Etna appeared to be *on top of* the limestone beds. So if Mt Etna was hundreds of thousands of years old, the fossil shellfish must have been much older than that. This all suggested things happened very slowly, on what is now called a geologic time scale. If things such as the building of volcanoes and fossilization of shells happened so slowly, then the Earth must have been at least tens of millions to hundreds of millions of years old. Much older than had been accepted.

Excellent abbreviated accounts of Lyell's studies on Etna can be found in *Four Centuries of Geological Travel: The Search for Knowledge on Foot* ..., edited by Patrick Wyse Jackson, and *The Day the Universe Changed*, by James Burke (episode/chapter 8).

1. Identify a geologic formation near where you live or that you have visited, e.g. a volcano, canyon, cutting, or cliff with visible layers (strata).

   (a) How old is the formation? (You may need to research this): _____

   (b) Is there any evidence that the formation has changed significantly in the past? For example, part of the structure may have collapsed, or there may be old lava flows nearby or evidence of old river beds, etc.

   _____

   _____

   (c) Has the formation changed in any way since you became aware of it? If so, did the change produce any noticeable effect on the formation, i.e. is it taller, shorter, or deeper, or has it changed in its orientation?

   _____

   _____

   _____

   (d) What does the change (or no change) you noted in (c) tell you about how long the formation took to appear or be produced? Does this tell you anything about the age of formations in the surrounding area or indeed the world?

   _____

   _____

   _____

©2018 **BIOZONE** International
ISBN: 978-1-927309-55-1
Photocopying Prohibited

SC EM ESS1.C

EXPLORE: Relative and absolute dating

Working out how old something is can be done in two ways. The first is by ranking an object's age compared to something else. This is called **relative dating**, e.g. A is older than B but is younger than C. It does not place a firm time or date on when a object was formed but relies on the **Principle of Superposition**, which states that in undisturbed strata, the oldest layers are always at the bottom. You can explore this further in the next chapter. The second method is **absolute dating**. This method uses a known measurement such as the rate at which a radioactive element decays into another. Absolute dating, as its name implies, is able to produce definitive times and dates for the formation of an object (within a known range based on the limits of the technique), e.g. hair from a frozen mammoth was radiocarbon dated to 9500 years ago ± 100 years.

One of the most common methods of absolute dating is by measuring the proportions of different elements in the object (e.g. fossil) or the rock in which it is found. This method uses the fact that the atoms of some elements are unstable and break down (**decay**) into other specific types of atoms over time at a rate known as a **half-life**. A half-life is the time taken for half the original number of atoms to decay into atoms of a different element. Using this method, the Earth has been dated at 4.5 billion years old.

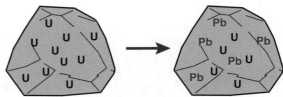

Unstable atoms, e.g. uranium change into stable atoms, e.g. lead, following a long but predictable decay series involving many radioactive elements.

Modeling half-lives

Half-lives can be modeled using a known number of M&M's® (e.g. 100).

Place the M&M's® in a lidded container and shake them up. Pour them out onto a plate and count, record and remove the M&M's® with the M upper most. These represent decayed atoms. Replace the rest of the M&M's® back into the container and repeat the procedure. Do this until there are no more M&M's®.

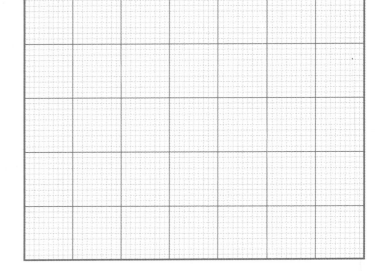

2. (a) Plot a line graph of the decay of M&M's® with the number of M&M's® remaining on the y axis, and the number of half-lives (throws) on the x axis. Join the points to form an approximate decay curve.
    Note: this experiment only approximates the spontaneous decay of atoms. The results obtained are slightly lower than occurs in reality. Can you explain why?

    (b) Compare your decay curve with other groups. Is it the same or different? Why?

    _____

    _____

    _____

    (c) Imagine you left your bag of 100 M&M's® with a friend who, after a while, became bored and decided to eat them using the procedure above. Every 5 minutes from when they started your friend carried out the procedure. When you came back there 6 M&M's® left.

    How long before your return did your friend start eating the M&M's®?

    _____

    _____

3. Study the image of rock strata, right. Using relative dating techniques:

    (a) Which layer was laid down first? _____

    (b) How old is sediment layer B? _____

    (c) How old is sediment D? _____

    (d) How old is fossil H? _____

    (e) Which layer was laid down last? _____

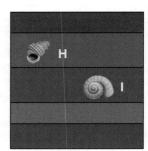

©2018 **BIOZONE** International
ISBN: 978-1-927309-55-1
Photocopying Prohibited

EXPLAIN: Dating the Earth

The very oldest material dated on Earth is a zircon crystal found in a metamorphosed sandstone from Western Australia. It is dated to 4.4 billion years old, just a 100 million years after the Earth formed. Zircon crystals often contain trace amounts of uranium because the zirconium and uranium atoms have similar electron structures and so can be interchanged in the crystal structure. The uranium in the crystal spontaneously decays to lead over time, so can be radiometrically dated.

The usefulness of radioisotopes for radiometric dating depends on the length of the half life and the material being dated. For example carbon isotope ^{14}C has a half life of 5740 years, and can only be used to date organic material up to 50,000 years old.

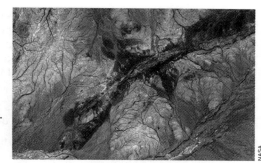

The Jack Hills formation (Australia) where the oldest minerals on Earth have been found.

Potassium-argon dating is a commonly used radiometric dating technique. Potassium decays into calcium 89% of the time and argon about 11% of the time. Argon is not contained in most minerals. Its mass in minerals is therefore related to the decay of potassium.

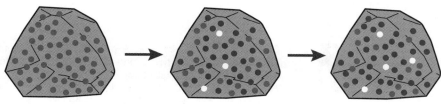

100 Potassium atoms
Time = 0 years

50 Potassium atoms
45 Calcium atoms
5 Argon atoms
Time = 1.28 billion years

25 Potassium atoms
68 Calcium atoms
7 Argon atoms
Time = 2.56 billion years

Evidence from meteorites

▶ Meteorites are pieces of rocky debris that have fallen to Earth and survived impact. Because meteorites originate from material that formed when the solar system formed they can be used to estimate the age of the solar system and therefore Earth.

▶ Meteorites can be dated using the ratios of lead-206, 207, and 208 to lead-204. Lead 206, 207, and 208 are all formed from the radioactive decay of uranium or thorium. Lead 204 is primordial, i.e. it is not formed from radioactive decay. Thus the ratio of lead-206, 207, or 208 to lead 204 can tell us how long ago the meteorite formed (for example, the ratio of lead-206 to lead-204 increases with time).

The Hoba meteorite, the largest known meteorite, weighs 60 tonnes and is 2.7 meters across.

Evidence from moon rocks

▶ Moon rocks here on Earth come from lunar meteorites and samples collected by the manned American Apollo and unmanned Soviet Luna missions. The Apollo missions brought back a total of 380.96 kg of moon rock from the various landing sites on the moon.

▶ The Moon has little active geology so most of the surface rocks are almost as old as the Moon. The Moon was formed very soon after the Earth. Radiometric dating places the age of moon rocks around 4.5 billion to about 3.16 billion years old.

4. How does potassium-argon dating help provide dates in rocks? _____

_____

_____

5. Why are objects, such as meteorites, useful for dating the age of the Earth? _____

_____

_____

6. Explain how dating rocks from the Moon can help in dating the age of the Earth: _____

_____

_____

_____

©2018 **BIOZONE** International
ISBN: 978-1-927309-55-1
Photocopying Prohibited

ELABORATE: Determining the age of rocks and fossils

When volcanic rocks cool, the elements in them form crystals. These crystals have known ratios of elements. Sometimes radioactive elements are incorporated into the crystals (e.g. uranium being incorporated into zircon crystals). When these radioactive elements decay, they produce specific daughter elements (e.g. uranium 238 produces lead 206). From the ratio of parent to daughter elements, the number of half lives can determined and therefore the age of the crystals and rock.

▶ The decay of uranium 238 (U-238) passes through a series of 15 radioactive elements that eventually ends at lead 206 (Pb-206). For simplicity it is said that uranium decays to lead with a half life of 4.5 billion years.

▶ For any number of half-lives (t) the proportion of radioactive atoms left (r) is equal to 0.5^t. The proportion of radioactive atoms left gives us the number of half-lives passed as $t = -1.4427 \, ln(r)$. Multiplying the number of half lives by the length of a half life gives us the number of years passed. You will need a scientific calculator.

▶ Sediments are difficult to date directly because sediments contain fragments of rock formed at earlier dates. However sediments can be dated when found between layers of volcanic material.

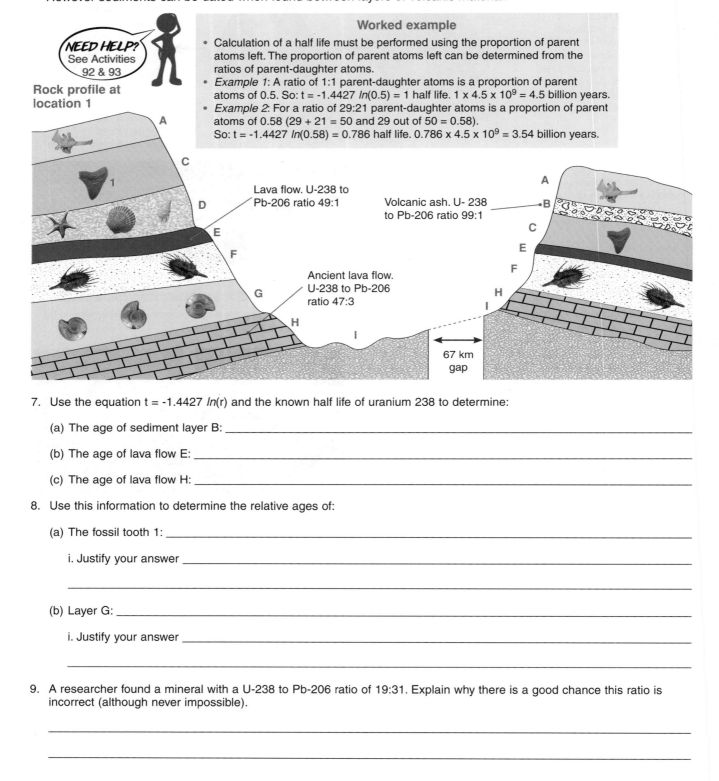

NEED HELP?
See Activities
92 & 93

Rock profile at location 1

Worked example

• Calculation of a half life must be performed using the proportion of parent atoms left. The proportion of parent atoms left can be determined from the ratios of parent-daughter atoms.
• *Example 1*: A ratio of 1:1 parent-daughter atoms is a proportion of parent atoms of 0.5. So: $t = -1.4427 \, ln(0.5) = 1$ half life. $1 \times 4.5 \times 10^9 = 4.5$ billion years.
• *Example 2*: For a ratio of 29:21 parent-daughter atoms is a proportion of parent atoms of 0.58 (29 + 21 = 50 and 29 out of 50 = 0.58). So: $t = -1.4427 \, ln(0.58) = 0.786$ half life. $0.786 \times 4.5 \times 10^9 = 3.54$ billion years.

Lava flow. U-238 to Pb-206 ratio 49:1

Volcanic ash. U- 238 to Pb-206 ratio 99:1

Ancient lava flow. U-238 to Pb-206 ratio 47:3

67 km gap

7. Use the equation $t = -1.4427 \, ln(r)$ and the known half life of uranium 238 to determine:

(a) The age of sediment layer B: _____

(b) The age of lava flow E: _____

(c) The age of lava flow H: _____

8. Use this information to determine the relative ages of:

(a) The fossil tooth 1: _____

   i. Justify your answer _____

   _____

(b) Layer G: _____

   i. Justify your answer _____

   _____

9. A researcher found a mineral with a U-238 to Pb-206 ratio of 19:31. Explain why there is a good chance this ratio is incorrect (although never impossible).

_____

_____

©2018 **BIOZONE** International
ISBN: 978-1-927309-55-1
Photocopying Prohibited

28 The Coevolution of Earth's Systems

ENGAGE: The Earth's lungs

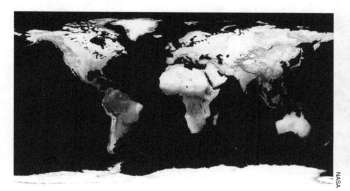 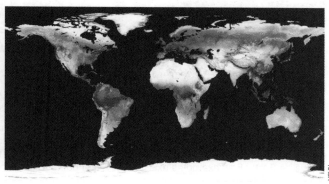

The images above are based on the amount of carbon dioxide (CO_2) over the land during the Northern Hemisphere winter (left) and summer (right). This can be translated into plant productivity (darker green = higher productivity). The majority of the land on Earth is situated in the Northern Hemisphere. Across North America and northern Eurasia there are vast tracts of conifer and deciduous forest. During winter, plants in these forests carry out little or no significant photosynthesis. During the spring the plants begin photosynthesizing again, using CO_2 to produce glucose.

1. What happens to the plant productivity in summer? _____

2. What happens to the plant productivity in the winter? _____

3. Suggest the effect of this changing productivity on the Earth's atmosphere: _____

_____

_____

EXPLORE: Present day seasonal atmospheric changes

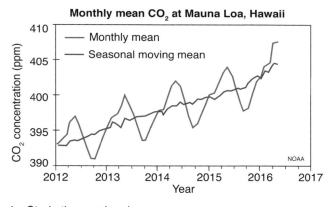 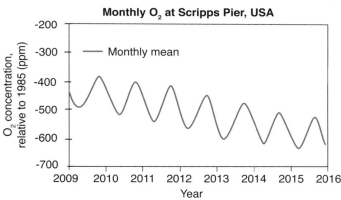

4. Study the graphs above:

   (a) What is the general trend of carbon dioxide concentration in the atmosphere over time? _____

   _____

   (b) What is the general trend of oxygen concentration in the atmosphere over time? _____

   _____

   (c) What pattern is common to the shape of these two graphs? _____

   _____

   (d) Explain how this pattern relates to photosynthesis and respiration in the biosphere: _____

   _____

   _____

   _____

©2018 **BIOZONE** International
ISBN: 978-1-927309-55-1
Photocopying Prohibited

SC   SSM   ESS2.E

EXPLORE: Photosynthesis, iron, and a snowball

Stromatolites (such as the ones above from Shark Bay, Western Australia), represent some of the most ancient living things on Earth. Few examples exist today, but fossil remains can be dated back to 3.7 billion years ago (bya). Stromatolites are rock-like structures formed from the build-up of sediment by microbes, particularly cyanobacteria (photosynthetic bacteria). Cyanobacteria flourished in

shallow waters throughout the world until ~1 bya after which their abundance and diversity declined. They were among the first bacteria to split water to produce oxygen during photosynthesis.
Banded iron formations (above) are mostly dated between 2.4 and 1.9 billion years old. They formed from the reaction of dissolved iron in the sea with oxygen to form iron oxides, which then settled out to

form vast tracts of sediments.
The Earth's early atmosphere was filled with methane, a potent greenhouse gas. Oxygen in the atmosphere reacted with CH_4 to produce CO_2, a much less potent greenhouse gas. This caused a decrease in the Earth's temperature, triggering the Huronian glaciation, a snowball Earth that lasted for 300 million years, from 2.4 to 2.1 billion years ago.

The rise of free oxygen

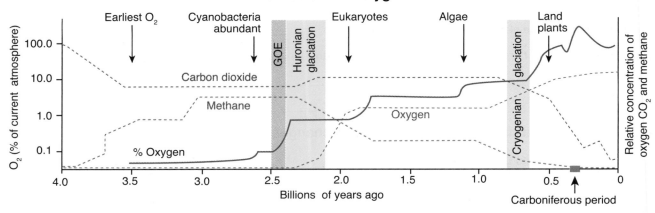

Atmospheric oxygen was very scarce for the first two billion years of Earth's history. During a time called the Great Oxygenation Event, oxygen content in the air began to rise. Oxygen levels in the atmosphere rose gradually over the next billion years, but soon after the Cryogenian glaciation oxygen levels quickly reached then surpassed today's level.

5. (a) What was the Great Oxygenation Event? _____

_____

_____

(b) Describe the evidence in the geological record for the occurrence of this event and its likely cause: _____

_____

_____

_____

_____

_____

_____

6. Use the graph above to explain the evidence for the events triggering the Huronian glaciation. _____

_____

_____

_____

©2018 **BIOZONE** International
ISBN: 978-1-927309-55-1
Photocopying Prohibited

EXPLORE: Swamps, coal, and more snow

During the Carboniferous period (358.9 to 298.9 million years ago) vast tracts of shallow-rooted swamp forests covered the land. The trees in these Carboniferous forests were high in lignin, a fiber largely resistant to the decomposers present at the time. When trees in the swamp forests died, they were buried without decompostion and the carbon they contained accumulated at rates faster than it could be recycled back to the atmosphere, becoming preserved as the coal deposits we see today. At the same time as atmospheric carbon was being removed, atmospheric oxygen rose to a peak of 35%. Around 300 mya, an extinction event called the Carboniferous Rainforest Collapse (**CRC**) saw the decline and fragmentation of the vast forests as the climate become drier and cooler.

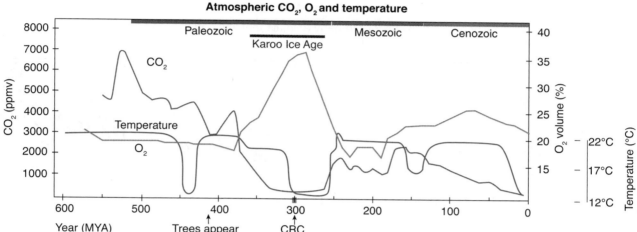

Atmospheric CO_2, O_2 and temperature

7. (a) When in Earth's history did oxygen peak and carbon dioxide and temperature both fall, all at the same time?

    _____

    (b) Describe the relationship between these atmospheric changes, events in the biosphere, and the formation of coal:

    _____

    _____

    _____

    _____

    _____

    _____

    _____

    _____

    (c) Carbon dioxide is a greenhouse gas, meaning it contributes to the ability of the atmosphere to capture and recycle energy emitted by Earth's surface, keeping it at a temperature capable of supporting life. Use the plot above to construct an explanation for the cooling of the planet 360-260 mya during the Karoo Ice Age.

    _____

    _____

8. The column chart right shows the extent of coal reserves in each geologic period.

    (a) Use the information on this page to account for the lack of coal deposits prior to the Carboniferous:

    _____

    _____

    (b) Coal is a fossil fuel. Suggest why it is called this and why it is regarded as a non-renewable resource:

    _____

    _____

    _____

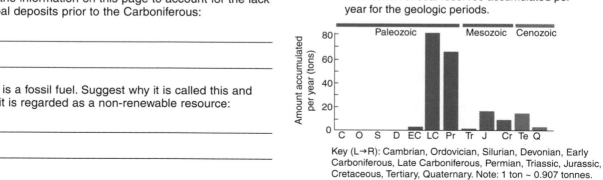

Distribution of coal reserves accumulated per year for the geologic periods.

Key (L→R): Cambrian, Ordovician, Silurian, Devonian, Early Carboniferous, Late Carboniferous, Permian, Triassic, Jurassic, Cretaceous, Tertiary, Quaternary. Note: 1 ton ~ 0.907 tonnes.

©2018 **BIOZONE** International
ISBN: 978-1-927309-55-1
Photocopying Prohibited

EXPLORE: Life makes soil

▸ Soils are a complex mixture of loosely arranged weathered rock and organic material. Before microbes invaded the land, the Earth had no soil as we know it now, but simply rock and earth produced by weathering and erosion. Rising oxygen levels after the GOE increased weathering rates and produced oxidized minerals.

▸ Soils as we think of them today didn't begin to appear until about 500 mya. Microorganisms used chemicals in the rocks to build their bodies and when they died they added their organic matter to the developing soils.

▸ Once soils developed, the first land plants could establish. This resulted in further development of the soils through biological weathering and the addition of dead material to accumulate organic matter. The rapid proliferation of plant life on land increased biological weathering and soil development, and removed carbon from the atmosphere.

▸ The developing soils provided new habitats for the early animals that invaded the land. They too helped soil development by churning the soil and adding organic matter.

1 **Disintegrating parent rock** — **Bedrock**

2 **Weathered parent rock** — **Bedrock**

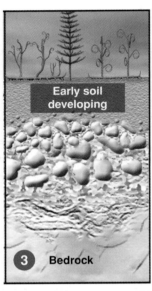

3 **Early soil developing** — **Bedrock**

4 **Developed soil** — Subsoil — **Bedrock**

The parent rock is broken down by weathering. Some minerals are available, others are locked up in rocks.

Microbes invade the land and make minerals locked in rocks available. Organic material from microbes builds up.

Early land plants appear ~470 mya, adding organic matter to soils. Microbes, including fungi, help plants access minerals.

Soils became more complex, supporting an increased plant diversity. Plants became larger and more complex.

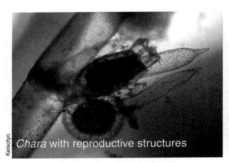

Chara with reproductive structures

Land plants evolved from a group of green algae, perhaps as early as 850 mya, similar to this Charophyte. These filamentous algae would have lived in shallow, seasonally dry pools.

The earliest land plants were small and flat similar to modern day liverworts (above). They produce tough spores called cryptospores, preserved as microfossils in Ordovician rocks ~450 mya.

The Devonian period (419.2-358.9 mya), saw a rapid diversification of plant life on land, including plants with vascular tissue (e.g. roots). These plants formed the vast forests that would form coal.

9. How does weathering contribute to soil formation? _____

_____

_____

10. How did invasion of the land by plants increase weathering and, in turn, increase the suitability of the land for plant life?

_____

_____

_____

©2018 **BIOZONE** International
ISBN: 978-1-927309-55-1
Photocopying Prohibited

EXPLORE: Interacting spheres

While much is known about the factors contributing to major events in Earth's history, these did not act in isolation. Instead, complex interactions and feedback loops between atmospheric conditions, biospheric activity, and geological processes were involved in shaping the history of Earth's spheres. For example, the Great Oxygenation Event triggered an increase in mineral diversity, with many elements appearing as oxidized weathering products at the surface. These transformations made elements locked up in rocks available to organisms, driving an increase in biological activity and further mineral diversity. Such is the interplay between the atmosphere, the Earth's crust, and life, that researchers at the Carnegie Institution Geophysical Laboratory documented a history of mineral diversity on Earth (below), linking ~70% of the Earth's minerals to the biological activity associated with changing atmospheric conditions. Clay minerals for example were formed by the weathering of rock, first by microbial mats and then by the root systems of plants.

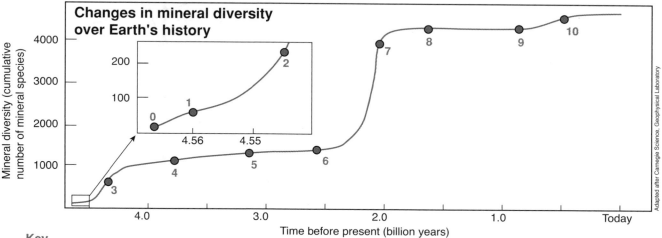

Key

1. Minerals in chondrite meteorites.
2. Planetary accretion ~ 250 minerals.
3. Earth differentiates into core, mantle and crust.
4. Granite forms the cores of continents.
5. Global scale plate tectonics.
6. Biological processes begin to affect surface mineralogy.
7. Rise in mineral diversity associated with biological changes in the atmosphere and oceans.
8. Formation of banded iron formations ceases. Oceans gradually become more oxygenated.
9. Snowball Earth events. Ice was Earth's most abundant surface mineral.
10. Biological innovations such as shells, teeth, and bones, and the rise of the terrestrial biosphere.

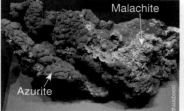

The copper(II) carbonate minerals azurite and malachite only form in an oxygen-rich environment.

Biominerals include calcium carbonate in shells and the eyes of this fossil trilobite…

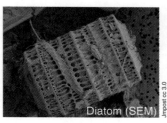

…silica in the cell walls of diatoms and calcium phosphate in bones and teeth.

11. (a) Use the information in this activity to annotate the graph above with significant events in the atmosphere, biosphere, and geosphere. You may work in pairs or small groups if you wish.

(b) What evidence is there for the interdependence of changes in the atmosphere, the biosphere, and geosphere? Use extra paper for your discussion if you wish and attach it to this page.

_____

_____

_____

_____

_____

_____

_____

©2018 **BIOZONE** International
ISBN: 978-1-927309-55-1
Photocopying Prohibited

EXPLAIN: The coevolution of Earth's systems

▶ Life began on Earth about 3.8 billion years ago. Life takes resources from its surroundings, modifies them, and then uses them to replicate itself. In doing this, life modifies the environment, either by modifying chemicals and structures in the environment directly or by adding chemicals from waste or through decay after death.

▶ The modification of the environment produces new conditions for the evolution of life and as life evolves so does the environment, forcing more change upon life.

▶ This reciprocal influence is termed **coevolution** and it can be observed between the biosphere and Earth's other systems. For example, Earth's current atmosphere is the result of activity by early photosynthetic organisms, which produced oxygen as a waste product of oxygenic photosynthesis. The presence of oxygen caused the extinction of many anaerobic bacteria but eventually led to the rise of multicellular life forms.

▶ The diagram right illustrates key stages of the coevolution of life and Earth's systems showing how changes in one influenced changes in the other.

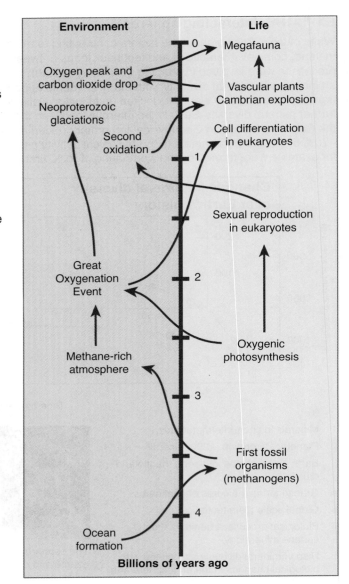

Mangroves have evolved to inhabit estuarine environments. They modify the estuary by collecting sediment and thus the environment evolves into a terrestrial instead of a marine one.

▶ The rise of free oxygen in the atmosphere had a profound effect on life. A 2009 study led by paleobiologist Jonathan L. Payne found that there were correlations between the evolution of eukaryotes and multicellular life, increase in body size, and the rise of oxygen in the atmosphere (right). The body size of organisms increased in two sharp jumps: at the evolution of eukaryotes and the evolution of multicellular organisms. Both these events correlate with sharp increases in the amount of available oxygen.

▶ Although the biodiversity of this planet has not increased at a steady rate since the evolution of life it has nevertheless become greater over time. Biodiversity is closely related to the environment. As biodiversity has increased over time, it has affected the environment and produced new conditions for organisms to exploit. One example is cyanobacteria producing the first oxygen atmosphere, which allowed the rise of eukaryotes. Another example is microbes in the ground, which helped to produce the first soils and enabled plants to colonize the land.

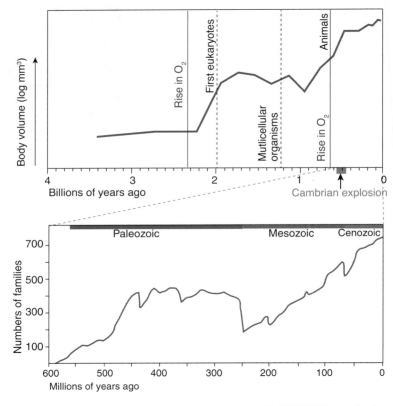

©2018 **BIOZONE** International
ISBN: **978-1-927309-55-1**
Photocopying Prohibited

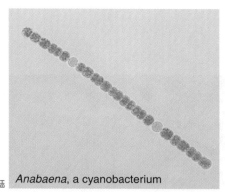

Multicellularity (being made of many cells) has evolved independently in eukaryotes at least 46 times and also in some prokaryotes such as cyanobacteria (left). One hypothesis for its origin is that it was a result of unicellular organisms (bacteria) failing to separate after dividing. Some cells can then take on different tasks, e.g. nitrogen fixation or reproduction, to the benefit of the colony as a whole. This mechanism has been observed in modern prokaryotes, such as the filamentous cyanobacteria, and in 16 different protistan phyla (e.g. *Volvox*, right), providing support for the hypothesis.

Anabaena, a cyanobacterium

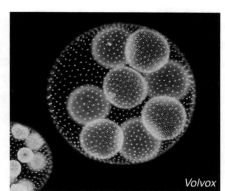

Volvox

Very high oxygen levels during the Carboniferous allowed invertebrates, including giant dragonflies and millipedes, to easily obtain the oxygen they needed to support a very large size.

The impact of a giant asteroid 65 mya produced a nuclear winter that marked the Cretaceous-Paleogene extinction event that ended the reign of the dinosaurs. In their place, mammals diversified.

A period of global warming in the Eocene, 53 to 47 mya, saw a rise in the biodiversity of mammals in North America. The number of genera increased to a record of 104 and many new plant forms also appeared.

12. Using the evidence in this activity and your understanding of photosynthesis and cellular respiration (including energy yields), present an argument (a reason or set of reasons) for the coevolution of life and Earth's other systems:

_____

_____

_____

_____

_____

_____

_____

_____

_____

_____

_____

_____

_____

_____

_____

_____

_____

_____

_____

_____

29 Carbon Cycling

ENGAGE: Biosphere 2 and its carbon problem

Biosphere 2 was built as a self contained ecosystem as a possible model of a future off-world colony. As such, it was designed to have its own nutrient cycles, including carbon, oxygen, and nitrogen cycles. However, almost as soon as the facility was sealed, the carbon dioxide level in the atmosphere began rising from 350 ppm to eventually 4000 ppm. At the same time oxygen levels were plummeting.

The problem with modeling a complex system is that it can be difficult to know just how complex it is and therefore difficult to predict the effect of small changes to the system. When building Biosphere 2, enriched soil was brought in to help the plants grow. It was assumed that the plants would absorb CO_2 and produce O_2. Instead, microbes in the soil quickly metabolized the nutrients, using up the O_2 and increasing the CO_2 levels. Interestingly, calculations showed that the CO_2 level should have been ten times greater at 40,000 ppm. Eventually, it was realized that the facility's concrete was acting as a carbon sink, soaking up the excess carbon dioxide and forming calcium carbonate.

1. (a) Biosphere 2's carbon cycle was out of balance. What was the immediate cause of the imbalance?

    _____

    _____

    (b) What was the effect of the imbalance? _____

    _____

    _____

    (c) Sinks are able to moderate chemical cycles by absorbing excess chemicals and storing them or releasing them slowly. What acted as a moderating sink in Biosphere 2?

    _____

EXPLORE: Carbon on Earth

Carbon is the fourth most abundant element in the universe and the 15th most abundant element in the Earth's crust. Its ability to form four stable covalent bonds with both other carbon atoms and other elements (especially non-metals such as oxygen) allows carbon to be the center for an almost unlimited number of molecules. The number of carbon based (organic) substances far outnumbers the number of inorganic (non carbon based) substances. On Earth, carbon is found bound in numerous different forms in all of the Earth's spheres.

Pure carbon is found in four different forms, which are all rare in nature. Carbon atoms in a tetrahedral shape form diamond, an extraordinarily hard substance. Other forms of pure carbon include graphite, amorphous carbon (e.g. charcoal and coal), and fullerenes.

Carbon is the basis of life. It is the "backbone" of molecules such as fats, proteins, and carbohydrates. The four covalent bonds of carbon allow it to form bonds with a wide range of elements, producing molecules that perform the biochemical reactions that sustain life.

On Earth, the vast majority of carbon is carried in rocks and in the mantle. Rocks in the Earth's crust that carry carbon include carbonates (e.g. limestone, above) formed by inorganic processes, and coal (and oil) formed from organic origins (burial of organic matter).

       LS2.B ESS2.D CE SSM EM

©2018 **BIOZONE** International
ISBN: 978-1-927309-55-1
Photocopying Prohibited

Only a small amount of the carbon on Earth is held in the atmosphere, as carbon dioxide (CO_2). CO_2 is transparent to incoming solar radiation, but absorbs radiation reflected from the Earth's surface. Some of this radiation is released back to the Earth's surface, causing the atmosphere to warm.

The oceans hold a huge amount of dissolved carbon as carbonate (CO_3^{2-}) and bicarbonate (HCO_3^{2-}) ions. Carbonate ions are available to living organisms to use, such as in the building of shells. Bicarbonate ions form when CO_2 dissolves in water. Bicarbonate ions are unavailable to livings organisms.

Carbon cycles between these reservoirs (also called sinks). For example, respiration moves carbon from organisms to the atmosphere. The speed at which carbon passes between reservoirs depends on the reservoirs involved. Carbon may remain in rocks for millions of years, but cycles out of the atmosphere after about ten years.

2. Identify four carbon reservoirs mentioned above and what form of carbon is found in them:

(a) _____

(b) _____

(c) _____

(d) _____

3. How might carbon move between each of these sinks (a)-(d) above?

(a) _____

_____

(b) _____

_____

(c) _____

_____

(d) _____

_____

EXPLAIN: The cycling of carbon

It is estimated there is over 1.5×10^{15} tonnes of carbon in the Earth. Most of this is in the mantle and may come to the surface during a volcanic eruption. Other reservoirs in the crust, atmosphere, and in the oceans hold small amounts of carbon in comparison, but they are significant in terms of their ability to move between reservoirs.

Reservoir	Amount in reservoir (gigatonnes (Gt))	Fluxes
Atmosphere	840	Movements included absorption and release by the ocean, and living organisms. Combustion of fossil fuels is adding carbon dioxide to the atmosphere.
Biomass	2610 (610 in plants and animals, ~ 2000 in soil)	Approximately 123 Gt per year is absorbed by photosynthesis, with 120 Gt per year being released by respiration.
Ocean	41,000	About 92 Gt per year is absorbed but 90 Gt per year is released.
Sedimentary rocks	60 million	Exchanges between rocks and other reservoirs and generally very small, in the order of 0.5 Gt per year
Fossil fuels	10,000	About 9 Gt per year into atmosphere from combustion of fossil fuels.

©2018 **BIOZONE** International
ISBN: 978-1-927309-55-1
Photocopying Prohibited

4. (a) Name two processes that remove carbon from the atmosphere: _____

   (b) Name two processes that add carbon to the atmosphere: _____

5. Using the information in this activity complete the diagram below by adding arrows and labels to show how carbon cycles through the environment (including the size of the carbon flux). Write the size of the reservoir in the white boxes.

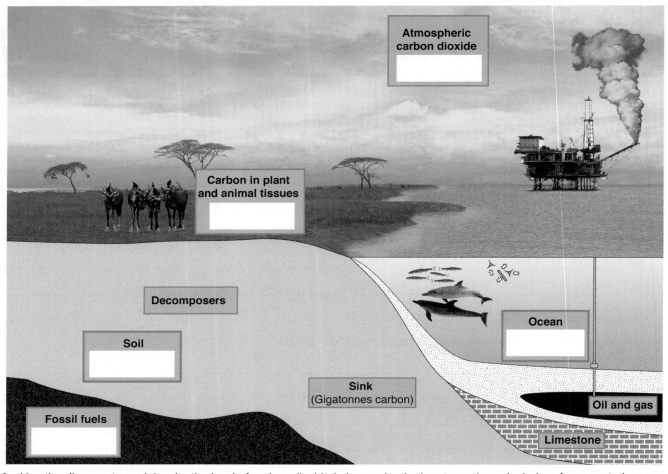

Atmospheric carbon dioxide

Carbon in plant and animal tissues

Decomposers

Soil

Ocean

Sink
(Gigatonnes carbon)

Fossil fuels

Oil and gas

Limestone

6. Use the diagram to explain why the level of carbon dioxide is increasing in the atmosphere. Include references to human activities and their effects on carbon reservoirs. Include any calculations to support your explanation:

_____

_____

_____

_____

_____

_____

_____

_____

_____

_____

_____

_____

_____

_____

_____

_____

_____

©2018 **BIOZONE** International
ISBN: 978-1-927309-55-1
Photocopying Prohibited

30 Modeling the Carbon Cycle

ENGAGE: Making a closed ecosystem

▶ In this activity, a simple model was used to mimic the carbon cycle in a small closed ecosystem.

▶ A large, clear soda bottle was used as the container. It was filled almost to the top with filtered pond water.

▶ Small pebbles or rocks were added to a depth of around 2-5 cm. A dead leaf was added as detritus.

▶ The living components of the system were added. (a few stems of an aquatic plant such as *Cabomba* and four small aquatic snails).

▶ The system was left to stand for a day before putting the lid on. The bottle was placed in an area that gets direct sunlight for a few hours each day.

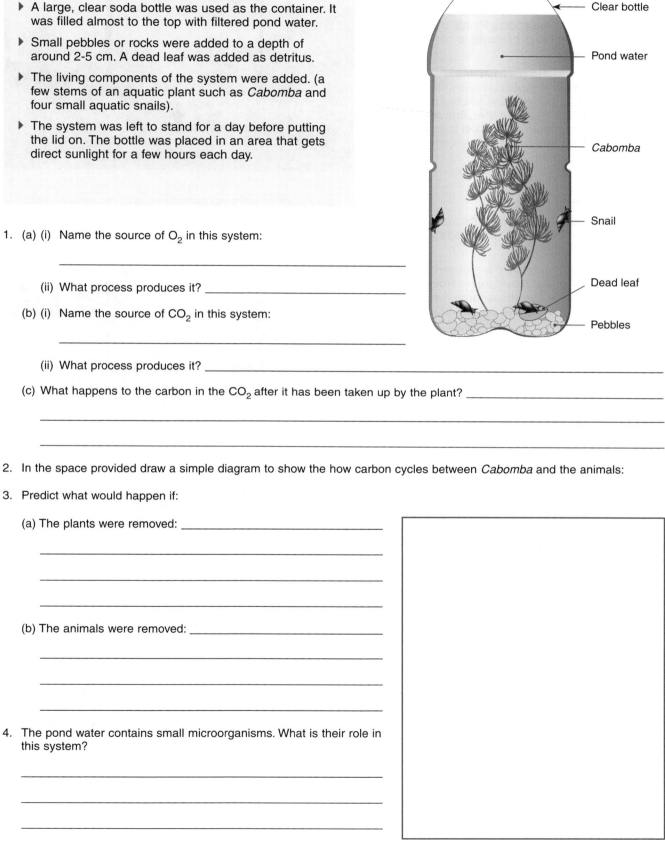

Air space
Clear bottle
Pond water
Cabomba
Snail
Dead leaf
Pebbles

1. (a) (i) Name the source of O_2 in this system:

    _____

    (ii) What process produces it? _____

    (b) (i) Name the source of CO_2 in this system:

    _____

    (ii) What process produces it? _____

    (c) What happens to the carbon in the CO_2 after it has been taken up by the plant? _____

    _____

    _____

2. In the space provided draw a simple diagram to show the how carbon cycles between *Cabomba* and the animals:

3. Predict what would happen if:

    (a) The plants were removed: _____

    _____

    _____

    _____

    (b) The animals were removed: _____

    _____

    _____

    _____

4. The pond water contains small microorganisms. What is their role in this system?

    _____

    _____

    _____

    _____

©2018 BIOZONE International
ISBN: 978-1-927309-55-1
Photocopying Prohibited

EM SSM CE ESS2.D LS2.B

EXPLAIN: Photosynthesis affects the carbon cycle

▶ The balance of photosynthesizing and respiring organisms can affect the amount of CO_2 in the atmosphere. If the biomass of photosynthesizing organisms vastly outweighs that of respiring organisms, CO_2 will be removed from the atmosphere and the carbon will be stored as biomass. Respiration returns carbon to the atmosphere.

Photosynthesis and carbon

▶ Photosynthesis removes carbon from the atmosphere by fixing the carbon in CO_2 into carbohydrate molecules. Plants use the carbohydrates (e.g. glucose) to build structures such as wood.

▶ Some carbon may be returned to the atmosphere during respiration (either from plants or from animals). If the amount or rate of carbon fixation is greater than that released during respiration then carbon will build up in the biosphere and be depleted in the atmosphere (diagram, right).

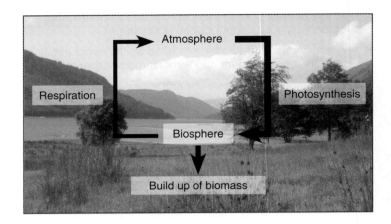

Respiration and carbon

▶ Cellular respiration releases carbon into the atmosphere as carbon dioxide as a result of the breakdown of glucose.

▶ If the rate of carbon release is greater than that fixed by photosynthesis then carbon may accumulate in the atmosphere over time (diagram bottom right). Before the Industrial Revolution, many thousands of gigatonnes (Gt) of carbon were contained in the biosphere of in the Earth's crust (e.g. coal).

▶ Deforestation and the burning of fossil fuels have increased the amount of carbon in the atmosphere.

Carbon cycling simulation

Plants move about 120 Gt of carbon from the atmosphere to the biosphere a year. Respiration accounts for about 60 Gt of carbon a year. A simulation was carried out to study the effect of varying the rates of respiration and photosynthesis on carbon deposition in the biosphere or atmosphere. To keep the simulation simple, only the effects to the atmosphere and biosphere were considered. Effects such as ocean deposition and deforestation were not studied. The results are shown in the tables right and below.

Table 1: Rate of photosynthesis equals the rate of cellular respiration.

Years	Gt carbon in biosphere	Gt carbon in atmosphere
0	610	600
20	608	600
40	608	600
60	609	598
80	612	598
100	610	596

Table 2: Rate of photosynthesis increases by 1 Gt per year.

Years	Gt carbon in biosphere	Gt carbon in atmosphere
0	610	600
20	632	580
40	651	558
60	671	538
80	691	518
100	710	498

Table 3: Rate of cellular respiration increases by 1 Gt per year.

Years	Gt carbon in biosphere	Gt carbon in atmosphere
0	610	600
20	590	619
40	570	641
60	548	664
80	528	686
100	509	703

©2018 **BIOZONE** International
ISBN: 978-1-927309-55-1
Photocopying Prohibited

5. Plot the data for tables 1,2, and 3 on the grid provided (above). Include a key and appropriate titles and axes.

6. (a) What is the effect of increasing the rate of photosynthesis on atmospheric carbon? _____

_____

(b) i. What is the effect of increasing the rate of photosynthesis on biospheric carbon? _____

_____

ii. How does this effect occur? _____

_____

_____

7. What is the effect of increasing the rate of cellular respiration on atmospheric and biospheric carbon? _____

_____

_____

8. In the real world, respiration is not necessarily increasing in comparison to photosynthesis and its effect is limited by CO_2 uptake by the oceans. However, many human activities cause the same effect as increased respiration.

(a) Name two human activities with the same effect on atmospheric carbon as increasing the rate of cellular respiration:

_____

(b) What effect does this extra atmospheric carbon have on the global climate? _____

_____

_____

©2018 **BIOZONE** International
ISBN: 978-1-927309-55-1
Photocopying Prohibited

31 How Carbon Dioxide Affects the Oceans

ENGAGE: Acids and bases

The pH scale (1 - 14) measures the concentration of hydrogen (H^+) ions in a solution. In a neutral solution the pH is 7, meaning the concentration of H^+ ions is 1×10^{-7} mol L^{-1} (the negative logarithm of which is 7). Acid substances have a pH below 7 and include fruit juice and vinegar, whereas basic substances have a pH above 7 and include baking soda and drain cleaner. Adding acids to bases (and vice versa) results in neutralization (the pH returning to 7).

The ocean is naturally slightly basic with a pH of about 8.1.

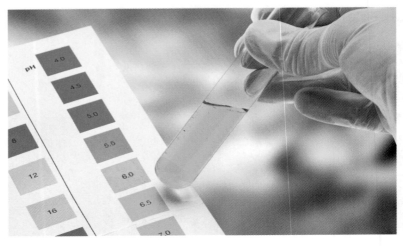

1. Write down some naturally occurring acidic substances you know of: _____

   _____

2. Write down some naturally occurring basic substances you know of: _____

   _____

EXPLORE: CO₂ and water

The images below show the effect of adding dry ice (solid CO_2) to a flask of water containing universal indicator. Universal indicator is blue in a basic solution, green in neutral, and red in acidic solutions. These products should be readily available in your school laboratory so you can repeat this demonstration for yourselves.

No solid CO₂ added   Solid CO₂ added

All photos Jobjabramon CC 4.0

3. (a) How does adding carbon dioxide affect the pH of the solution? _____

   _____

   (b) Why does this happen? _____

   _____

   _____

EXPLORE: pH and the ocean

The image below shows the change in pH of the ocean surface waters since 1700. Note areas with the most change.

pH is a logarithmic scale so even a small change in pH represents a large change in H^+ concentration. Some areas of the ocean (e.g. areas of increased human activity or underwater volcanic eruptions) are more affected by pH change than others. Temperate waters are also more affected than tropical waters because cold water absorbs more CO_2 than warm water.

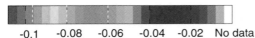

-0.1   -0.08   -0.06   -0.04   -0.02   No data
Change in ocean pH

Atlantic Ocean

Pacific Ocean

Indian Ocean

Change of -0.09 pH units

  LS2.A LS2.B CE

©2018 **BIOZONE** International
ISBN: 978-1-927309-55-1
Photocopying Prohibited

The pH of the oceans has fluctuated throughout geologic history but has always remained at around pH 8.1 - 8.2. Recent studies have measured current ocean pH at around 8.0. Projections show that if CO_2 continues to dissolve into the ocean at its current rate, the pH of the ocean could fall to 7.8 in a hundred years time.

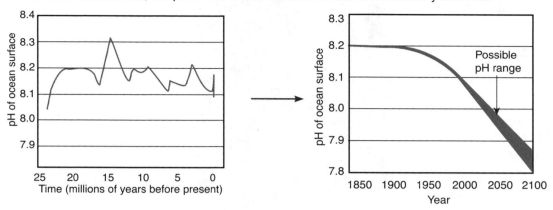

4. (a) Which part of the ocean has had the greatest change in pH? _____

_____

(b) What do you think might be the reason for this? _____

_____

_____

(c) How many units has the pH of the ocean reduced by since 1850? _____

EXPLAIN: CO_2 and shellfish

The decrease in pH of the oceans is often called "ocean acidification". This is a somewhat misleading term, as the oceans are not becoming acid. A better term might be "ocean neutralization", as the pH of the oceans moves closer to a pH of 7. It is often portrayed that the shells of marine organisms (including corals and planktonic pteropods) are "dissolved by acid" but this is also misleading. Shell-building organisms use carbonate ions from the sea to build their shells. As ocean pH falls, these ions in the seawater become locked up as bicarbonate ions. This makes it harder for many animals to build and maintain their shells, causing them to deform and slowly become thinner ("dissolve").

The oceans act as a carbon sink, absorbing much of the CO_2 produced from burning fossil fuels. When CO_2 reacts with water it forms carbonic acid (H_2CO_3). H_2CO_3 dissociates (separates) into HCO_3^- and H^+ ions. The increase in H^+ ions decreases the pH of the oceans. Naturally occurring carbonate ions (CO_3^{2-}) in the ocean waters react with the extra H^+ ions to form more HCO_3^- ions. This process lowers the CO_3^{2-} ions available to shell-making organisms. In addition, because H^+ interferes with carbonate availability, shell-forming organisms use proton pumps to pump H^+ out of their bodies. The more H^+ is in the water, the more energy they have to expend to obtain carbonate ions.

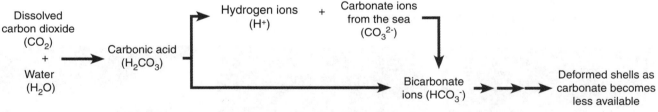

5. Explain how the burning of fossil fuels leads to the disintegration of the shells of shell-forming organisms:

_____

_____

_____

_____

_____

_____

_____

_____

©2018 **BIOZONE** International
ISBN: 978-1-927309-55-1
Photocopying Prohibited

32 Fossil Fuels and the Environment

ENGAGE: CO_2 and cars

The vast majority of cars use petroleum based fuels, either gasoline (petrol) or diesel. Some use CNG or LPG. The combustion of these fuels in the car's engine produces water and carbon dioxide (with some toxic gases such as carbon monoxide and nitrogen oxides, which are mostly removed by the car's catalytic converter).

The fuel used in these cars originated tens to hundreds of millions of years ago and contains carbon that was removed from the environment by burial. Returning it to the environment by combustion to form CO_2 has added a new "adding" pathway to the carbon cycle, which does not yet have a balancing "removal" pathway, resulting in a build up of CO_2 in the atmosphere.

CA EP&Cs III: Human practices can alter the cycles and processes the operate within natural systems. (III c)

Burning fossil fuels (oil, gas, coal) alters the carbon cycle.

▶ Fossil fuels contain carbon buried hundreds of millions of years ago. In the case of coal, the carbon was removed from the atmosphere by plants by photosynthesis.

▶ The conditions under which coal formed do not exist today.

▶ Burning fossil fuels returns carbon to the air, but at much faster rates than it was removed.

1. An average car uses about 45 liters of gasoline a week. This produces about 90 kg of CO_2 from the exhaust system.

   (a) If your household uses a car, find out how many liters of gasoline it uses per week: _____

   (b) i. One liter of gasoline produces about 2 kg of CO_2. How many kg of CO_2 does the car produce per week?

   _____

   ii. How many kg of CO_2 is produced per year? _____

   (c) A report in the journal *Nature* in 2009 stated that each metric tonne (1000 kg) of CO_2 released leads to an increase in atmospheric temperature of 1.5×10^{-12} °C. What temperature increase does the car in (a) and (b) produce in a year?

   _____

   (d) There are around 184 million private cars on the road in the USA. Assuming average car values:

   i. How many tonnes of CO_2 do they emit in a year in total? _____

   ii. How many °C does this CO_2 increase the atmospheric temperature by? _____

EXPLORE: Fossil fuels

Most of the world's energy for transport and domestic use comes from fossil fuels. These include coal, oil, and natural gas. In 2016 the world consumed 132,051 TWh (teraWatt hours) worth of fossil fuels.

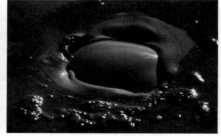

Fossil fuels are important for electricity production. They are an easy to store, cheap, convenient, high energy fuel. Coal is important as coke in steel production, while oil is also used to make plastics.

Oil is the source of products such as gasoline, diesel, and liquid fuels. These are important in the transport industry due to their energy density, transportability, and ease of use in engine systems.

Natural gas is becoming more important in generating electricity due to its "clean burning". This simply means it produces only CO_2 and H_2O during combustion, with no soot or carbon monoxide.

2. Fossil fuels contain hydrocarbons. When they burn in oxygen they produce carbon dioxide and water. The shorter the hydrocarbon chain, the more cleanly it tends to burn as each molecule more easily reacts with oxygen.

   Write a balanced chemical equation illustrating the statement above, using methane (CH_4) as the hydrocarbon:

   _____

   LS1.C EM SC

©2018 **BIOZONE** International
ISBN: 978-1-927309-55-1
Photocopying Prohibited

EXPLORE: Fossil fuels and electricity

3. The graph right shows the source of the electricity generated in California in August, 2017.

    (a) What percentage of electricity was generated from fossil fuels?

    _____

    (b) What percentage of electricity was generated from renewables?

    _____

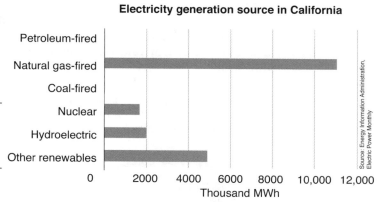

Electricity generation source in California

Petroleum-fired
Natural gas-fired
Coal-fired
Nuclear
Hydroelectric
Other renewables

0   2000   4000   6000   8000   10,000   12,000
Thousand MWh

Source: Energy Information Administration, Electric Power Monthly

EXPLORE: Fossil fuels are ancient

All fossil fuels are formed by similar processes but both the contributing material, duration and depth of burial, and extent of heat and pressure determine the final product.

Coal is formed from the remains of terrestrial plant material buried in vast shallow tropical swamps during the Carboniferous period (358.9 to 298.9 mya). Deep burial creates heat and pressure and over time transforms the plant layers into a hard black rock. The first stage of coal formation is peat (partly decomposed plant material). Deeper burial and increasing temperature and pressure produce lignite (brown coal), bituminous coal (soft black coal), and finally anthracite (hard black coal).

Oil and natural gas are formed from the remains of algae and zooplankton that settled to the bottom of shallow seas and lakes about the same time as the coal-forming swamps. These remains were buried and compressed under layers of non-porous sediment. The amount of pressure and heat, together with the composition of the biomass, determined if the material became oil or natural gas. Greater heat or mainly algal biomass produced natural gas. The process, although continuous, occurs so slowly that oil and natural gas (like coal) are essentially non-renewable.

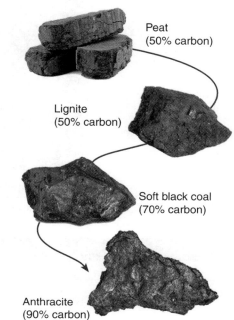

Peat
(50% carbon)

Lignite
(50% carbon)

Soft black coal
(70% carbon)

Anthracite
(90% carbon)

4. Working in pairs, devise sequences to show how (1) coal and (2) oil and natural gas are formed and how their formation differs. Draw your sequences below:

Coal

Oil and natural gas

©2018 **BIOZONE** International
ISBN: 978-1-927309-55-1
Photocopying Prohibited

EXPLAIN: Why use fossil fuels?

▸ The energy in a barrel of oil is approximately 6.1 gigajoules (about six times the amount of solar energy falling on an average house roof per day in winter). A tonne of coal contains about 31 gigajoules. The energy gained in the production of coal or oil or the **energy return on energy invested** (ERoEI) can be expressed as a ratio of energy expended to energy gained.

▸ In the early 1900s, the ratio for a barrel of oil was around 100:1 (100 barrels of oil were produced from 1 barrel of oil invested). The ratio has reduced over time to between 30:1 and 10:1, as the resources become increasingly difficult to harvest and process. Crude oil has a return of about 10:1, open pit oil about 7:1, and *in situ* (in place) oil extraction from oil sands returns about 5:1.

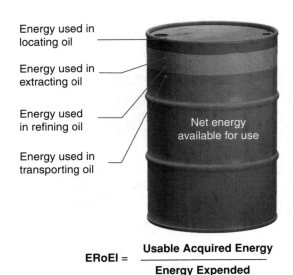

Energy used in locating oil

Energy used in extracting oil

Energy used in refining oil

Net energy available for use

Energy used in transporting oil

$$ERoEI = \frac{\text{Usable Acquired Energy}}{\text{Energy Expended}}$$

CA EP&Cs V: The spectrum of what is considered in making decisions about resources and natural systems and how those factors influence decisions (V a)

Is it worth it?

▸ The effort in and return from exploiting a resource must be considered. Put simply, "Is it worth it?"

▸ This question must consider not only the energy return but the economical, social, and environmental aspects of that return.

▸ Is it worth removing 100 hectares of forest for 10,000 barrels of oil? Does the type of forest matter? What value do we place on the services provided by the forest, including recreation and aesthetics, shelter, air and water purification, climate moderation, and protection from erosion.

Using oil	
Benefits	**Costs**
Large supply	Many reserves are offshore and difficult to extract
High net energy gain	High CO_2 production
Can be refined to produce many different fuel types	Potential for large environmental damage if spilled
Easy to transport	Rate of use will use up reserves in near future

Using coal	
Benefits	**Costs**
Huge supplies (billions of tonnes)	High CO_2 production when burned
High net energy yields.	High particle pollution from soot
Can be used to produce syngas and converted to other fuels.	Low grade coals produce high pollution and contribute to acid rain
Relatively easy to extract when close to the surface	High land disturbance through mining, including subsidence
Important in industry as coke (reducer)	Noise and dust pollution during mining

5. Use the information above to explain why coal, oil, and gas are the most commonly used fuels for both electricity production and as a transport fuel:

_____

_____

_____

_____

6. In groups, discuss the following scenario and evaluate the costs and benefits associated with it: "*In situ* extraction of oil uses enough clean burning natural gas in one day to heat 3 million homes. It has a ERoEI of 5:1, but produces large amounts greenhouse gases compared to other extraction methods". Summarize your discussion below, including reference to energy inputs and outputs, environmental issues, and pros and cons of energy intensive oil extraction.

_____

_____

_____

_____

©2018 **BIOZONE** International
ISBN: 978-1-927309-55-1
Photocopying Prohibited

EXPLAIN: The cost of extraction

▶ Because fossil fuels formed underground they require mining. As these resources become increasingly scarce, the mining and extraction processes have become more complex. As the complexity of the extraction increases so does the risk of environmental damage due to a failure in the process.

▶ When these extraction processes are near water there is potential for chemical leaks to rapidly spread, causing considerable harm.

▶ Oil extraction, for example, is associated with a number of environmental issues. Drilling for oil on land risks groundwater pollution. Oil spills from drilling offshore can affect vast areas of ocean and coastline. Mining oil sands and oil shales destroys thousands of hectares of forests.

CA EP&Cs II & IV:

Methods used to extract, harvest, transport, and consume natural resources influence the geographic extent, composition, biological diversity, and viability of natural systems (II b).

The effects of human activities on natural systems are directly related to the quantities of resources consumed and to the quantity and characteristics of the resulting products. (IV a)

To extract fossil fuels, the landscape must be modified.

▶ Mining requires large areas of land for tailings and overburden, and contaminates surface and groundwater.

▶ Some mines cover huge areas, causing major disruption to natural systems and landscapes. The world's largest coal mine, the North Antelope Rochelle Mine, in Wyoming, has already disturbed more than 17,000 hectares.

In situ extraction

In situ extraction is a method of removing oil from oil sands that are too deep to mine. It uses large amounts of energy and water. The water must be stored in tailings ponds for decontamination. Extraction produces up to three times as much CO_2 as the same quantity of conventional oil.

Offshore oil platform

Oil platforms disrupt the seabed when wellheads and pipelines are laid down. There is the potential for oil spills to affect large areas of the ocean. The flaring of the gases contributes to global warming.

Hydraulic fracturing well

In this method, oil-containing rock is fractured by high pressure fluid. The fracking fluid is mostly water with numerous additives to enhance oil mobility. The water must be stored and storage ponds may contaminate groundwater. Groundwater can be contaminated by additives and oil if the well casings are not sealed correctly or fissures link fractured layers to groundwater.

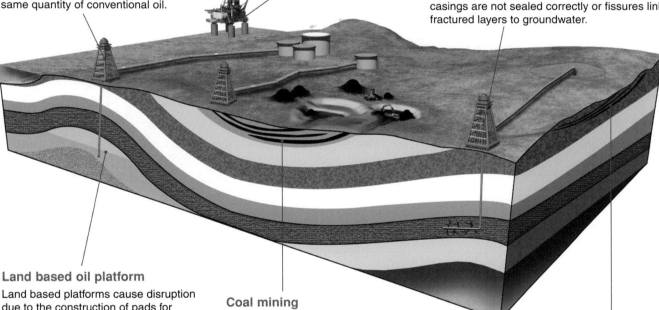

Land based oil platform

Land based platforms cause disruption due to the construction of pads for pumps, storage and pipelines. Runoff and leaks from wells can contaminate ground and surface water. Accidental release of air pollutants, such as methane, contributes to global warming. There is the possibility of land subsidence above oil or gas fields.

Coal mining

Coal mining produces both dust and noise pollution. Even high quality coal can contain many heavy metals, making its processing and use potentially damaging. When compared to even some of the most complex oil extraction techniques, coal use is many times more toxic.

Oil sands and oil shale mining

This requires the removal of vast tracts of forest. It produces large volumes of solid and liquid toxic tailings which occupy huge tracts of land. Leakages of tailings ponds contaminate water.

7. Explain why there is such potential for environmental damage from fossil fuel extraction when water is involved:

_____

_____

_____

_____

©2018 **BIOZONE** International
ISBN: 978-1-927309-55-1
Photocopying Prohibited

33 Revisiting Model Worlds

The anchoring phenomenon for this chapter was two model worlds and one real one. Now that you have completed this chapter you should be able to discuss the relationships between the biosphere, the atmosphere, and the geosphere in these worlds, and explain why life on Earth appears sustainable while the other two "worlds" are less so.

An EcoSphere®

Biosphere 2

The Earth

1. Discuss the theory behind the sustainability of the shrimp/algae EcoSphere®, and suggest why this may not be a fully sustainable system:

_____

_____

_____

_____

_____

_____

2. Discuss how Biosphere 2 is a scaled up version of the EcoSphere®, and why it too faced long term sustainability issues:

_____

_____

_____

_____

_____

3. Describe the interactions between the Earth's organisms, atmosphere, and geosphere. Explain how human activity may be altering the balance between these systems. You may use extra paper if you wish:

_____

_____

_____

_____

_____

_____

_____

_____

_____

©2018 **BIOZONE** International
ISBN: 978-1-927309-55-1
Photocopying Prohibited

34 Summative Assessment

1. Photosynthesis is the process in which carbon dioxide is fixed to form glucose. In the space below, draw a diagram to show the process of photosynthesis, including reactants and products and the location of the reactions involved:

2. Carbon dioxide is increasing in the atmosphere. This is causing a slow increase in the Earth's average surface temperature. The graphs below shows the effect of carbon dioxide concentration on plant growth, and the effect of temperature on the rate of photosynthesis.

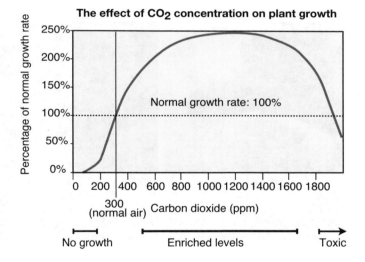

The effect of CO$_2$ concentration on plant growth

Percentage of normal growth rate

Normal growth rate: 100%

300 (normal air) Carbon dioxide (ppm)

No growth — Enriched levels — Toxic

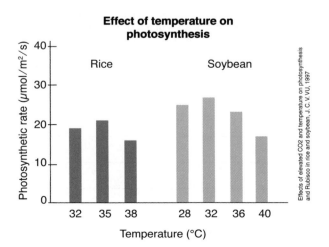

Effect of temperature on photosynthesis

Photosynthetic rate (μmol/m^2/s)

Rice   Soybean

Temperature (°C)

Effects of elevated CO2 and temperature on photosynthesis and Rubisco in rice and soybean, J. C. V. VU, 1997

Use the graphs to describe how plant growth may be affected by global warming caused by an increase in the level of CO$_2$ in the atmosphere.

_____

_____

_____

_____

_____

_____

_____

_____

_____

_____

_____

_____

_____

_____

_____

_____

_____

©2018 **BIOZONE** International
ISBN: 978-1-927309-55-1
Photocopying Prohibited

   SC EM SSM ESS2.D ESS1.C LS2.B LS1.C

3. Cellular respiration is a continuous, integrated process. A simple diagram of the process in a eukaryote is shown below.
   (a) In the diagram, fill in the rectangles with the process and the ovals with the substance used or produced. Use the following word list (some words can be used more than once): *pyruvate, glycolysis, glucose, oxygen (O₂), link reaction, electron transport chain (ETC), Krebs cycle, ATP, carbon dioxide (CO₂), water (H₂O)*
   (b) Add in a pathway to show fermentation. Write the two possible products of this pathway in eukaryotes.

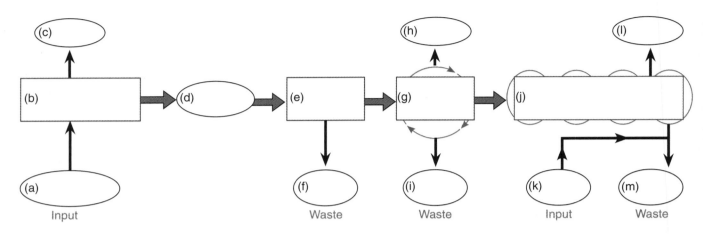

(c) Use the completed diagram to explain the difference in ATP yield between aerobic and anaerobic pathways:

_____

_____

_____

4. The three case studies below present some evidence for the fate of glucose in producer and consumer organisms. Use the information provided and your knowledge from the earlier activities to explain how the carbon, hydrogen, and oxygen from sugar molecules are used to make the many other molecules found in organisms:

Isotope experiments with plants

¹³C isotopes were used to trace the movement of glucose in plant leaves. It was found that some glucose is converted to fructose (an isomer of glucose in which the atoms are arranged into a 5-sided ring). Fructose molecules can be joined together with a terminal sucrose to be stored as fructan in plant vacuoles. Fructose is also added to glucose to produce sucrose. Sucrose is transported out of the leaf.

Isotope experiments with animals

Experiments using ¹³C isotopes to identify the fate of glucose in guinea pigs showed that 25% of the glucose intake was used as fuel for cellular respiration. The rest of the glucose was incorporated into proteins, fats, and glycogen. Corals live in a symbiotic relationship with algae, which transfer sugars to the coral in return for a safe environment to live. Experiments with ¹³C showed that the major molecule transferred to the coral was glucose.

_____

_____

_____

_____

_____

©2018 **BIOZONE** International
ISBN: 978-1-927309-55-1
Photocopying Prohibited

5. Some species of yeast preferentially use fermentation (an anaerobic process) to produce ATP when glucose is in excess, even in the presence of oxygen. They only use aerobic respiration when glucose concentrations are very low.

(a) Suggest why some yeast might preferentially use fermentation when glucose levels are high, but respire aerobically when glucose is limited:

_____

_____

_____

(b) The table below shows the oxygen (O_2) uptake and the carbon dioxide (CO_2) production of various yeast species and the ratio of fermented glucose to respired glucose.

Complete the table by writing the dominant ATP production method in the final column:

Organism	µl of gas per 10^7 cells per 10 min		Ratio fermented glucose: respired glucose	Respiration or fermentation dominant
	Respiration (O_2 uptake)	Fermentation (CO_2 production)		
Saccharomyces cerevisiae	4.8	78.0	49.0	
Saccharomyces chevalieri	1.2	90.8	250.0	
Saccharomyces fragilis	24.5	1.9	0.23	
Canidida tropicalis	27.7	0.9	0.1	
Debrayomyces globosus	12.3	22.2	5.4	
Torulopsis dattila	0.0	52.0	∞	
Torulopsis sphaerica	25.7	3.5	0.4	

The Crabtree Effect: A Regulatory System in Yeast. R. H. DE DEKEN, 1965

(c) The graph right shows the effect of glucose concentration on the biomass yield from aerobic cellular respiration in the yeast *Brettanomyces bruxellenisis*. What is the effect of glucose concentration >55 g/L on biomass yield from respiration in this yeast? Biomass yield is the ratio of biomass produced to substrate consumed (gram per gram).

_____

_____

(d) Explain the result: _____

_____

_____

_____

_____

_____

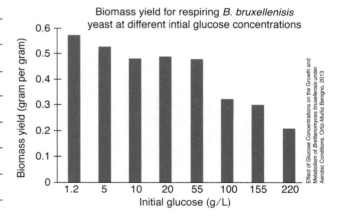

Biomass yield for respiring *B. bruxellenisis* yeast at different intial glucose concentrations

Effect of Glucose Concentrations on the Growth and Metabolism of Brettanomyces bruxellensis under Aerobic Conditions. Ortiz-Muñiz Benigno, 2013

6. The following two equations summarize important processes in the cycling of matter between the Earth's systems:

i) Denitrification (an anaerobic process carried out by bacteria): $2NO_3^- + 12H^+ \rightarrow N_2 + 6H_2O$

ii) Photosynthesis (anaerobic) and cellular respiration (aerobic): $6CO_2 + 12H_2O \rightleftarrows C_6H_{12}O_6 + 6O_2$

Use these equations to help you to explain how matter cycles and energy flows in the Earth's systems. You may use more paper if you wish:

_____

_____

_____

_____

©2018 BIOZONE International
ISBN: 978-1-927309-55-1
Photocopying Prohibited

7. Using evidence you have studied in this chapter, discuss the importance of photosynthesis and respiration in the evolution of the Earth's atmosphere:

_____

_____

_____

_____

_____

_____

_____

_____

_____

_____

_____

8. Uranium 238 has a half-life of approximately 4.5 billion years during which time it decays to lead 206. Under certain circumstances it can be used to date rocks and minerals.

   (a) Use the grid to draw a graph showing the decay curve of uranium 238 over four half-lives, assuming a starting radioactivity of 100%.

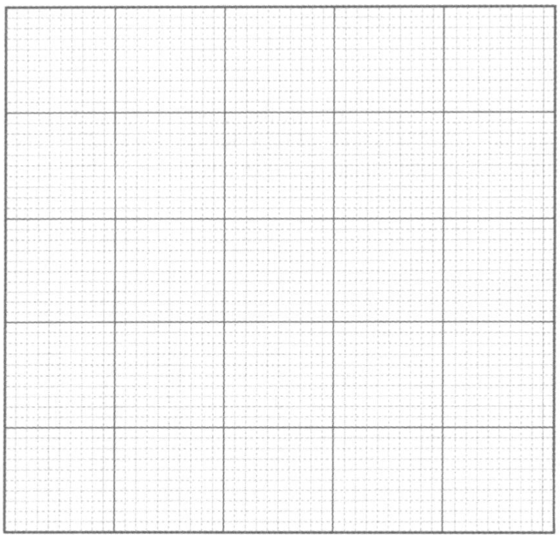

   (b) Explain how the radioactive decay of uranium (and other radioisotopes) can be used to date rock samples to determine the age of the Earth and the age of fossils in rock strata:

   _____

   _____

   _____

   _____

   _____

   _____

   _____

©2018 **BIOZONE** International
ISBN: 978-1-927309-55-1
Photocopying Prohibited

9. (a) Use the space below to complete a model to show how carbon cycles through the biosphere, geosphere, hydrosphere, and atmosphere. Include quantities where relevant:

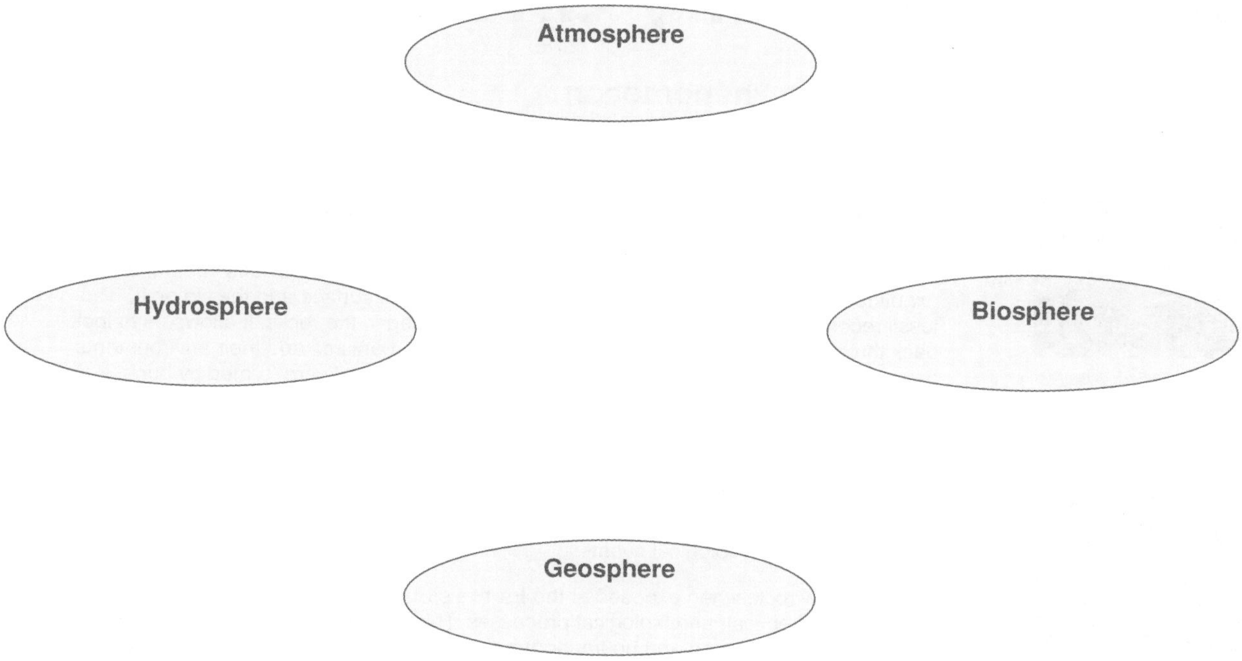

Atmosphere

Hydrosphere

Biosphere

Geosphere

(b) If you have not already done so, add arrows and labels to your model to indicate where humans intervene in the cycle. Use the model to discuss the effect of human use of fossil fuels on the carbon cycle:

_____

_____

_____

_____

_____

_____

_____

_____

_____

_____

_____

(c) Use the following equations to explain the effect of fossil fuel use on the pH of the oceans. Explain the impact of this on shell-building marine organisms. **Equation 1**: $CO_2 + H_2O \rightarrow H_2CO_3$ **Equation 2**: $H^+ + CO_3^{2-} \rightarrow HCO_3^-$

_____

_____

_____

_____

_____

_____

_____

©2018 **BIOZONE** International
ISBN: 978-1-927309-55-1
Photocopying Prohibited

Instructional Segment 3

Evidence of Common Ancestry and Diversity

Anchoring Phenomenon

The rise of the tyrants: Fossils and diversity over time.

35 46

How do layers of rock form and how do they contain fossils?

☐ 1 You have already explored aspects of the Earth's history in the previous chapter and examined the evidence for the coevolution of the Earth's surface and the life on it. The fossil record provides a history of life on Earth recorded in the rocks. It allows us to look back through time in order to better understand how organisms and their environments, have changed over time. Use a model to describe how fossils are formed by burial in an environment that prevents decay, how they become part of the geosphere, and how they provide a record of the physical and biotic environment of the time.

36

☐ 2 Recall how rock strata are dated by both absolute and relative dating methods. Apply basic principles of stratigraphy to interpret sedimentary rock sequences and explain how the ordering of layers and arrangement of materials can reveal clues about past environments and geological events.

36 37

☐ 3 Describe how rock, when exposed at the Earth's surface, is subjected to weathering by physical, chemical, and biological processes. This weathered material may then be removed through erosion and be transported and deposited elsewhere. Use a stream table to investigate the role of erosion and deposition in producing California's fertile Central Valley. Use your findings to explain how erosion and deposition have provided the basis for California's agricultural economy.

37 47

☐ 4 Use your stream table to investigate the role of water in producing the Earth's distinctive surface features, such as the Grand Canyon and Table Mountain in Tuolumne county. Use your understanding to explain why there are layers of rock and how these layers are deposited and accumulate over time. How does each layer preserve a record of the environment at the time?

37 47

☐ 5 Use your stream table to model changing rates of erosion and deposition and use this to construct an explanation for the rapid burial and preservation of organic material.

37

☐ 6 Investigate coastal erosion in California's Pacifica region using the material provided on BIOZONE'S resource hub. Using coastal erosion as an example, explain how the rates at which land-forming processes occur can vary over different time scales.

37

☐ 7 ENGINEERING CONNECTION: COASTLINE EROSION
Coastline erosion that affects humans is a natural hazard. Explain some of the common impacts of erosion in California and work in groups to come up with design solutions for the control of coastal erosion in California. Compare and evaluate your design solutions based on your prioritized criteria and tradeoffs that account for a range of constraints (e.g. cost, safety, longevity, and aesthetics). What impact might your design have on the natural environment? How could you test this? What modifications could you make to reduce the impact of your design solution?

37 47

What evidence shows that different species are related?

☐ 8 Multiple lines of evidence provide support for common ancestry and biological evolution. Evaluate and communicate the contribution of each of these lines of evidence to the scientific understanding of the evolution of life throughout Earth's history. Communicate the information using writing, visual displays (such as diagrams, charts, and annotated photographs), or oral presentations. Construct evidence-based arguments for common ancestry and biological evolution including:

38 39 47

☐ 8a **Fossil evidence**, including transitional fossils (e.g. *Archaeopteryx*) and fossil sequences documenting morphological change in a taxon over time (e.g. horses, whales).

☐ 8b **Anatomical evidence**, e.g. limb homology in vertebrates. How do analogous structures provide evidence for the role of adaptation in the evolution of structures with a similar function in unrelated groups?

☐ 8c **Molecular evidence**, e.g. from DNA sequences and protein homologies.

☐ 8d **Developmental evidence**, e.g. the order of appearance of structures in embryos.

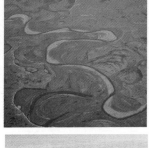

Activity number

Jeff Podos

Jeff Podos

Why do we see similar fossils across the world from each other but living organisms that are very different?

NATURAL SELECTION AND ADAPTATION   40

☐ 9 Describe variation in natural populations and recognize its origin in the differences in both the genetic makeup of individuals and the environment in which they develop. Collect, analyze, and interpret data from humans (e.g. classmates, extended family) to provide evidence for variation in natural populations. Do you see any patterns in the data you have collected?

☐ 10 Analyze and interpret data from natural populations to show that some of the variation   40
in individuals is inherited and that individuals within the population produce many more offspring than survive (because resources are limited).

☐ 11 Analyze and interpret data to show that those individuals within the population with   40
particular combinations of traits have higher fitness (survival and reproduction) under certain environmental conditions than other individuals.

☐ 12 Develop and use a model to demonstrate how differential survival of favorable   40
phenotypes (natural selection) can lead to phenotypic change over time. Use your model and the evidence you have analyzed from natural populations to explain the process of evolution (change in the genetic makeup of a population over time).

☐ 13 Analyze shifts in the numerical distribution of phenotypic traits (e.g. beak size, coat color)   40 59
to support your explanations that organisms with a favorable heritable trait increase in   also IS4
proportion to individuals without that trait. (Also IS4).

☐ 14 Construct an evidence-based explanation for how differences in selective environments   40 41 43
contribute to genetic change in populations and result in the adaptations we see in organisms. Evidence could include examples of convergence in different taxa as well as modern examples such as insecticide resistance in pest insects and antibiotic resistance in bacteria.

☐ 15 Explain how the study of organisms with short generation times can record observable   43
change in the genetic makeup of populations. Create and use a computational model to explain how quickly antibiotic resistance can spread in a bacterial population given a specific selective environment and argue the case for completing prescription medication for treatable bacterial infections.

☐ 16 Evaluate the evidence for claims that natural or human-induced changes in the   41-44 47
environment may contribute to the expansion of some species, the emergence of new species (speciation), or species extinction.

CONTINENTAL DRIFT AND THE CURRENT LOCATION OF FOSSILS: 45 47

☐ 17 Evaluate the fossil evidence for plate tectonics and the movement of continents. How is the timing of these events tracked? Use a model to help you to explain why the fossil evidence for continental drift is so compelling, i.e. what would the fossil record have shown if populations had been somehow transported to separate continents and existed in isolation there?

How did modern day humans evolve?

☐ 18 Throughout this chapter, you have explored how natural selection favors some traits over   44
others in the prevailing environment. Isolation of populations and genetic divergence in response to different selection pressures in different environments can lead to the emergence of new species. While changing environments can lead to the emergence of new species, existing species may become extinct if they are unable to adapt. Now you will ask questions and evaluate the evidence for the role of different selection pressures in the evolution of the human family.

☐ 19a The epicenter of human evolution was eastern Africa. Study a model of hominin evolution (which is less like a linear timeline and more like a bushy tree), noting the large number of species, often coexisting in time and space. One species, *H. sapiens*, thrived, while the others, hugely successful in their own right at the time, did not persist.

☐ 19b Divide into groups of four. Within each group, each individual will investigate the relative importance of one of four aspects of human evolution: climate change, physical changes in the evolution of humans, molecular evidence of human ancestry and evolution, and human cultural evolution (behavior and communication).

For your aspect of choice, analyze data and evaluate and communicate information about its role in the emergence of *Homo sapiens* and the success of our species. Report back to your group on what you have found and then construct an evidence-based explanation for the relative importance of each of the four aspects in the enduring success of humans.

35 The Rise of the Tyrants

ANCHORING PHENOMENON: Fossils and diversity over time

The map below shows the distribution of some of the fossils found belonging to the dinosaur group Tyrannosauroidea, of which *Tyrannosaurus rex* (North America) is the most famous. The group also includes giants such as *Tarbosaurus* (Asia), and *Albertosaurus* (North America). Tyrannosaurs and Tarbosaurs are very closely related, with some paleontologists arguing they should be in the same genus. Members of Tyrannosauroidea first appear in the fossil record in what are now England and Siberia. There are no swimming or flying members. In the middle Jurassic the group consisted of about 3 genera. By the end of the Cretaceous the group had about 13 genera.

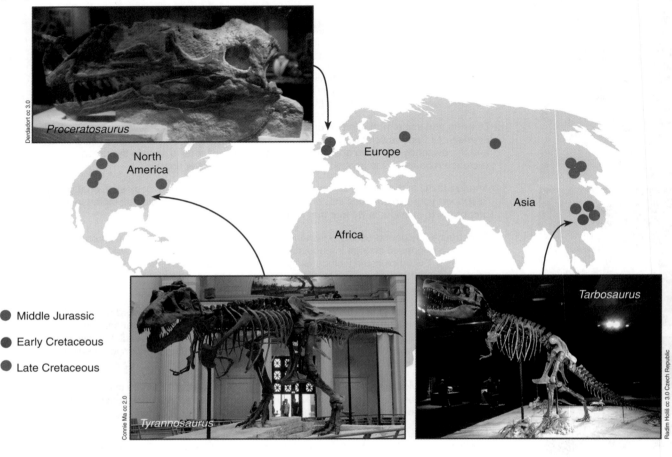

1. (a) How old is the oldest fossil you have ever seen? _____

    (b) What was this fossil? _____

    (c) Where was it found? _____

2. The photos above show two closely related dinosaur genera, *Tyrannosaurus* and *Tarbosaurus*. Their fossils have been found in areas separated by thousands of kilometers.

    (a) In groups or as a class discuss how these two closely related dinosaurs might have become separated by such a large distance:

    _____

    _____

    _____

    (b) *Tyrannosaurus* and *Tarbosaurus* share many features but differ very subtly. For example, the eyes of *Tyrannosaurus* point more forward than the eyes of *Tarbosaurus*, much like today's predators. What do you think might have caused this slight difference in the position of the eyes in what appear to be extremely powerful and capable predators?

    _____

    _____

    _____

©2018 **BIOZONE** International
ISBN: 978-1-927309-55-1
Photocopying Prohibited

36 What are Fossils and How Do They Form?

ENGAGE: Fossils are part of the geosphere

Fossils are the remains of long-dead organisms that have escaped decay and have, after many years, become part of the Earth's crust. For fossilization to occur, the organism must be buried rapidly (usually in water-borne sediment). This is followed by chemical alteration, whereby minerals are added or removed. Fossilization requires the normal processes of decay to be permanently stopped. This can occur if the organism's remains are isolated from the air or water and decomposing microbes are prevented from breaking them down. Fossils record the appearance and extinction of organisms, from species to entire taxonomic groups. Once this record is calibrated against a time scale (by using a range of dating techniques), we can build up a picture of the evolutionary changes that have taken place.

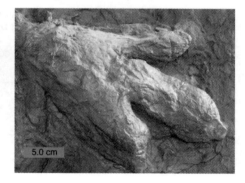

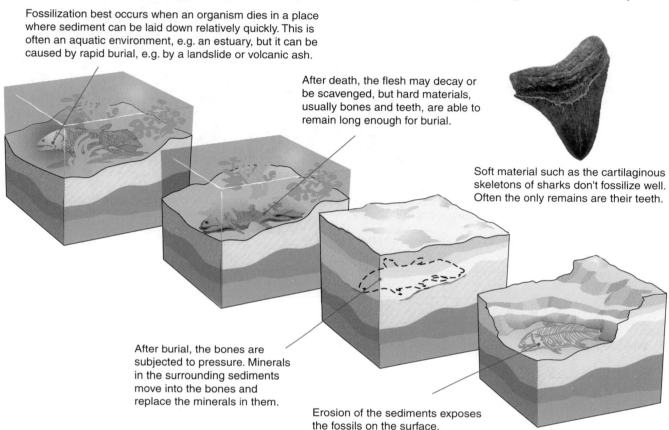

Fossilization best occurs when an organism dies in a place where sediment can be laid down relatively quickly. This is often an aquatic environment, e.g. an estuary, but it can be caused by rapid burial, e.g. by a landslide or volcanic ash.

After death, the flesh may decay or be scavenged, but hard materials, usually bones and teeth, are able to remain long enough for burial.

Soft material such as the cartilaginous skeletons of sharks don't fossilize well. Often the only remains are their teeth.

After burial, the bones are subjected to pressure. Minerals in the surrounding sediments move into the bones and replace the minerals in them.

Erosion of the sediments exposes the fossils on the surface.

A fossil may be the preserved remains of the organism itself (as in the ammonite below), the impression of it in the sediment (a mold and compression-impression fossils), or marks made by it during its lifetime (trace fossils).

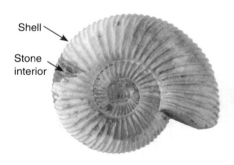

Shell

Stone interior

Ammonite: This ammonite still has a layer of the original shell covering the stone interior, Jurassic (Madagascar).

Compression fossil: This fern frond shows traces of carbon and wax from the original plant, Carboniferous (USA).

5.0 cm

Trace fossil: A dinosaur footprint in Lower Jurassic rock, SW Utah.

1. In what way are fossils part of the geosphere? _____

_____

_____

_____

©2018 BIOZONE International
ISBN: 978-1-927309-55-1
Photocopying Prohibited

 SC SF CE LS4.A

EXPLORE: Understanding past events and the Principle of Superposition

Our interpretation of sedimentary rock layers (strata) and the fossils they contain, including relative age, the environment in which the sediments were deposited, and the events after their deposition (folding, faulting, uplift, erosion) is based on a few basic principles (below):

The Principle of Superposition
In otherwise undisturbed strata, the youngest layer is at the top and oldest is at the bottom.

Principle of Original Horizontality
Layers of rocks, deposited from above under the influence of gravity, are laid down horizontally.

Principle of Lateral Continuity
Within the depositional basin in which they form, strata extend in all directions (laterally continuous) until they thin out.

Principle of Cross-Cutting Relationships
Deformation events, such as volcanic intrusions or faults, that cut across rocks are younger than the rocks they cut across.

Principle of Faunal Succession
Unique fossil assemblages in strata can be used to correlate rocks of the same age in widely separated geographical locations.

Photo: Strata in the Colorado Plateau area of southeastern Utah.

2. There are seven layers of sedimentary rock represented in the photograph above. The oldest rocks are Permian white sandstones (indicated). The youngest rocks are Jurassic. Mark the boundaries of the other 6 layers on the photograph.

3. What three principles of stratigraphy are evident in this example? _____

_____

_____

_____

4. Why are the stratigraphic principles outlined above helpful when interpreting a rock sequence and the fossils it contains?

_____

_____

_____

_____

EXPLAIN: Applying stratigraphic principles

5. (a) Working in groups, apply the stratigraphic principles above to construct a plausible explanation of the events that gave rise to the profile represented by the age ordered sequence A-F below. Attach your explanation to this page:

   (b) Which layer(s) could you expect to contain fossils? Why?

   _____

   _____

   _____

   _____

   _____

   _____

   (c) What unidentified process is currently working on the land? How can you tell?

   _____

   _____

   _____

©2018 **BIOZONE** International
ISBN: 978-1-927309-55-1
Photocopying Prohibited

37 Erosion Shapes the Landscape

ENGAGE: **Water as an agent of erosion**

Bluff Cove, California

Water, including water as snow and ice, is able to erode (remove) rock and soil. Water moves weathered rock to the lowest elevation possible, and transports it either in solution (dissolved), in suspension, or carried along the base of water courses as bed load. The erosive force of water on land is clearly seen where soil is removed along drainage lines by surface water runoff or where the sea meets coastal cliffs, as along California's coastline (above). Erosion does not always proceed at the same rates. River banks and coastal cliffs may be very gradually undercut and then, once unstable, may slump, forming a flatter topography as a large volume of material is removed.

1. (a) Is there an area affected by erosion near where you live? Describe its features: _____

    _____

    _____

    _____

    (b) What was the agent of erosion? How do you know?_____

    _____

    _____

2. Do you know what the area looked like before being eroded? How has it changed? _____

    _____

    _____

EXPLORE: **Modeling the effect of water on the landscape**

Stream trays or tables are a simple way of modeling and observing how rivers develop, and change the land by erosion and deposition of sediment. Any long tray can be used as long as there is a water supply and an outlet for the water is drilled at the lower end.

You will use your stream tray set-up to explore how water affects the landscape and what features of the landscape influence the land forms that result.

▸ Set up the tray by placing it on a slight angle with the outlet at the lower end. Place your substrate, e.g. gravel, silt, or sand in the tray and work the sediment so that it becomes thinner near the lower end.

▸ The simplest set up is to make a sediment "mountain" near the upper end of the tray to initially block water flow, forming a "lake".

©2018 **BIOZONE** International
ISBN: 978-1-927309-55-1
Photocopying Prohibited

SC SPQ CE ETS1.B ESS2.C

3. Before you do the modeling, record your predictions for the following:

(a) The effect of flow velocity on erosion: _____

_____

(b) Where most of the erosion will occur in a meandering stream: _____

_____

(c) The effect of layering materials of different hardness: _____

_____

(d) How and where the sediment that is moved will be deposited: _____

_____

Observe how the lake overflows and forms a channel (below). The variation of particle sizes in the sediment and shape of the mountain range influences how the channel will form.
Try this several times with different shaped mountain ranges. Use different materials to simulate the effect of a river meeting different types of rock. Softer layers of gravel and sand could be topped with a layer of harder clay and vice versa.

Another simulation is to create a river meander (sinuous track) and observe how it changes over time as water moves at different velocities around the bends, depositing and eroding material at different places. Other investigations could include adding larger rocks and vegetation to observe their effects on erosion and river channel formation. Investigate the effect of increasing water velocity when banks are already undercut by erosion.

Overflow channel

4. Draw the river flow through the different river models you have made. Include notes about the changes and how features such as large rocks and vegetation affect the river's shape. What is the effect of increasing the velocity of water flow?

©2018 **BIOZONE** International
ISBN: 978-1-927309-55-1
Photocopying Prohibited

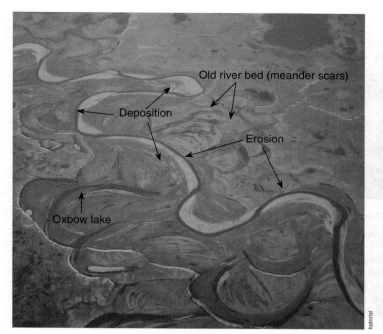

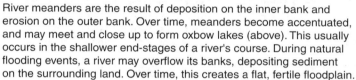

Deposition — Old river bed (meander scars) — Erosion — Oxbow lake

katorisi

River meanders are the result of deposition on the inner bank and erosion on the outer bank. Over time, meanders become accentuated, and may meet and close up to form oxbow lakes (above). This usually occurs in the shallower end-stages of a river's course. During natural flooding events, a river may overflow its banks, depositing sediment on the surrounding land. Over time, this creates a flat, fertile floodplain.

California's Central Valley is a flat, fertile valley between the coastal mountain ranges and the Sierra Nevada. The Sacramento River (above) drains its largest basin and is heavily modified to control flooding and to provide water for cities and agriculture. It joins the San Joaquin River to form a large delta, which drains into San Francisco Bay.

5. Build the gravel and sand into a mountain at the top end of the tray. Let the water flow, forming a new stream. Near the bottom of the tray, in the water flow, place a small animal or plant figurine, lying it down. Observe the build up of sediment on the figurine. This process models part of the process of fossilization:

(a) Which step in fossilization is being modeled? _____

(b) What other process needs to occur for fossilization to proceed? _____

6. (a) Use the information provided and your models to explain how river meanders arise: _____

_____

_____

(b) How would this pattern of deposition change during a large flood?_____

_____

_____

7. Based on your models and your understanding, explain how erosion and deposition form the basis of California's agricultural economy:

_____

_____

_____

_____

_____

_____

_____

_____

_____

_____

©2018 **BIOZONE** International
ISBN: 978-1-927309-55-1
Photocopying Prohibited

EXPLORE: Shaping the Earth

The processes that shape the surface of the Earth can be divided into two main categories: **constructive factors**, which build surface features, and **destructive factors**, which break down surface features.

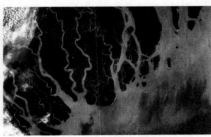

Weathering

Weathering is caused by physical, chemical, and biological processes. Together they break down rocks into finer particles. Physical weathering includes the effects of changes in heat and pressure. Chemical weathering involves chemical reactions that may dissolve minerals in the rock. Biological weathering occurs when rocks are exposed to the actions of living organisms, such as the organic acids and enzymes produced by microorganisms.

Tectonic activity

New land can be formed by uplift during earthquakes. The seabed rising out of the sea extends the beach and adds to the land. For example the seabed near Cape Cleare, in Alaska, was lifted 10 m by a large earthquake in 1964 creating a new beach (white area). Orogenies are long term uplifts of large areas of land, which may result from many earthquakes. The Laramide Orogeny in North America, ending about 35 million years ago, formed the Rocky Mountains.

Sediment deposition

Deposition is the process of sediments being added to a land mass. Sediments, such as sand and silt, may be carried by streams or rivers, blown by wind, carried in ice, or slide down hills as landslides. Deposition during floods adds sediments to floodplains. New land can be formed when sediments are deposited in river deltas or near the shore. If the sediments are held in place by tree roots, e.g. mangroves, then the land may become permanent and be extended out to sea.

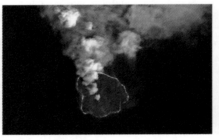

Mass wasting

Mass wasting, or mass movement, is the sudden movement of large volumes of rock and material, as in landslides. High cliffs and steep slopes are particularly susceptible to mass wasting. A high cliff undercut by a river or wave action can quickly collapse, causing the edge of the cliff to retreat.

Volcanic activity

Volcanic activity builds land. **Magma** (hot fluid or semi-fluid rock) welling up from deep below the crust bursts out of cracks in the crust as a volcano. When magma reaches the surface it is called **lava**. Pulverized rock and lava thrown into the air by the eruption of a volcano is called ash. Together lava and ash can build new land.

Erosion

Erosion is the loosening and removal of weathered material. A key part of erosion is the transport of materials away from their origin so it lowers the average level of the land. Erosion may occur through the action of water, wind, or glaciers. It is often linked to deposition as transported materials are deposited elsewhere (e.g. in deltas).

8. On the photograph below, identify, label, and describe the factors that are changing the landscape:

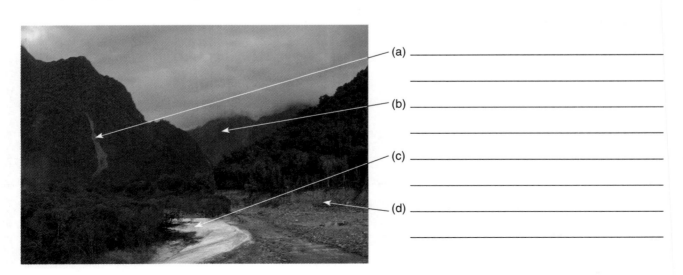

(a) _____

_____

(b) _____

_____

(c) _____

_____

(d) _____

_____

©2018 **BIOZONE** International
ISBN: 978-1-927309-55-1
Photocopying Prohibited

9. Place the processes that shape the Earth in the following categories:

    (a) Constructive factors: _____

    (b) Destructive factors: _____

10. Use the information presented in this activity to draw and label a model of land formation and destruction:

[blank drawing box]

EXPLAIN: Weathering and erosion

▶ **Weathering** and **erosion** are important processes in shaping the Earth's surface. They usually work closely together and are often confused as the same thing. It is important to remember weathering and erosion are quite different and separate processes.

▶ **Weathering** is the chemical, physical, and biological process of breaking rocks and minerals down into smaller pieces, e.g. dissolving limestone by rain.

▶ **Erosion** is the loosening, removal, and transport of the weathered materials. Erosion is followed by deposition of the material elsewhere (e.g. in a river delta). The processes of erosion combine to lower the Earth's surface.

Frost wedging

Weathering

Physical weathering

Physical weathering occurs when rocks break apart without any change to their chemical structure. Physical weathering includes changes in pressure and temperature affecting the rock. These combine to put constant physical stress on the rock until it shatters. One way this occurs, especially in high mountains, is the process of freeze-thaw, causing frost wedging (above right and right).

Chemical weathering

Chemical weathering is the breakdown of rock by chemically changing the minerals in the rock. This includes processes such as dissolving and oxidation. Rain water is slightly acidic (pH of about 6 due to dissolved carbon dioxide forming carbonic acid). Chemical reactions occur when it comes in contact with the minerals in rocks. An example is the weathering of limestone. The calcium carbonate in the limestone reacts with the excess hydrogen ions in the water forming bicarbonate ions, which are soluble and so are washed away in the water. Another form of chemical weathering is oxidation. For example, oxygen in the air or water reacts with iron in rocks forming oxides and rust, which can slowly break down a rock.

Biological weathering

Biological weathering is any weathering process carried out by a living organism. It can therefore also be chemical (e.g. organic acids and enzymes produces by organisms) or it can be physical (e.g. tree roots lifting pavements). Lichens and algae growing on the surface of rocks can slowly etch the surface, producing a greater surface area and slowly allowing other processes to take hold.

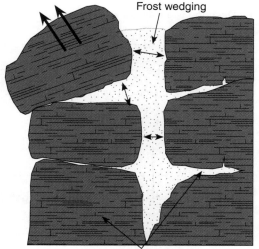
Frost wedging

Fracturing

Frost wedging occurs when water seeps into cracks in rocks. As it freezes, it expands and forces the cracks open a tiny bit more. When the ice thaws, water seeps into the newly widened cracks, ready to freeze again. This continuous freeze-thaw cycle can crumble whole mountainsides over the course of thousands of years.

©2018 **BIOZONE** International
ISBN: 978-1-927309-55-1
Photocopying Prohibited

Erosion

Canyonlands National Park, Utah

Wind-carved alcove, Utah

Glacier, Alaska

Water

Water is a major contributor to erosion. The continual flow of water over a stony river bed or the continual action of waves can shape rocks into rounded smooth shapes. The transport of sediment by rivers plays a major role in layering sediment and debris according to size and weight. In a delta, where water flow slows, heavier debris drops out of the water flow first with the finer sediment being carried further out to sea.

Wind

Much like water, wind has the power to carry sediment and shape rocks. In deserts, where there is a lack of vegetation to cover the dry ground, sand is picked up by the wind and transported across continents and hurled at cliffs and mountains, physically weathering their rocks and helping to carve them into the most extraordinary shapes. Grinding by wind-borne particles creates grooves or depressions, and can cut and polish rocks.

Glaciers

Glaciers form in areas where snow and ice accumulate year round and flow slowly downhill. Currently about 10% of Earth's surface is covered by glaciers. At the height of the last glaciation this figure was about 30%. Glaciers have enormous erosive power, and are capable of moving massive boulders and scouring out huge U shaped valleys. The rock frozen in the ice acts as an effective abrasive as it moves down the valley, removing soil and earth as it goes.

11. Use the boxes below to distinguish between weathering and erosion, including examples:

Weathering

Definition

Examples:

Erosion

Definition

Examples:

12. Describe two ways in which chemical weathering breaks down rocks:

(a) _____

(b) _____

13. Explain how biological weathering can incorporate both physical and chemical weathering: _____

_____

_____

_____

14. Explain how weathering and erosion together can flatten a mountainous landscape: _____

_____

_____

_____

_____

_____

_____

©2018 **BIOZONE** International
ISBN: 978-1-927309-55-1
Photocopying Prohibited

Evaluate: Surface area affects weathering

Surface area plays an important role in the process of weathering. The information below steps you through a controlled investigation into the effect of surface area on the chemical weathering of rocks. You will analyze and graph data and draw conclusions about chemical weathering of rocks. If you have time, you may wish to repeat this experiment in groups and collect your own data for analysis.

Background

Rainwater is very slightly acidic. When it comes in contact with limestone it reacts with it, forming carbon dioxide gas and calcium ions in solution. This causes the limestone to dissolve.

Limestone that is already partially eroded presents a greater surface area for rain to fall on and therefore a greater surface area for reaction.

Aim

To investigate the effect of surface area on the reaction of acid with calcium carbonate (limestone).

Hypothesis

If reaction rate is dependent on surface area, an increase in the surface area of limestone (calcium carbonate) will produce an increase in reaction rate.

Experimental method

▶ A roughly cubic, one gram piece of calcium carbonate ($CaCO_3$) was placed into excess 1 mol/L hydrochloric acid (HCl) at room temperature, 21°C (the reaction with rainwater is very slow due its weak acidity, so a stronger more concentrated acid was substituted).

▶ A stopwatch was started and the time taken for the $CaCO_3$ to completely dissolve recorded.

▶ One gram of $CaCO_3$ was then roughly crumbled and again placed into excess HCl and the time for it to completely dissolve recorded.

▶ Finally one gram of powered $CaCO_3$ was added to HCl and the time taken for it to completely dissolve recorded. The whole experiment was then repeated twice more.

Time to dissolve (minutes)			
Surface area of sample	Trial 1	Trial 2	Trial 3
Low (Single chip)	20	21	19.5
Medium (Crumbled)	5	5.5	4.8
High (Powdered)	1	1.5	1.2

NEED HELP?
See Activity 96

Controlling variables

To carry out an investigation fairly, all the **variables** (factors that could be changed) are kept the same, except the factor that is being investigated. These are called **controlled variables**. The variable that is being changed by the investigator is called the **independent variable**. The variable being measured is called the **dependent variable**.

15. Identify the independent variable for the experiment: _____

16. Identify the dependent variable for the experiment: _____

17. What is the sample size for each surface area of calcium carbonate? _____

18. What assumption is this experiment based on? _____

_____

19. Research and write out an equation for the weathering of calcium carbonate (limestone) by water: _____

_____

20. Complete the table below:

NEED HELP? See Activity 99

SA	Time to dissolve (minutes)			
	Sample 1	Sample 2	Sample 3	Mean
Low	20	21	19.5	
Medium	5	5.5	4.8	
High	1	1.5	1.2	

21. Use your calculations to plot a column graph on the grid below:

22. Write a conclusion based on the findings: _____

_____

_____

_____

23. Discuss how this experiment relates to chemical erosion and how the experiment could be improved:

_____

_____

_____

_____

_____

24. In cities where there is particularly high air pollution (usually caused by burning fossils fuels in cars and industry) there is often rapid weathering of sculptures and carvings, especially those made from limestone. Explain why this happens:

_____

_____

_____

_____

©2018 **BIOZONE** International
ISBN: 978-1-927309-55-1
Photocopying Prohibited

EXPLAIN: Table Mountain, Tuolumne County

Table mountains form when an area of hard material, e.g. a lava flow or hard sandstone, covers part of a larger area of softer ground. The uncovered softer ground then erodes away and leaves a flat topped mountain.

The numerous buttes and mesas of Monument Valley were formed when surrounding shales were eroded away leaving the hard sandstone monuments standing above the valley floor.

The rocks making up Mt. Roraima in Venezuela are 2 billion year old sandstones uplifted from the bottom of an ancient shallow sea. The softer surrounding rock has been eroded away.

North Table Mountain, part of the Rocky Mountains, is capped by lava flows dating to the early Paleocene. The underlying sedimentary rock has eroded away.

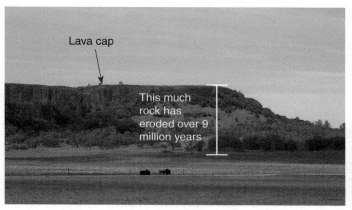

Lava cap

This much rock has eroded over 9 million years

Table Mountain (above) in Tuolumne County east of San Francisco is a striking example of a table formation. The long sinuous mountain is capped by a lava flow that follows an ancient river course (right).

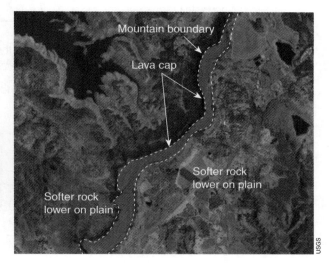

Mountain boundary

Lava cap

Softer rock lower on plain

Softer rock lower on plain

25. What activity produced Table Mountain's lava cap? _____

26. Suggest why Table Mountain has an elongated meandering shape: _____

_____

_____

27. Why does Table Mountain stand high above the lower plain? _____

_____

_____

28. Draw a series of diagrams to show how Table Mountain formed:

©2018 **BIOZONE** International
ISBN: 978-1-927309-55-1
Photocopying Prohibited

Elaborate: Interpreting past environments

Studying rock strata can tell us about the geological history of a particular area. Different land forms and environments leave particular formations in rocks that give us clues to the habitat, environment, and life that once lived in the area. The rock strata below shows a hypothetical example of this:

Relatively thick volcanic ash layer, indicating volcanic eruption.

Mud layer with ripple marks preserved in mud layer. Alternate thin layers. Indicates a beach.

Mud layers showing shallow ripple marks. Some marine fossils such as young fish, but also evidence of non marine organisms. Indicates possible lagoon with brackish water.

Youngest layer of fine grained mudstone indicating a lake.

Thick layer of rounded pebbles and ripple marks preserved in the arrangement of the pebbles. These indicate a river flowed here.

Layer of volcanic ash indicating volcanic activity.

Silt and sand layers with fossils of exclusively marine animals indicate a shallow sea.

29. The interpretation of rock strata varies depending on the data, and there may be a need to review other climate data for an accurate interpretation. For example the strata above could be caused by land being uplifted or by falling sea levels.

Write an interpretation of the history of the area around where the rock strata above were found. Justify your interpretation with evidence in the rocks:

_____

_____

_____

_____

_____

_____

_____

30. How can fossils help us determine the type of environment in which the fossils formed? _____

_____

_____

_____

Evaluate: Coastal erosion

Erosion is a constantly occurring process. It is especially so when water is involved, such as along a river or on the ocean's shoreline. Coastal erosion is more rapid and severe where waves can crash directly into land formations or where the shoreline is very steep, such as the base of cliffs. Here, the constant action of waves undermines the cliff or bank, which eventually collapses. Coastal erosion is a constant phenomenon along much of California's coastline, e.g. at Pacifica (right), where loss of land threatens dwellings and infrastructure. Scientists can study the action of waves on different coastal defense structures and design the best structures for preventing erosion and storm damage. The type of structure used depends on many factors. One important factor is the direction of the waves and currents along the beach. Waves coming in at an angle have a different erosive effect that those coming directly at the beach.

Bluff erosion Pacifica, CA, 2009

Brocken Inaglory cc 3.0

▶ Go to **BIOZONE's Resource Hub** to see video clips and examine coastal erosion at Pacifica, CA, 2016-2017.

©2018 **BIOZONE** International
ISBN: 978-1-927309-55-1
Photocopying Prohibited

Aim

To investigate the efficiency of different coastal erosion defenses.

Method

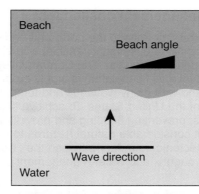

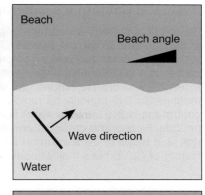

- Using a tray to make a small wave tank, set up the two scenarios, top right, making a sloped beach. A range of beach material could be tested (e.g. sand, gravel, rocks).

- Carry out a control for each type of beach material and wave direction, testing the effect of the wave action on an unprotected beach. Record the results (either diagrams or descriptions).

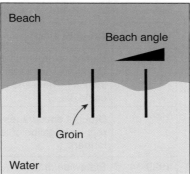

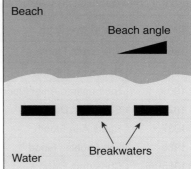

- Two types of erosion defense are shown below right, groins (bottom left) and breakwaters (bottom right). Test the effect of each of these types on preventing erosion for each beach material and wave direction. Record the results.

- What other types of erosion defense can you think of? Research other types and test them in your wave tank.

- Repeat your tests for different slopes of a beach (e.g. steep or vertical banks).

21. Record your results on a separate page, either as diagrams or descriptions, and staple them to this activity:

CA EP&Cs III: Human practices can alter the cycles and processes operating in natural systems (III c)

- Humans can provide engineering solutions to coastline erosion. Soft solutions involve replenishing beach sand (beach nourishment). Hard solutions involve creating anti-erosion structures.

- Once erosion control is started, it has to continue. Engineering solutions improve over time.

CA EP&Cs V: Decisions affecting resources and natural systems are based on a wide range of considerations and decision-making processes (V a,b)

- Engineering solutions to coastline erosion involve trade-offs among constraints, including cost, reliability, and aesthetics.

- What is being preserved by the erosion control? Is it worth it? Can buildings be relocated?

31. A settlement near a sandy beach wants to build coastal erosion defenses to prevent erosion. Studies show a slight longshore current parallel to the beach. The wave angle is perpendicular to the beach but the seafloor and beach have a relatively gentle slope. In small groups, discuss the efficiency of coastal defenses and determine which is the best defense in this situation. Summarize your main points here:

_____
_____
_____
_____
_____
_____
_____
_____
_____

32. Coastal erosion defenses can affect the habitat of shore-living species. Refine your beach erosion defense design to reduce its impact on the habitat. You may need to do some extra research for this. Summarize your changes here:

_____
_____
_____
_____

©2018 **BIOZONE** International
ISBN: 978-1-927309-55-1
Photocopying Prohibited

Evaluate: California's water management challenges

California, situated between temperate rainforests to the north and arid deserts to the south, boasts a diverse range of habitats and biological diversity. The climate is highly variable, with significant precipitation (as rain, snow, and ice) only in winter and prolonged droughts every year. The Central Valley, which dominates California's geographical center, is a sediment-filled valley and (like all floodplains) is highly productive. The Central Valley was once an extensive and diverse natural grassland with many wetlands and temporary water bodies, but today it is one of the most productive agricultural regions of the United States. To achieve this, those who farmed there had to intervene to control the highly variable water flows, preventing flooding and providing water for agriculture through prolonged dry periods. Floods and droughts present considerable natural hazards to California, which also has a high earthquake risk as a result of its position on the tectonic boundary between the Pacific and the North American Plates. A brief history of California's major floods, droughts, and water management policy is outlined in the tables below.

Table 1: Flood events in California

Date	Flood event
1861-62	The Great Flood
1905	Salton Sea Filling
1909	Sacramento River Flood
1928	St Francis Dam
1937	Santa Ana Flood
1938	Los Angeles Flood
1955-56	Christmas Flood
1964	North Coast Flood
1986	St Valentine's Day Flood
1996-97	Central Valley Flood
2017	Oroville Dam Emergency

Dark blue = statewide floods
Green = Southern California floods
Purple = Northern California floods

The latest filling of the Salton Sea was as a result of failed water management.

Table 2: Drought events in California

Date	Drought event
1862-65	Extensive droughts destroy California's cattle ranching industry. Thousands of cattle die.
1924	Drought encourages regular irrigation of land by farmers.
1929-34	Statewide drought coinciding with the Dust Bowl period.
1947-50	Multi-year droughts.
1959-60	Multi-year droughts.
1976-77	1977 was the driest in California's history to that point, prompting urban water conservation efforts.
1986-92	A 6 year drought ended by an El Niño in the Pacific.
2007-2009	Severe drought prompting a statewide emergency.
2011-17	One of the worse drought periods on record. More than 100 million trees died and low river levels prevented fish populations from spawning. The 2014 drought was the most severe in 1200 years.

Table 3: Water policy in California

Date	Water policy
1860	Formation of Levee and Reclamation Districts
1922	Colorado River Compact
1928	Reasonable and Beneficial Use Doctrine
1931	County of Origin Law
1933	Central Valley Project (CVP)
1959	Delta Protection Act and State Water Project (SWP)
1969	Porter-Cologne Water Quality Act
1970	California Environmental Quality Act and California Endangered Species Act
1972	Wild and Scenic Rivers Clean Water Acts
1974	Federal Safe Drinking Water Act
1985	Pesticide Contamination Prevention Act
2009	Delta Reform Act
2014	Sustainable Groundwater Management Act

Folsom Dam and Folsom Lake July 20, 2011 before the 2011-2017 drought (left) and January 16, 2014, during 2014 (right)

33. Divide the class into working groups and together construct a timeline of California's 'water history', based on the information presented above. Each group should select different events of interest, find out more about them, and add any relevant information to the timeline. Is there a pattern to the floods and droughts? Did any water management policies coincide with particular events or follow them closely? Why?

©2018 **BIOZONE** International
ISBN: 978-1-927309-55-1
Photocopying Prohibited

The Oroville Dam crisis

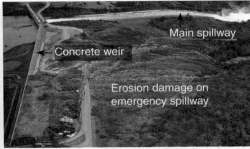

Main spillway

Concrete weir

Erosion damage on emergency spillway

Collapse of the main spillway causes erosion

Debris dam below main spillway

William Croyle, California Dept of Water Resources, Public Domain

The Oroville Dam is situated on the Feather River upriver and east of the city of Oroville, CA. The dam impounds Lake Oroville. The lake water is used to produce hydroelectric power and is a key facility in the State Water Project. Lake Oroville has two spillways used to prevent overflow of the dam during high rainfall. The main spillway is lined with concrete whereas the emergency spillway is topped with a concrete weir, with a bare earth embankment below.

In February 2017, heavy rainfall led to water from the lake being let into the main spillway. A crater began to form on the south-eastern side of the spillway so flows into the spillway were reduced to prevent further damage. This allowed the lake to rise. The rising water overtopped the emergency spillway (as per design). However, the upper bank eroded faster than expected, leading to fears of the weir collapsing. Release of water into the main spillway was increased to lower the lake level. This damaged the spillway further, so that the south-eastern side collapsed, allowing water to flow out of the spillway and erode the adjacent hillside (above left). A large debris dam formed below the damaged spillway, blocking the outflow of the power stations and forcing them to be shut down (above right).

34. Discuss the effect of erosion on the main spillway and emergency spillway of the Oroville Dam, including why this could have led to uncontrolled water flow out of Lake Oroville and the hazards this might have caused. Also include a discussion of the effect of deposition below the spillway.

_____

_____

_____

_____

_____

_____

_____

_____

_____

_____

_____

_____

_____

_____

_____

35. Form small groups to discuss the environmental and engineering challenges associated with managing water flows in a highly seasonal floodplain environment such as California's Central Valley. How do the properties that make the Central Valley such an agriculturally productive region also present challenges to the development and settlement of the region. Present a summary of your group's discussion to the class (e.g. as poster, flow chart, or slideshow). You may have to investigate some of the events identified in Tables 1-3 further and use them to illustrate your arguments.

©2018 **BIOZONE** International
ISBN: 978-1-927309-55-1
Photocopying Prohibited

38 Evidence for Evolution

ENGAGE: A horse is a horse, of course, of course

Horse evolution is well documented. There are many transitional fossils that document the evolution, over many generations, of a three toed animal about the size of a medium sized dog into the large single toed horse (and related species) we know today. Although each fossil type is a separate species, the similarity of skeletal structures and clear cases of features derived from earlier species, clearly connects these species in a evolutionary sequence.

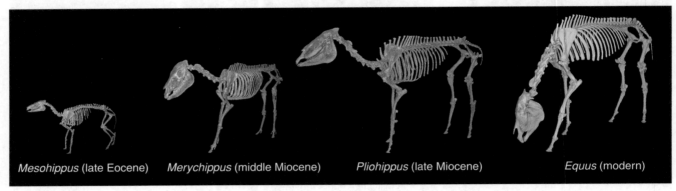

Mesohippus (late Eocene)   *Merychippus* (middle Miocene)   *Pliohippus* (late Miocene)   *Equus* (modern)

1. (a) How many modern species of horse (genus *Equus*) can you think of? _____

_____

   (b) What feature do you think connects these species as *Equus*? _____

_____

   (c) What special condition(s) or pressure(s) do you think acted on the horse's ancestors to produce the equids of today?

_____

_____

EXPLORE: Homologous structures

Homologous structures are structures found in different organisms that are the result of their inheritance from a common ancestor. Their presence indicates the evolutionary relationship between organisms. Homologous structures have a common origin, but they may have different functions.

For example, the forelimbs of birds and sea lions are homologous structures. They have the same basic skeletal structure, but have different functions. A bird's wings have been adapted for flight and a sea lion's flippers are modified as paddles for swimming.

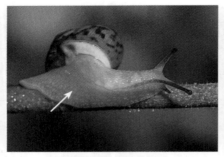

Hardened forewing (elytron)

Duck wing (top) and sea lion flipper.   Snail foot (top) and squid tentacles.   Butterfly forewing (top) and beetle elytron.

LS4.A   P   CE   SF

©2018 **BIOZONE** International
ISBN: 978-1-927309-55-1
Photocopying Prohibited

2. Describe the purpose of the homologous structure below and the adaptation to provide this purpose:

(a) (i) Duck wing: _____

_____

(ii) Sea lion flipper: _____

_____

(b) (i) Snail foot: _____

_____

(ii) Squid tentacles: _____

_____

(c) (i) Butterfly forewing: _____

_____

(ii) Beetle elytron: _____

_____

EXPLORE: The pentadactyl limb

A **pentadactyl limb** is a limb with five fingers or toes (e.g. hands and feet), with the bones arranged in a specific pattern (below, left).

Generalized pentadactyl limb

The forelimbs and hind limbs have the same arrangement of bones. In many cases bones in different parts of the limb have been modified for a specialized locomotory function.

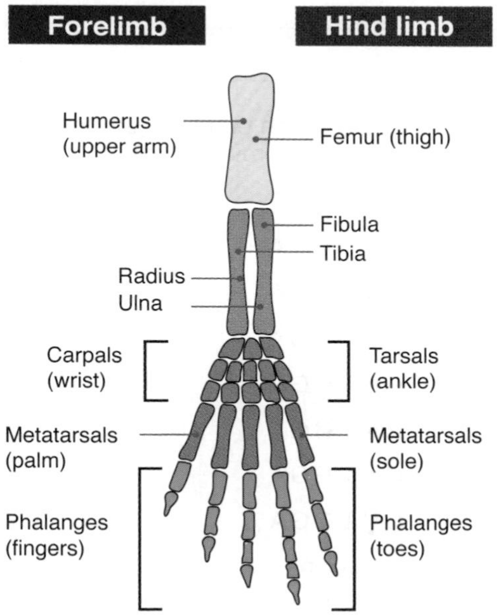

Forelimb	Hind limb

Humerus (upper arm) — Femur (thigh)

Fibula
Tibia

Radius
Ulna

Carpals (wrist) — Tarsals (ankle)

Metatarsals (palm) — Metatarsals (sole)

Phalanges (fingers) — Phalanges (toes)

Specializations of pentadactyl limbs

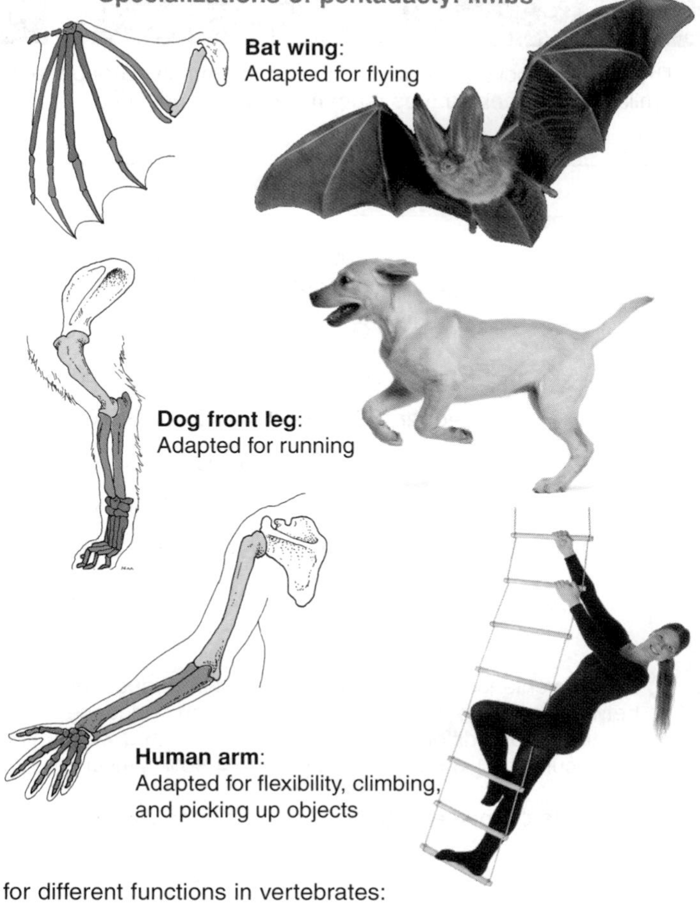

Bat wing: Adapted for flying

Dog front leg: Adapted for running

Human arm: Adapted for flexibility, climbing, and picking up objects

3. Describe how the pentadactyl limb has been adapted for different functions in vertebrates:

_____

_____

_____

_____

©2018 **BIOZONE** International
ISBN: 978-1-927309-55-1
Photocopying Prohibited

EXPLORE: How does DNA show evolutionary relationships?

▶ The DNA (genetic material) of mammals is very similar. They all share the same kinds of genes (e.g. protein-coding sequences that code for hair production, muscle fibers, milk production, bone, etc). However, there is clearly some difference in these genes because different groups of mammals look different to other groups.

▶ By comparing the DNA sequences in genes we can determine how closely related the groups are and, within limits, how long ago they diverged from each other. Consider the following examples of mammals:

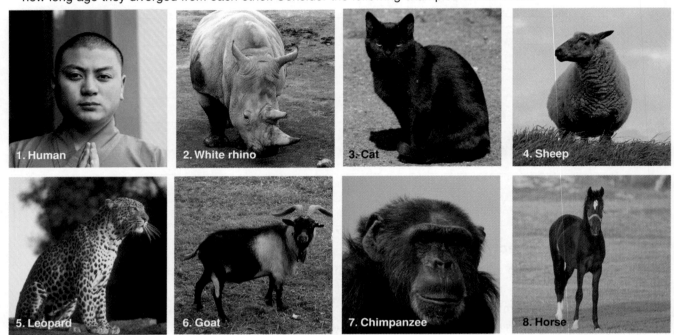

1. Human   2. White rhino   3. Cat   4. Sheep
5. Leopard   6. Goat   7. Chimpanzee   8. Horse

4. The images above show four groups of mammals. Group the images together to produce four groups of mammals with similar anatomical features (summarize these briefly):

A: _____   C: _____

B: _____   D: _____

We can now compare the similarity of DNA sequences for the mammalian gene FLNA, which produces an actin binding protein. The aligned sequences between bases 465 and 495 of the FLNA gene are provided below.

Human: GCGACGGGCTGCGGCTTATCGCGCTGTTGGA

Rhino: GCGACGGGCTGCGGCTCATCGCGCTGCTCGA

Leopard: GCGATGGGCTGCGGCTCATCGCACTGCTGGA

Cat: GCGATGGGCTGCGGCTCATCGCACTGCTGGA

Sheep: GCGACGGGCTGCGGCTTATCGCACTGCTCGA

Goat: GCGACGGGCTGCGGCTTATCGCACTGCTCGA

Chimp: GCGACGGGCTGCGGCTCATCGCGCTGTTGGA

Horse: GCGACGGGCTGCGGCTCATTGCGCTGCTCGA

5. (a) For convenience we will compare the DNA sequences above to the human sequence. By counting the difference between any one sequence and the human sequence we can estimate the percentage similarity. For example there are three base differences between humans and white rhinos (shown in red above) out of 31 bases shown. To calculate: 31-3 = 28. 28 ÷ 31 x 100 = 90% similarity.

Calculate the percentage similarity of human DNA to the other six mammals in the list:

(i) Leopard: _31-4 = 27 ÷ 31 x 100 = 87%_   (iv) Cat: _87%_

(ii) Sheep: _90%_   (v) Goat: _90%_

(iii) Chimpanzee: _31-3 = 30 ÷ 31 x 100 = 96%_ (vi) Horse: _87%_

(b) We can use the similarities to group the mammals as in (4). Now regroup them according to % DNA similarity:

_____

_____

©2018 BIOZONE International
ISBN: 978-1-927309-55-1
Photocopying Prohibited

6. Now look at the actual differences in the DNA sequences (base change and position) rather than the overall percentage similarity. These can also be used to group the animals. Regroup the animals based on these differences.

_____

_____

7. How well do all these comparisons agree? How does having three lines of evidence help confirm relationships between the groups of mammals?

_____

_____

_____

_____

EXPLORE: Proteins can show evolution

In much the same way that DNA sequences can be used to determine relationships between species so can the proteins that the DNA codes for. Researchers look for homologies (similarities) between the proteins of similar **genes** (protein-coding sequences) in different species. These similar proteins are often highly conserved, meaning they change (mutate) very little over time. This is because they have critical roles (e.g. in cellular respiration) and mutations are likely to be detrimental to their function.

Evidence indicates that these highly conserved proteins are homologous and have been derived from a common ancestor. Because they are highly conserved, differences between them are likely to represent major divergences between groups during the course of evolution.

Hemoglobin homology

Hemoglobin is the oxygen-transporting blood protein found in most vertebrates. The beta chain hemoglobin sequences from different organisms can be compared to determine evolutionary relationships.

As genetic relatedness decreases, the number of amino acid differences between the hemoglobin chains of different vertebrates increases (below). For example, there are no amino acid differences between humans and chimpanzees, indicating they recently shared a common ancestor. Humans and frogs have 67 amino acid differences, indicating they had a common ancestor a very long time ago.

Human – chimpanzee 0 Gorilla 1 Gibbon 2 Rhesus monkey 8 Dog 15 Horse 25 Mouse 27 Kangaroo 38 Chicken 45 Frog 67

Increasing difference in amino acid sequence

Primates   Placental mammals   Marsupial   Non-mammalian vertebrates

8. Compare the differences in the hemoglobin sequence of humans, rhesus monkeys, and horses. What do these tell you about the relative relatedness of these organisms?

_____

_____

_____

9. The fossil record shows that amphibians diverged from the lineages that evolved into mammals before the birds did. How is this record supported by protein homologies?

_____

_____

_____

©2018 **BIOZONE** International
ISBN: 978-1-927309-55-1
Photocopying Prohibited

EXPLORE: Shared stages of development reflect common ancestry

Developmental biology studies the process by which organisms grow and develop. In the past, it was restricted to the appearance (morphology) of a growing fetus. Today, developmental biology focuses on the genetic control of development and its role in producing the large differences we see in the adult appearance of different species.

During development, vertebrate embryos pass through the same stages, in the same sequence, regardless of the total time period of development. This similarity is strong evidence of their shared ancestry. The stage of embryonic development is identified using a standardized system based on the development of structures, not by size or the number of days of development. In humans, the Carnegie stages (right) cover the first 60 days of development.

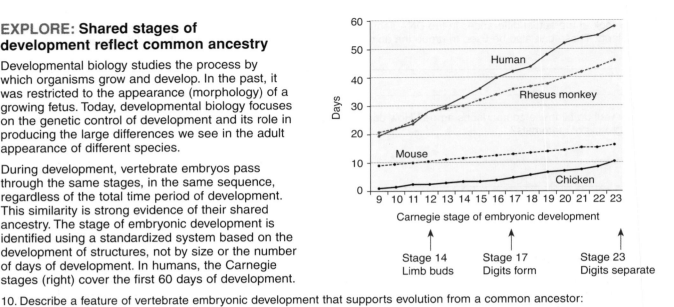

Carnegie stage of embryonic development

Stage 14
Limb buds

Stage 17
Digits form

Stage 23
Digits separate

10. Describe a feature of vertebrate embryonic development that supports evolution from a common ancestor:

_____

_____

11. How many days does it take for the following to reach Carnegie stage 23?

(a) Chicken: _____ (b) Rhesus monkey: _____

EXPLORE: The fossil record documents stages in evolutionary history

Horse evolution

The evolution of the horse from the ancestral *Eohippus* to modern *Equus* is well documented in the fossil record. The rich fossil record, with at least 24 known genera, including numerous transitional fossils, has enabled scientists to develop a robust model of the evolutionary history (phylogeny) of horses. It is a complex lineage with many divergences (right) and a diverse array of species (often coexisting). The environmental transition from forest to grasslands drove many of the changes observed in the fossil record. These include reduction in toe number, increased size of cheek teeth, and increasing body size.

12. How do relatively complete fossil records, such as the horse fossil record, provide evidence for evolution and common ancestry of organisms?

_____

_____

_____

_____

_____

_____

_____

_____

_____

_____

_____

_____

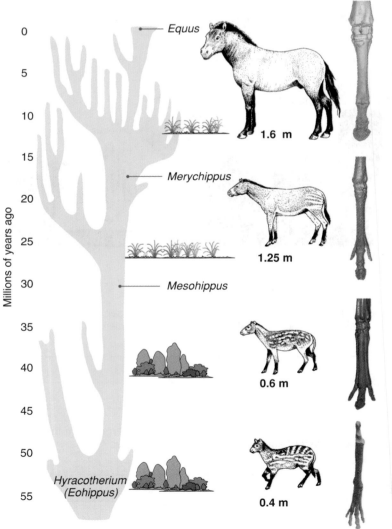

©2018 **BIOZONE** International
ISBN: 978-1-927309-55-1
Photocopying Prohibited

Whale evolution

▶ The evolution of modern whales from an ancestral land mammal is well documented in the fossil record. The fossil record of whales includes many transitional forms, which has enabled scientists to develop an excellent model of whale evolution. The evolution of the whales (below and opposite) shows a gradual accumulation of adaptive features that have equipped them for life in the open ocean.

▶ The fossil evidence is well supported by the molecular evidence. DNA analysis has played an important role in determining the origin of whales and their transition from a terrestrial (land) form to a fully aquatic form. The evidence shows hippopotamuses to be the closest living ancestor of whales and both whales and hippos evolved from a family of water-loving extinct, artiodactyl (even-toed) ungulates (hoofed mammals) more than 50 million years ago. The toothed and baleen whales are thought to have diverged some 28-33 million years ago.

Pakicetus (mainly terrestrial):

Early Eocene (50 mya)

Kevin Guertin cc 2.0

Rodhocetus (mainly aquatic):

Mid Eocene (45 mya)

Dorudon (fully aquatic):

Mid-Late Eocene (40 mya)

Cetotherium (fully aquatic):

1 m

Mid-Late Miocene (18 mya)

Hind limbs are vestigial (functionless remnants) and absent from the fossil

All photos: Nobu Tamura;

Pavel Gol'din, Dmitry Startsev, and Tatiana Krakhmalnaya

13. The sequence above shows just four of the species in a very comprehensive fossil record documenting the evolution of whales. Study the skeletons and reconstructions as well as the material on **BIOZONE's Resource Hub** and, working in pairs, make notes to the left of the images about the features you notice. Think about the limbs, vertebrae, tail flukes (present or absent), and position of the nasal openings. Can you document any new (derived) features in each species?

14. (a) What adaptations for an aquatic life have evolved in whales over time? _____

_____

_____

_____

(b) What features of the environment do you think would have been involved in the evolution of these adaptations?

_____

_____

_____

_____

©2018 **BIOZONE** International
ISBN: 978-1-927309-55-1
Photocopying Prohibited

EXPLORE: Transitional fossils provide evidence of evolution

Transitional fossils are fossils that have a mixture of features found in two different, but related, groups. Transitional fossils provide important links in the fossil record and provide evidence to support how one group may have given rise to the other by evolutionary processes.

Important examples of transitional fossils are found in the fossil record of horses, whales, and birds, e.g. *Archaeopteryx* (below) is a transitional form between birds and non-avian dinosaurs.

Archaeopteryx was crow-sized (50 cm length) and lived about 150 million years ago. It is regarded as the first primitive bird and had a number of birdlike (avian) features, including feathers. However, it also had many non-avian features, which it shared with theropod dinosaurs of the time (theropods were a group of flesh-eating dinosaurs).

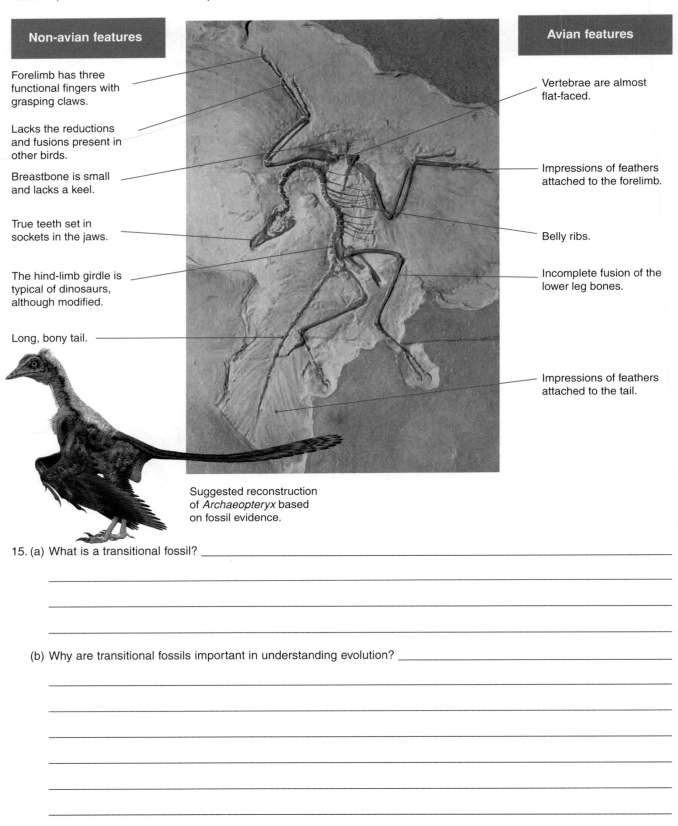

Non-avian features

Forelimb has three functional fingers with grasping claws.

Lacks the reductions and fusions present in other birds.

Breastbone is small and lacks a keel.

True teeth set in sockets in the jaws.

The hind-limb girdle is typical of dinosaurs, although modified.

Long, bony tail.

Avian features

Vertebrae are almost flat-faced.

Impressions of feathers attached to the forelimb.

Belly ribs.

Incomplete fusion of the lower leg bones.

Impressions of feathers attached to the tail.

Suggested reconstruction of *Archaeopteryx* based on fossil evidence.

15. (a) What is a transitional fossil? _____

_____

_____

_____

(b) Why are transitional fossils important in understanding evolution? _____

_____

_____

_____

_____

_____

_____

©2018 **BIOZONE** International
ISBN: 978-1-927309-55-1
Photocopying Prohibited

EXPLAIN: Common ancestry

The three mammals below are all primates. Certain features of their biology can be used to show that chimpanzees are more closely related to gorillas than to macaques. Some of these features are presented below.

Gorilla	**Chimpanzee**	**Macaque**

Andre Chalmers CC 3.0

Skeleton	**Skeleton**	**Skeleton**
	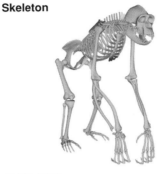	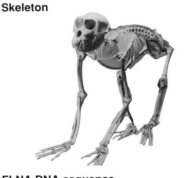

NoraSmb CC4.0

FLNA DNA sequence
GCC CTG GTG GAC AGC TGT GCC
CCG GGC CTG TGT

$32 \div 33 \times 100 = 96\%$

FLNA DNA sequence
GCC CTG GTG GAC AGC TGT GCC
CCG GGT CTG TGT

FLNA DNA sequence
GCC CTG GTC GAC AGC TGT GCC
CCG GGC CTG TGT

$31 \div 33 \times 100 = 93\%$

16. Use the evidence presented to describe the relationship between the three primates. Present evidence to support the argument that chimpanzees share a more recent common ancestor with gorillas than with macaques:

_____

_____

_____

_____

_____

_____

_____

_____

_____

_____

_____

_____

_____

_____

_____

_____

_____

17. "Cattle and whales share a more recent common ancestor than cattle and horses". Using what you have learned in this chapter and revisiting **BIOZONE's Resource Hub** where necessary, work in pairs to construct an evidence-based explanation for this statement. Summarize your argument on a separate sheet together with any relevant drawings, photographs, or molecular evidence supporting specific points. Attach all the material to this page.

©2018 **BIOZONE** International
ISBN: 978-1-927309-55-1
Photocopying Prohibited

39 All Life is Related

ENGAGE: There is a universal genetic code

DNA encodes the genetic instructions of all life. The form of these genetic instructions, the **genetic code**, is effectively universal, i.e. the same combination of three DNA bases code for the same amino acid in almost all organisms. The very few exceptions in which there are minor coding alternatives occur only in some bacteria and mitochondrial DNA.

EXPLORE: How do we know about the relatedness of organisms?

▸ Traditionally, the phylogeny (evolutionary history) of organisms was established using morphological comparisons. In recent decades, molecular techniques involving the analysis of DNA, RNA, and proteins have provided more information about how all life on Earth is related.

▸ These newer methods have enabled scientists to clarify the origin of the eukaryotes and to recognize two prokaryote domains. The universality of the genetic code and the similarities in the molecular machinery of all cells provide powerful evidence for a common ancestor to all life on Earth.

All known living organisms use ATP. It is the universal energy carrying molecule in cells.

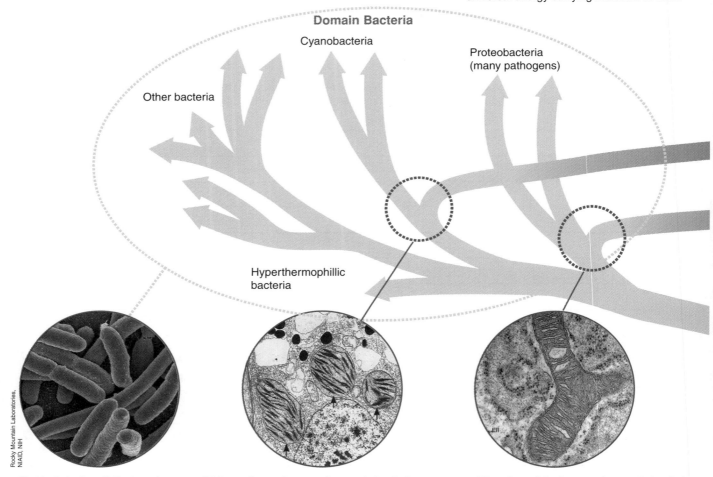

Domain Bacteria

Cyanobacteria

Proteobacteria (many pathogens)

Other bacteria

Hyperthermophillic bacteria

Rocky Mountain Laboratories, NIAID, NIH

Bacteria lack a distinct nucleus and cell organelles. Features of the cell wall are unique to bacteria and are not found among archaea or eukaryotes. Typically found in less extreme environments than archaea.

Chloroplasts have a bacterial origin

Cyanobacteria are considered to be the ancestors of chloroplasts. The evidence for this comes from similarities in the ribosomes and membrane organization, as well as from genetic studies. Chloroplasts were acquired independently of mitochondria, from a different bacterial lineage, but by a similar process.

Mitochondria have a bacterial origin

Evidence from mitochondrial gene sequences, ribosomes, and protein synthesis indicate that mitochondria came from prokaryotes. Mitochondria were probably symbiotic inclusions in an early eukaryotic ancestor.

    LS4.A P SF

©2018 **BIOZONE** International
ISBN: 978-1-927309-55-1
Photocopying Prohibited

1. Identify three features of the metabolic machinery of cells that support a common ancestry of life:

(a) _____

(b) _____

(c) _____

2. Using the diagram below, describe the evolutionary sequence from which animals arose. Alternatively you could draw an annotated diagram and attach it to this page:

_____

_____

_____

Eukarya (the eukaryotes) are characterized by complex cells with organelles and a membrane-bound nucleus. This domain is made up of the four kingdoms recognized under a traditional classification scheme.

Archaea resemble bacteria but membrane and cell wall composition and aspects of metabolism are very different. Many live in extreme environments similar to those on primeval Earth.

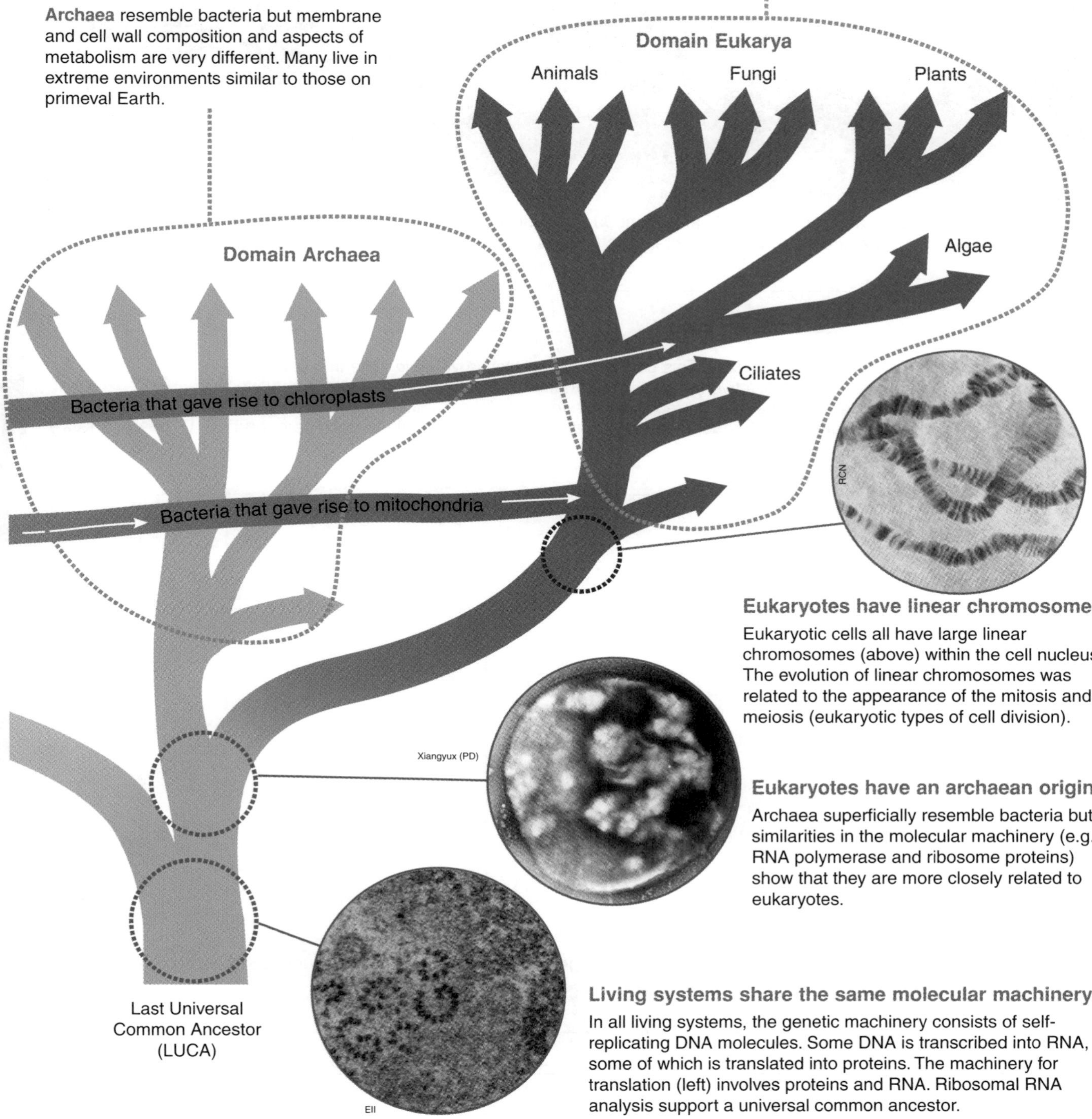

Domain Eukarya

Animals   Fungi   Plants

Algae

Domain Archaea

Ciliates

Bacteria that gave rise to chloroplasts

Bacteria that gave rise to mitochondria

RCN

Eukaryotes have linear chromosomes

Eukaryotic cells all have large linear chromosomes (above) within the cell nucleus. The evolution of linear chromosomes was related to the appearance of the mitosis and meiosis (eukaryotic types of cell division).

Xiangyux (PD)

Eukaryotes have an archaean origin

Archaea superficially resemble bacteria but similarities in the molecular machinery (e.g. RNA polymerase and ribosome proteins) show that they are more closely related to eukaryotes.

Last Universal
Common Ancestor
(LUCA)

Living systems share the same molecular machinery

In all living systems, the genetic machinery consists of self-replicating DNA molecules. Some DNA is transcribed into RNA, some of which is translated into proteins. The machinery for translation (left) involves proteins and RNA. Ribosomal RNA analysis support a universal common ancestor.

EII

©2018 **BIOZONE** International
ISBN: 978-1-927309-55-1
Photocopying Prohibited

40 Natural Selection

ENGAGE: Variability in populations

In any population there is some variability. The interaction of genes and the environment during development results in variations in the phenotypes (physical and behavioral characteristics) we see in a population, as can be easily seen in this pack of gray wolves. Not all of these phenotypic variants are equally successful at leaving offspring. This differential survival and reproduction of individuals due to differences in phenotype is called **natural selection**.

1. In the case of humans, any two individuals are about 99.9% genetically identical. Look at the people in your class and record some of the variability in various characteristics (e.g. height, eye color, etc):

   _____

   _____

2. Identify a phenotypic variable in your class. This could be a quantitative variable (one that can be measured, such as height) or a qualitative (categorical) variable to which you can assign a ranking (e.g. skin color). Record a value for the variable for each person in your class. Tabulate the data in the space provided and plot a histogram of the frequency of the variable (number of people in each category):

NEED HELP?
See Activities
91 & 97

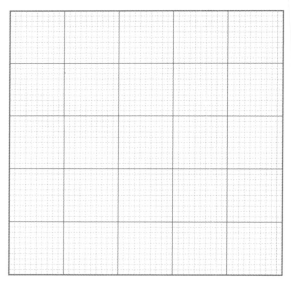

3. For your chosen phenotypic variable, suggest how a change in the environment might influence the success of the phenotypes at one extreme of the phenotypic range:

   _____

   _____

   _____

©2018 **BIOZONE** International
ISBN: 978-1-927309-55-1
Photocopying Prohibited

EXPLORE: Populations and the production of offspring

All populations must produce offspring, or they will die out (become extinct). The data below examines the production and survival of offspring from several populations. These populations remained relatively steady from year to year. A separate study showed that 39.2% of seedlings survived until at least the 9th year of growth.

Fate of common ash (*Fraxinus excelsior*) seeds (%) 1966-1969

Fate of seed	Year			
	1966	1967*	1968	1969
Undeveloped	2	NA	11	7
Seed infested by insects	23	NA	35	14
Seed eaten by small mammals	70	NA	47	76
Seed germination	5	0	7	3

* Very small crop produced. No seeds germinated.

Population data on three species of Australian seabirds

	Adult mortality (%)	Young reared per pair per year	Yearling survival (%)	Young returned to breed (%)
Yellow eyed penguin	14	0.36	50	26
Little penguin	14	0.71	>50	2.8
Short tailed shearwater	7	NA	44	44

4. (a) What percentage of common ash seeds germinated in 1968? _____

(b) What percentage of seeds produced in 1968 might have survived to their 9th year? _____

(c) Common ash begins fruiting in its 30th year. Given that this is many years after it germinated, what might we expect about the number of seedlings from any particular year that survive to reproductive age?

_____
_____

5. (a) For yellow eyed penguins, how many years does it take to rear a single chick? _____

(b) How many surviving chicks survive to return to breed? _____

6. (a) From the data above, what can be said about the number of offspring produced by a population compared to the number needed to maintain that population?

_____
_____

(b) Why is this number of offspring produced? _____
_____

EXPLORE: Variation

Populations have variation. The ladybug *Harmonia axyridis* has a wide range of elytra (hardened forewings) colors and patterns.

7. How many variations in elytra can you see in the photo?

Color variation: _____

Number of spots: _____

8. Why might variation be important?

_____
_____
_____
_____

©2018 **BIOZONE** International
ISBN: 978-1-927309-55-1
Photocopying Prohibited

EXPLORE: Natural selection and inheritance

▸ **Natural selection** acts on the phenotypes of individuals. Individuals with phenotypes that increase their fitness (survival and successful reproduction) leave more offspring, increasing the proportion of the genes corresponding to that phenotype in the next generation.

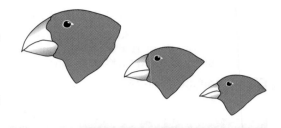

▸ Numerous population studies have shown that natural selection can cause phenotypic changes in a population relatively quickly.

▸ The finches on the Galápagos Island (Darwin's finches) are famous in that they are commonly used as examples of how evolution produces new species.

In this activity you will analyze data from the measurement of beak depths of the medium ground finch (*Geospiza fortis*) on the island of Daphne Major near the center of the Galápagos Islands. The measurements were taken in 1976 before a major drought hit the island and in 1978 after the drought (survivors and survivors' offspring).

Beak depth (mm)	No. 1976 birds	No. 1978 survivors	Beak depth of offspring (mm)	Number of birds
7.30-7.79	1	0	7.30-7.79	2
7.80-8.29	12	1	7.80-8.29	2
8.30-8.79	30	3	8.30-8.79	5
8.80-9.29	47	3	8.80-9.29	21
9.30-9.79	45	6	9.30-9.79	34
9.80-10.29	40	9	9.80-10.29	37
10.30-10.79	25	10	10.30-10.79	19
10.80-11.29	3	1	10.80-11.29	15
11.30+	0	0	11.30+	2

9. Use the data above to draw two separate sets of histograms:

   (a) On the left hand grid draw side-by-side histograms for the number of 1976 birds per beak depth and the number of 1978 survivors per beak depth (you can use different colors for the different years).

   (b) On the right hand grid draw a histogram of the beak depths of the offspring of the 1978 survivors.

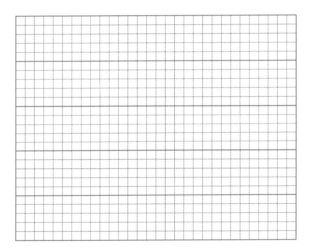
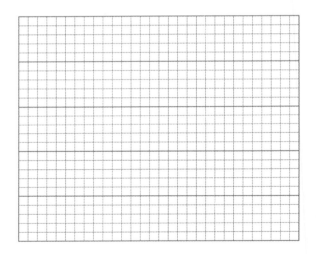

10. (a) Mark on the graphs of the 1976 beak depths and the 1978 offspring the approximate mean beak depth.

    (b) How much has the mean moved from 1976 to 1978? _____

    (c) Is beak depth heritable? _____

    (d) How do the data sets show evidence for this? _____

    _____

    _____

    _____

    _____

    _____

©2018 **BIOZONE** International
ISBN: 978-1-927309-55-1
Photocopying Prohibited

EXPLORE: Selection for insecticide resistance

Insecticides are pesticides used to control insects considered harmful to humans, their livelihood, or environment. Insecticide use has increased since the use of synthetic insecticides began in the 1940s.

▸ The widespread but often ineffective use of insecticides can lead to chemical resistance in insects, meaning the same level of an insecticide no longer controls the pest. Effective control then requires higher dosage rate or a different chemical.

▸ Random mutations (DNA changes) may also produce characteristics in individuals that are favored in the environment in which insecticides are being used.

▸ Ineffective application may include applying chemicals at the wrong dosage or at the wrong time, e.g. before rain, and applying contact sprays that may be avoided by hiding under leaves.

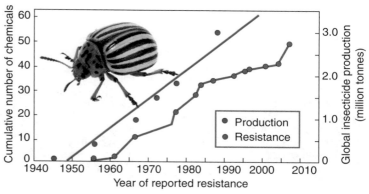

The Colorado potato beetle is a major potato pest that was originally found living on buffalo-bur (*Solanum rostratum*) in the Rocky mountains. Since synthetic insecticides began to be produced, it has become resistant to more than 50 different types.

How does resistance become more common?

The application of an insecticide can act as a selective agent for chemical resistance in pest insects. Insects with a low natural resistance die from an insecticide application, but a few (those with a naturally higher resistance) will survive, particularly if the insecticide is not applied properly. These individuals will reproduce, giving rise to a new generation which will, on average, have a higher resistance to the insecticide.

11. (a) Using the blue and red circles for susceptible and resistant individuals, draw a sequence in the boxes right to show what you think will happen to the proportion of susceptible and resistant individuals over time.

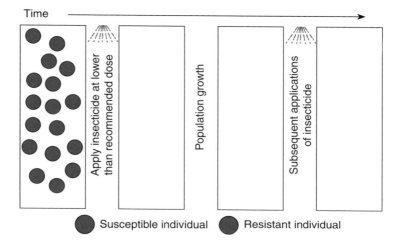

(b) What did you think would happen to the relative number of resistant and susceptible individuals over time in environments where insecticide is being regularly applied:

_____

_____

_____

(c) How does this help to explain the graph above showing increase in insecticide resistance in Colorado potato beetles and trends in global production of insecticides?

_____

_____

_____

12. There are various mechanisms by which insecticide resistance arises. Some insects may produce stronger physical barriers so that the chemical cannot penetrate their cuticle or they may detoxify the chemical using enzymes. These mechanisms come at an energetic cost (there is no free lunch). Predict what would happen next in your time sequence above if the farmer stopped using the insecticide to control the insect population. Explain your reasoning:

_____

_____

_____

©2018 **BIOZONE** International
ISBN: 978-1-927309-55-1
Photocopying Prohibited

EXPLORE: Selection in M&M's®

A simple example of selection can be carried out in class or on your own using M&M's®. M&M's® have several colors, but all taste the same. Therefore the only selection pressure is color. We can see the effect of selection pressure when people choose to eat the candy that is their favorite color. After one round of eating the candy, the relative amounts of different colors remaining can be calculated and the candy replaced to the original population number at the new proportions. After doing this several times, one or two colors will become more common than the others.

#1

#2

#3

In a bag of 90 M&M's®, there are many colors, which represents the variation in a population. You and a friend each eat 10 M&M's®, picking your favorite color first, but leave the blue ones, which you both dislike, and return them to bag, along with any uneaten M&M's®.

You count the number of different colors left and calculate their proportions. From a second bag of M&M's® you replace the colors in the proportions you calculated (i.e. the surviving M&M's® replace themselves in the proportions in which they survive in the population).

After doing this several times you are eventually left with a bag of blue M&M's®. Your selective preference for the other colors changed the make-up of the M&M's® population. This is the basic principle of selection that drives evolution in natural populations.

13. Try carrying out the activity above. You could try any colored candy. For a particularly interesting (and possibly difficult) variation, liquorice all-sorts could be tried. For this you would need a rigorous selection and replacement method. Write you results below, including the number of rounds (generations) required to produce a generation with only one color.

EXPLAIN: How natural selection works

Natural selection is the term for the mechanism by which better adapted organisms survive to produce a greater number of viable offspring. This has the effect of increasing their proportion in the population so that they become more common. This is the basis of Darwin's theory of evolution by natural selection.

1. Variation through mutation and sexual reproduction:
In a sexually reproducing population of brown beetles, changes in the DNA independently produce red coloration and 2 spot marking on the elytra. The individuals compete for limited resources.

2. Selective predation:
Brown mottled beetles are eaten by birds but red ones are avoided.

3. Change in the genetics of the population:
Red beetles have higher fitness and become more numerous with each generation. Brown beetles have poor fitness and become rare.

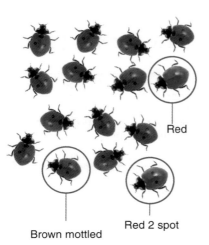
Red
Brown mottled
Red 2 spot

©2018 **BIOZONE** International
ISBN: 978-1-927309-55-1
Photocopying Prohibited

14. While developing his theory of evolution by natural selection, Charles Darwin noticed **four important aspects** that every naturally occurring population shares (you have been exploring these on previous pages). These are listed in the boxes below. Complete the boxes with an explanation of each of the four aspects and how they lead to evolution. You could use diagrams to illustrate your ideas:

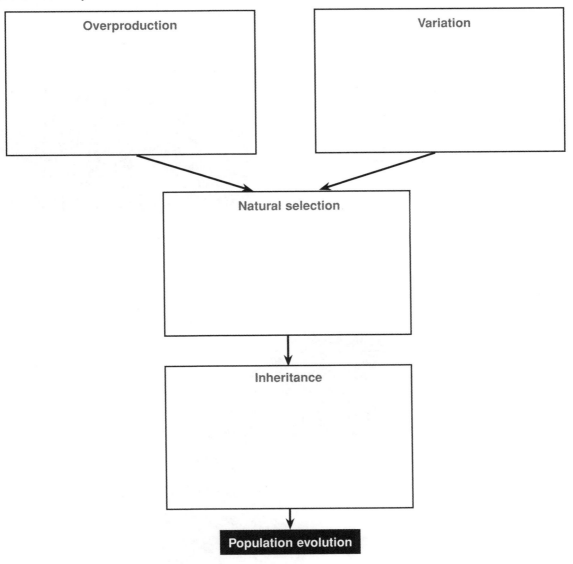

Overproduction

Variation

Natural selection

Inheritance

Population evolution

15. Using your answers above, explain how the genetic make-up of a population can change over time: _____

_____

_____

_____

_____

_____

_____

_____

16. The example of the beetles (opposite) shows selection by predation. Describe some other possible mechanisms for selection in a population:

_____

_____

_____

_____

©2018 **BIOZONE** International
ISBN: 978-1-927309-55-1
Photocopying Prohibited

EXPLAIN: Adaptations and fitness

An adaptation is any **heritable** trait that equips an organism for its niche, enhancing its exploitation of the environment and contributing to its survival and successful reproduction. Adaptations are a **product of natural selection** and can be structural (morphological), physiological, or behavioral characteristics. Adaptation is important in an evolutionary sense because adaptive features promote **fitness**. Fitness is a measure of an organism's reproductive success (genetic contribution to the next generation).

Adaptations of the North American beaver

North American beavers (*Castor canadensis*) are semi-aquatic herbivores, feeding on leaves, bark, twigs, roots, and aquatic plants. They do not hibernate. They live in domelike homes called lodges, which they build from mud and branches. Lodges are usually built in the middle of a pond or lake, with an underwater entrance, making it difficult for predators to capture them. Their adaptations enable them to exploit both aquatic and terrestrial environments.

Ears and nostrils

Valves in the ears and nose close when underwater. These keep water out.

Lips

Their lips can close behind their front teeth.
This lets them carry objects and gnaw underwater, but keeps water out and stops them drowning.

Front feet

Front paws are good at manipulating objects.
The paws are used in dam and lodge construction to pack mud and manipulate branches.

Teeth

Large, strong chisel-shaped front teeth (incisors) grow constantly.
These let them fell trees and branches for food and to make lodges with.

Waterproof coat

A double-coat of fur (coarse outer hairs and short, fine inner hairs). An oil is secreted from glands and spread through the fur.
The underfur traps air against the skin for insulation and the oil acts as a waterproofing agent and keeps the skin dry in the water.

Eyes

A clear eyelid (nictitating membrane). This protects the eye and allows the beaver to still see while swimming.

Oxygen conservation

During dives, beavers slow their heartbeat and reduce blood flow to the extremities to conserve oxygen and energy.
This enables them to stay submerged for 15 minutes even though they are not particularly good at storing oxygen in the tissues.

Thick insulating fat

Thick fat layer under the skin. Insulates the beaver from the cold water and helps to keep it warm.

Large, flat paddle-like tail

Assists swimming and acts like a rudder. Tail is also used to slap the water in communication with other beavers, to store fat for the winter, and as a means of temperature regulation in hot weather because heat can be lost over the large unfurred surface area.

Large, webbed, hind feet

The webbing between the toes acts like a diver's swimming fins, and helps to propel the beaver through the water.

17. In pairs, discuss how adaptations arise as a result of natural selection and how they increase fitness. Summarize your discussion as a list of points below:

_____

_____

_____

_____

_____

©2018 **BIOZONE** International
ISBN: 978-1-927309-55-1
Photocopying Prohibited

Adaptations for diving in air-breathing animals

Air breathing animals that dive must cope with lack of oxygen (which limits the length of the dive) and pressure (which limits the depth of the dive). Many different vertebrate classes, including birds, reptiles, and mammals, have diving representatives, which have evolved from terrestrial ancestors and become adapted for an aquatic life. Diving air-breathers must maintain a supply of oxygen to the tissues during dives and can only stay under water for as long as their oxygen supplies last. Their adaptations enable them to conserve oxygen and so prolong their dive time.

Species for which there is a comprehensive fossil record, e.g. whales (below), show that adaptations for a diving lifestyle accumulated slowly during the course of the group's evolution.

Penguin

Humpback whale

Weddell seals

Diving birds

Penguins show many of the adaptations typical of diving birds. During dives, a bird's heart rate slows, and blood is diverted to the head, heart, and eyes. This response is called the **diving reflex** and it is present in all air-breathing vertebrates to some extent. It is highly developed in diving mammals.

Diving mammals

Dolphins, whales, and seals are among the most well adapted diving mammals. They exhale before diving, so that there is no air in the lungs and nitrogen does not enter the blood. This prevents them getting the bends when they surface (a condition in which dissolved gases come out of solution at reduced pressures and form bubbles in the tissues). During dives, oxygen is conserved by reducing heart rate dramatically and redistributing blood only to critical organs. Diving mammals have high levels of muscle myoglobin, which stores oxygen, but their muscles also function efficiently using anaerobic metabolism.

18. The following list identifies some adaptations in a beaver. Suggest a selection pressure that might have been acting on the beaver to produce the adaptations below:

    (a) Large front teeth: _____

    (b) Thick layer of insulating fat: _____

    (c) Oil secreting glands in skin: _____

    (d) Large webbed hind feet: _____

19. Why would animals with a longer dive time have an adaptive advantage? _____

20. (a) All air breathing vertebrates show the diving reflex to some extent when the nostrils are immersed in water, regardless of whether or not they are diving animals. What could this suggest about their evolution?

    (b) From what you understand about natural selection and adaptation, explain why this reflex is highly developed in mammals that have adopted a fully aquatic lifestyle:

©2018 **BIOZONE** International
ISBN: 978-1-927309-55-1
Photocopying Prohibited

ELABORATE: Adaptations and environment

Sometimes, genetically unrelated (or very distantly related) organisms evolve similar adaptations in response to the particular environmental challenges they face. This may result in very different species with a very similar appearance (**convergence**). Although the organisms are not closely related, evolution has produced similar solutions to similar ecological problems. This phenomenon is called **convergent evolution**. Convergent evolution also accounts for the appearance of analogous structures, i.e. structures that serve the same function in unrelated taxa and are not the result of shared ancestry (e.g. the wings of insects and birds and the eyes of cephalopod mollusks and vertebrates).

Similar adaptations in unrelated animals

Gliding mammals

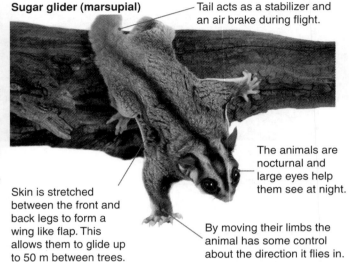

Sugar glider (marsupial) — Tail acts as a stabilizer and an air brake during flight.

Flying squirrel (rodent)

Colugo (related to lemurs)

The ability for mammals to glide between trees has evolved independently in unrelated animals. The characteristics listed right for the sugar glider are typically found in the gliding mammals shown here.

Skin is stretched between the front and back legs to form a wing like flap. This allows them to glide up to 50 m between trees.

The animals are nocturnal and large eyes help them see at night.

By moving their limbs the animal has some control about the direction it flies in.

Echolocating mammals

Microbats use echolocation (sonar) for navigation and foraging. They produce a series of species-specific high frequency calls and listen to the variation in the echo in order to pick out obstacles and locate prey (flying insects).

Dolphins and toothed whales use sonar for navigation and foraging. Outgoing sound is focused and modified by the fatty melon in the dolphin's forehead. Incoming echoes are detected by the lower jaw bone and directed to the ear.

Oilbirds are nocturnal (night-active) and nest in caves. They have good night vision, but use echolocation in caves, where there is no light at all. The echolocation is not as advanced as in bats or dolphins and is only used as a navigation tool.

21. (a) Suggest why gliding between trees (rather than walking) is an advantage to the gliding mammals described above:

_____

_____

(b) What features of the gliding mammals show convergence? _____

_____

_____

22. Oilbirds probably developed echolocation later than microbats or dolphins. Why might we suggest this? _____

_____

_____

©2018 **BIOZONE** International
ISBN: 978-1-927309-55-1
Photocopying Prohibited

Similar adaptations in unrelated plants

Cholla cactus (SW USA)

Euphorbia baioensis (Kenya)

Guaraná

Citrus

Tea

Caffeine

A C Moraes cc 2.0

Ellen Levy Finch cc 3.0

▶ The North American cactus and African *Euphorbia* species shown above are both xerophytes. They have evolved similar structural adaptations to conserve water and survive in a hot, dry, desert environment. Although they have a similar appearance, they are not related. They provide an excellent illustration of how unrelated organisms living in the same environment have independently evolved the same adaptations to survive.

▶ Their appearance is so similar at first glance that the *Euphorbia* is often mistaken for a cactus. Both have thick stems to store water and both have lost the presence of obvious leaves. Instead, they have spines or thorns to conserve water (a leafy plant would quickly exhaust its water reserves because of losses via transpiration). In cacti, spines are highly modified leaves. In *Euphorbia*, the thorns are modified stalks. It is not until the two flower that their differences are obvious.

▶ Convergent evolution is often seen in similarities of body shape or behavior. But convergent evolution can also occur in the production of chemicals in the body. In plants, several unrelated species have all evolved quite different pathways for the production of the molecule caffeine.

▶ The molecule is identical in the cacao, citrus, guaraná, coffee, and tea lineages. In some cases, the molecule is produced in the flowers, enhancing bee pollination, while in others the caffeine is found in the leaves, which may be to deter browsing animals.

23. Using the North American cactus and African *Euphorbia* as examples, explain why organisms show convergence:

_____

_____

_____

_____

24. Caffeine is manufactured in different plants in different ways, but seems to have the same purposes in different plants. What does this tell us about the adaptive value of caffeine manufacture?

_____

_____

_____

25. How does natural selection lead to adaptation? _____

_____

_____

_____

_____

_____

©2018 **BIOZONE** International
ISBN: 978-1-927309-55-1
Photocopying Prohibited

EVALUATE: Modeling natural selection:

Natural selection acts on phenotypes. Those individuals better suited to an environment will have a greater chance of reproductive success. Natural selection can be modeled in a simple activity based on predation as a selective pressure. You can carry out the following activity by yourself, or work with a partner to increase the size of the population. The black, gray, and white squares on the opposite pages represent phenotypes of a population. Cut them out and follow the **steps 1-10** below to model natural selection. You will also need a sheet of white paper and a sheet of black paper.

1. Cut out the squares on the opposite page and record the number of black, gray, and white squares. Work out the proportion of each phenotype in the population (e.g. 0.33 black 0.34 gray, 0.33 white) and place these values in the table below. This represents your starting population (you can combine populations with a partner to increase the population size for more reliable results).

2. For the first half of the activity you will also need a black sheet of paper or material that will act as the environment (A3 is a good size). For the second half of the activity you will need a white sheet of paper.

3. Place 14 each of the black, gray, and white squares in a bag and shake them up to mix them. Keep the others for making up population proportions later.

4. Now take the squares out of the bag and randomly distribute them over the sheet of black paper (this works best if your partner does this while you aren't looking).

5. For 20 seconds, pick up the squares that stand out (are obvious) on the black paper. These squares represent animals in the population that have been preyed upon and killed (you are acting the part of a predator on the snails). Place them to one side and pick up the rest of the squares. These represent the population that survived to reproduce.

6. Count the remaining phenotype colors and calculate the proportions of each phenotype. Record them in the table below in the proportions row of generation 2. Use the formula: Proportion = number of colored squares /total number of squares remaining. For example: for one student doing this activity: proportion of white after predation = 10/30 = 0.33.

7. Before the next round of selection, the population must be rebuilt to its original total number using the newly calculated proportions of colors and the second half of the squares from step 3. Use the following formula to calculate the number of each color: number of colored squares required = proportion x number of squares in original population (42 if you are by yourself, 84 with a partner). For example: for one student doing this activity: 0.33 x 42 = 13.9 = 14 (you can't have half a phenotype). Therefore in generation 2 there should be 14 white squares. Do this for all phenotypes using the spare colors to make up the numbers if needed. Record the numbers in the numbers row of generation 2. Place generation 2 into the bag.

8. Repeat steps 4 to 7 for generation 2, and 3 more generations (5 generations in total or more if you wish).

9. On separate graph paper, draw a line graph of the proportions of each color over the five generations. Which colors have increased, which have decreased?

10. Now repeat the whole activity using a white sheet background instead of the black sheet. What do you notice about the proportions this time?

Generation		Black	Gray	White
1	Number			
	Proportion			
2	Number			
	Proportion			
3	Number			
	Proportion			
4	Number			
	Proportion			
5	Number			
	Proportion			

©2018 **BIOZONE** International
ISBN: 978-1-927309-55-1
Photocopying Prohibited

The set of all the versions of all the genes in a population (its genetic make-up) is called the **gene pool**. Cut out the squares below and use them to model the events described in opposite.

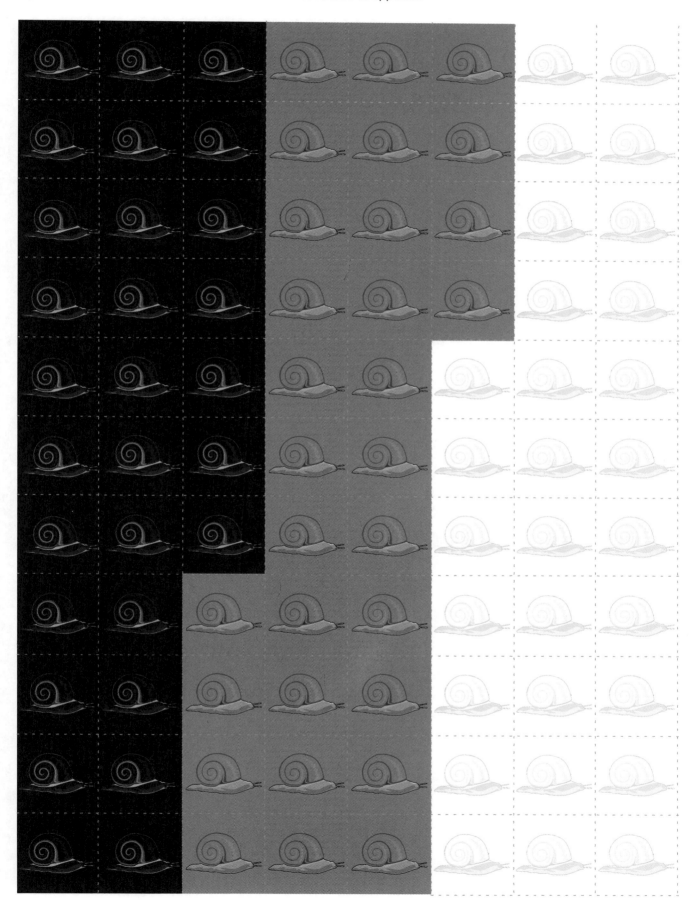

©2018 **BIOZONE** International
ISBN: 978-1-927309-55-1
Photocopying Prohibited

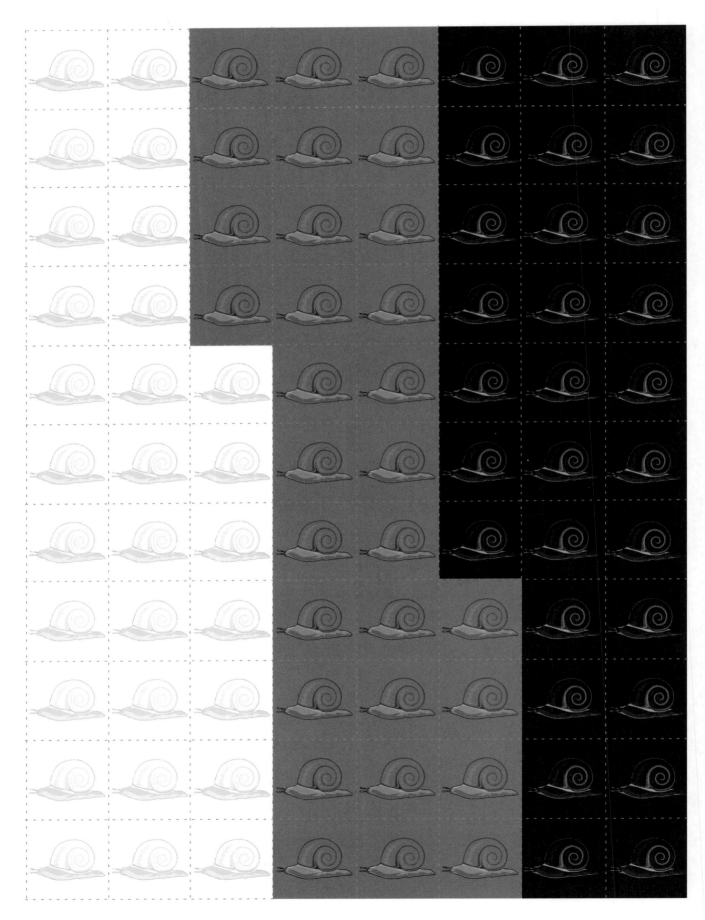

©2018 **BIOZONE** International
ISBN: 978-1-927309-55-1
Photocopying Prohibited

41 The Emergence of Species

ENGAGE: What is a species?

Examine the photographs below showing a range of different animals. The images below show three breeds of one single species (closely related), and three species of one genus (less related).

A

B

C

D

E

F

1. Group the photos above (using the letters A-F) belonging to one species and to the three separate species. Include a reason as to why you have place the images in those groups:

   (a) A single species: _____

   _____

   _____

   (b) Three species: _____

   _____

   _____

2. What other types of organisms can you think of that have breeds with very different appearances but are all one species, or are defined as separate species, but look very similar or can interbreed?

   _____

   _____

   _____

3. How would you define a species? _____

   _____

   _____

4. Breeds of domestic animals can differ a great deal in their appearance and behavior. Why are they not considered separate species?

   _____

   _____

©2018 **BIOZONE** International
ISBN: 978-1-927309-55-1
Photocopying Prohibited

SC CE P LS4.C

EXPLORE: The species is the basic unit of classification

Biological species

▶ A biological species is defined as a group of organisms capable of interbreeding to produce fertile offspring. Reproductive isolation is the cornerstone of this definition (all members of a species share the same gene pool).

▶ In practice, some closely related species can interbreed to produce fertile hybrids (e.g. species of *Canis*, which includes wolves, coyotes, domestic dogs, and dingoes).

▶ The concept of a biological species is difficult to apply to extinct organisms and those that reproduce asexually. Plants hybridize easily and can reproduce vegetatively (asexually). Some organisms, such as bacteria can also transfer genetic material to unrelated species.

▶ Increasingly, biologists are using DNA analyses to clarify relationships between closely related populations.

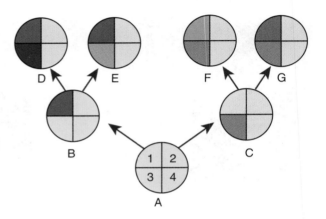

These breeds of domestic dog and the wolf (a different species) can all interbreed to produce viable offspring.

Defining species by their evolutionary history

Another way of defining a species is by using evolutionary history (**phylogeny**). This is determined on the basis of **shared derived characteristics** (physical or molecular characteristics that evolved in an ancestor and are present in all its descendants).

▶ In the model right, you can see how new characteristics arise in the descendants of a common ancestor.

▶ Defining a species as the smallest group of populations that can be distinguished by a unique set of characteristics can lead to splitting existing species into many 'new' species, especially as increasingly sophisticated genetic techniques are able to detect small differences between individuals.

5. The gray wolf (*Canis lupus*) is found in North America and Eurasia. These populations have been separated since the Bering Strait formed about 12,000 years ago, yet they are still genetically the same. What does this tell us about the process of species formation?

_____

_____

6. *Ensatina eschscholtzii* is a species of lungless salamander found on the west coast of North America. There are a number of subspecies, which probably expanded southwards from an single ancestral population in Oregon along either side of California's Central Valley. Adjacent populations interbreed to produce fertile offspring (◄—►) except just south of San Francisco, where interbreeding is rare, and in southern California, where eastern and western populations overlap but do not interbreed successfully (■—■). Studies of protein variation and mitochondrial DNA suggest that the subspecies are genetically distinct. In pairs, discuss the species status of the *Ensatina* complex and present your argument for *Ensatina eschscholtzii* being one species (with a number of subspecies), or seven separate species. Are there arguments for both points of view? Use more paper if you wish and attach it here.

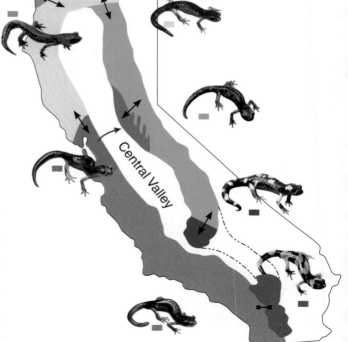

_____

_____

_____

_____

_____

_____

©2018 **BIOZONE** International
ISBN: 978-1-927309-55-1
Photocopying Prohibited

EXPLAIN: How do new species emerge?

▶ Species evolve in response to selection pressures in the environment. These may be naturally occurring or caused by humans. The diagram below represents a possible sequence for the evolution of two hypothetical species of butterfly from an ancestral population. As time progresses (from top to bottom) the amount of genetic difference between the populations increases, with each one becoming increasingly genetically isolated from the other. The greater the genetic difference between individuals, the less likely it is that they will interbreed successfully.

▶ The isolation of two populations and their 'pool' of genes often begins with a 'parent' population being split into two smaller populations by a geographical barrier. Food and habitat preferences may then diverge in the different environments. This may be followed by mechanisms that prevent successful reproduction, either before mating (such as differences in behavior) or after mating (e.g. hybrid sterility). As the two populations become increasingly different from each other, they are progressively labeled: population, race, subspecies, and finally species.

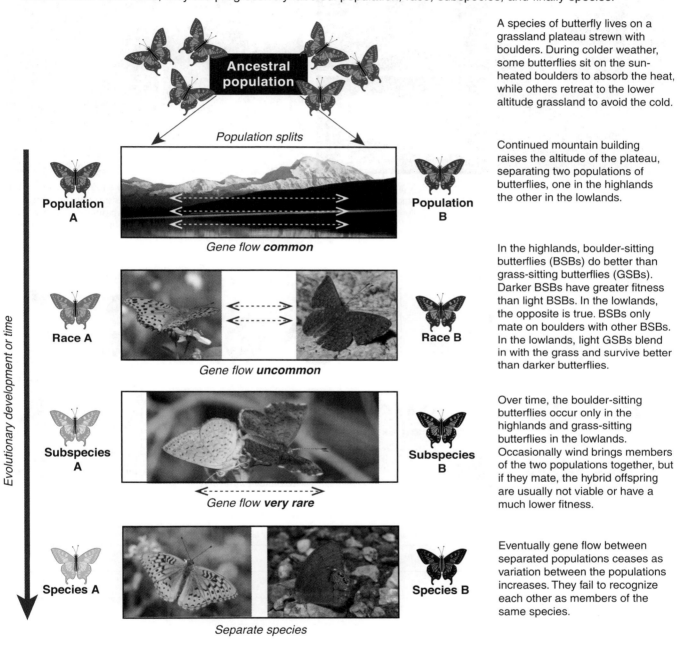

Ancestral population

Population splits

Evolutionary development or time

Population A — **Population B** — *Gene flow common*

Race A — **Race B** — *Gene flow uncommon*

Subspecies A — **Subspecies B** — *Gene flow very rare*

Species A — **Species B** — *Separate species*

A species of butterfly lives on a grassland plateau strewn with boulders. During colder weather, some butterflies sit on the sun-heated boulders to absorb the heat, while others retreat to the lower altitude grassland to avoid the cold.

Continued mountain building raises the altitude of the plateau, separating two populations of butterflies, one in the highlands the other in the lowlands.

In the highlands, boulder-sitting butterflies (BSBs) do better than grass-sitting butterflies (GSBs). Darker BSBs have greater fitness than light BSBs. In the lowlands, the opposite is true. BSBs only mate on boulders with other BSBs. In the lowlands, light GSBs blend in with the grass and survive better than darker butterflies.

Over time, the boulder-sitting butterflies occur only in the highlands and grass-sitting butterflies in the lowlands. Occasionally wind brings members of the two populations together, but if they mate, the hybrid offspring are usually not viable or have a much lower fitness.

Eventually gene flow between separated populations ceases as variation between the populations increases. They fail to recognize each other as members of the same species.

7. (a) What was the initial condition that separated the ancestral population above into populations A and B?

_____

(b) What selection pressure(s) caused the formation of races (and subspecies) A and B? _____

_____

_____

_____

©2018 **BIOZONE** International
ISBN: 978-1-927309-55-1
Photocopying Prohibited

EXPLORE: Isolation can lead to speciation

In a sexually reproducing population, genetic divergence is unlikely to occur when all the individuals in the population are in contact. In most animal populations, the formation of a new species (called speciation) often begins with the geographical isolation of a subset of the population. This allows populations to diverge in other ways, e.g. through different habitat preferences and through the evolution of different structural and behavioral features. Differences that become part of a species biology are called reproductive isolating mechanisms. Most operate before mating and fertilization, but those acting after this (such as hybrid infertility) are important in preventing offspring between closely related species (e.g. horses and donkeys) and are often the result of incompatible chromosome numbers.

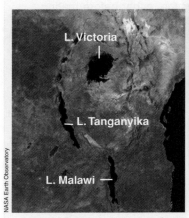

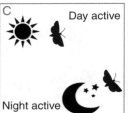

The cichlids in the rift lakes of East Africa are a striking example of speciation as a result of geographic isolation. Geologic changes to the lake basins split populations and resulted in a diversification into hundreds of species in the different lakes.

The western and eastern meadowlarks look very similar but their songs are very different. The eastern species sings pure whistles, while the song of the western species is more flute-like. Where their ranges overlap, they do not interbreed.

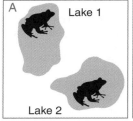
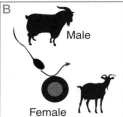
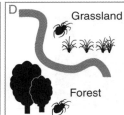
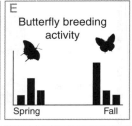
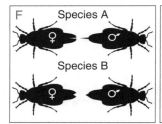
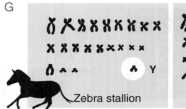

8. For the pictographs A-H, identify how the species represented are prevented from breeding:

(a) A: _____

(b) B: _____

(c) C: _____

(d) D: _____

(e) E: _____

(f) F: _____

(g) G: _____

(h) H: _____

9. Using the letters, put the pictographs in a sequence that you think might represent a hierarchy of reproductive isolation, from the least degree of isolation to the most (i.e. from the mechanisms that arise first to those that arise last):

_____

EXPLAIN: Patterns of evolution

▶ The diversification of an ancestral group into two or more species is called **divergent evolution**. This is shown right, where two species diverge from a **common ancestor**. Note that another species (W) arose, but became extinct.

▶ When divergent evolution involves the formation of a large number of species to occupy different niches, it is called an **adaptive radiation**.

▶ The evolution of species may not necessarily involve branching. A species may accumulate genetic changes that, over time, result in a new species. This is known as **sequential evolution**.

▶ Evolution of species does not always happen at the same pace. Two models describe the pace of evolution.

1) Phyletic gradualism proposes that populations diverge slowly by accumulating adaptive features in response to different selective pressures.

2) Punctuated equilibrium proposes that species are stable (in stasis) for most of their existence and the formation of new species is rapid. The stimulus for evolution is a change in the environment, e.g. an increase in mean temperature.

It is likely that both mechanisms operate at different times for different taxonomic groups.

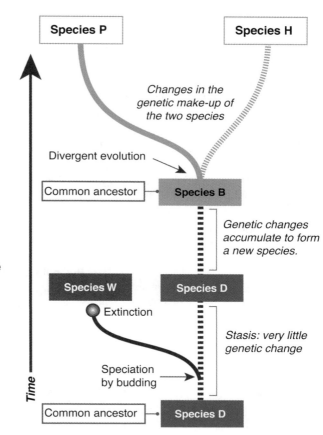

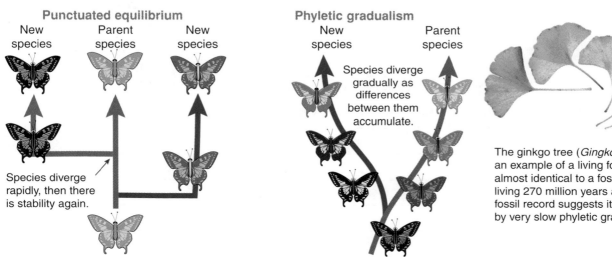

The ginkgo tree (*Gingko biloba*) is an example of a living fossil. It is almost identical to a fossil species living 270 million years ago. The fossil record suggests it evolved by very slow phyletic gradualism.

10. In the hypothetical example of divergent evolution illustrated at the top of the page:

(a) Identify the type of evolution that produced species B from species D: _____

(b) Identify the type of evolution that produced species P and H from species B: _____

(c) Name all species that evolved from: Common ancestor D: _____ Common ancestor B: _____

11. What might provide the stimulus for the adaptive radiation of an ancestral group of organisms: _____
_____
_____

12. When do you think punctuated equilibrium is most likely to occur and why? _____
_____
_____
_____

©2018 **BIOZONE** International
ISBN: 978-1-927309-55-1
Photocopying Prohibited

ELABORATE: Do changes in the environment lead to evolution?

▶ Evidence for evolution is also provided by experimental studies. These studies use organisms with very short generation times (e.g. *Drosophila*, bacteria) so that many generations can be observed in a short amount of time.

▶ A 2013 study observed 20 generations of mites with a generation time of 5 weeks to determine the effect of harvesting regime (simulating deaths) on the age and size at maturity of remaining individuals. Three 'populations' of mites were prepared. Each population had a different harvesting regime: no harvesting, juvenile harvesting, and adult harvesting. Measurements of the remaining (unharvested) individuals were made at 18, 37, 63, and 95 weeks to assess age and size at maturity. The data presented below show the results from 18 and 95 weeks only. Harvesting (selection pressure) occurred from weeks 13 to 83, with the last measurements being taken at 95 weeks. There were four to five harvesting events per generation.

▶ To obtain information about the effect of food level, replicate treatments of high food and low food availability were established for each harvesting regime.

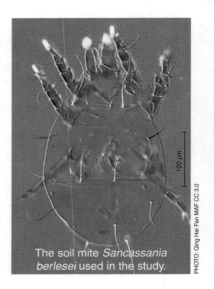

The soil mite *Sancassania berlesei* used in the study.

PHOTO: Qing Hai Fan MAF CC 3.0

Soil mite age and size at maturity

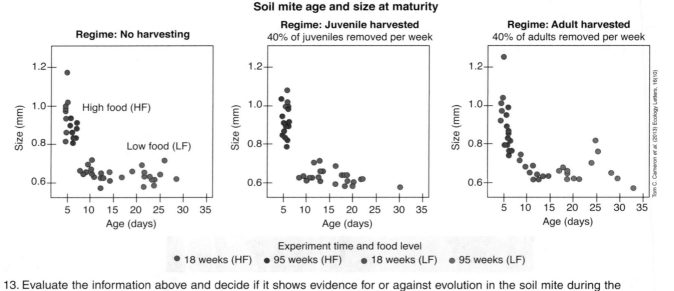

Tom C. Cameron et al. (2013) Ecology Letters, 16(10)

Experiment time and food level
● 18 weeks (HF)   ● 95 weeks (HF)   ● 18 weeks (LF)   ● 95 weeks (LF)

13. Evaluate the information above and decide if it shows evidence for or against evolution in the soil mite during the experimental period. Justify your answer with examples:

_____

_____

_____

_____

_____

_____

_____

14. Using the understanding you have developed and information in this activity explain how and why new species arise:

_____

_____

_____

_____

_____

_____

_____

_____

©2018 **BIOZONE** International
ISBN: 978-1-927309-55-1
Photocopying Prohibited

42 The Extinction of Species

EXPLORE: Extinction is a natural process

Extinction is the death of an entire species, so that no individuals are left alive. More than 98% of species that have ever lived are now extinct, most of these before humans were present.

▶ A **mass extinction** describes the widespread and rapid (in geologic terms) decrease in life on Earth and involves not only the loss of species, but the loss of entire families (which are made up of many genera and species). There have been five previous mass extinctions (below).

Graptolite

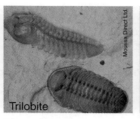
Trilobite

Coral

Conodonts (a type of jawless fish) disappeared in the Triassic extinction.

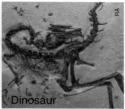
Dinosaur

Ordovician extinction (458-440 MYA). Second largest extinction of marine life: >60% of marine invertebrates died. One of the coldest periods in Earth's history.

Devonian extinction (375-360 MYA). Marine life affected especially brachiopods, trilobites, and reef building organisms.

Permian extinction (252 MYA). Nearly all life on Earth died. 57% of families and 83% of genera were wiped out. 96% of marine species became extinct.

Triassic extinction (201.3 MYA). At least half of all species present became extinct, vacating niches and ushering in the age of the dinosaurs.

Cretaceous extinction (66 MYA). Marked by the extinction of nearly all dinosaur species (their descendants, the birds, survive).

▶ The diagram below shows how the diversity of life has varied over the history of life on Earth and aligns this with major geologic and climatic events.

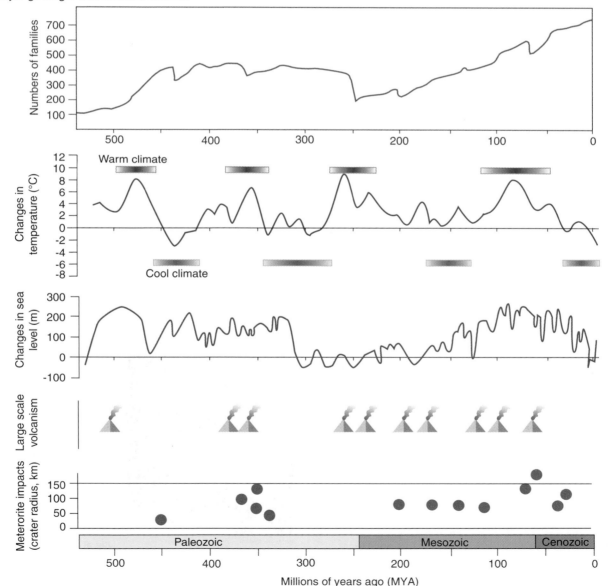

©2018 **BIOZONE** International
ISBN: 978-1-927309-55-1
Photocopying Prohibited

CE LS4.C

1. Study the data on the previous page carefully and answer the following questions:

   (a) What is the evidence that extinction is a natural process? _____

   _____

   _____

   (b) What is the general cause of a species becoming extinct? _____

   _____

   _____

   (c) Mark and label the five mass extinctions on the top graph on the previous page:

   (d) Is there reason to believe that there is any one cause for any of these mass extinctions? Explain your answer:

   _____

   _____

   _____

   _____

   (e) What has happened to the diversity of life soon after each mass extinction?_____

   _____

   _____

   _____

2. Why do you think extinction is an important process in evolution? _____

   _____

   _____

EXPLAIN: How have humans affected extinction rates?

▶ Human activity is the cause of many recent extinctions. One very famous extinction was the extermination of the dodo within 70 years of its first sighting on the island of Mauritius in 1598. Another was the extinction of the passenger pigeon, which went from an estimated 3 billion individuals before the arrival of Europeans in North America to extinction by 1914. More recent extinctions include the Yangtze River dolphin in 2006 and the Pinta Island tortoise, a giant tortoise, in 2012 (with the death of the last individual, Lonesome George).

▶ Human activities such as hunting and destruction of habitat (by pollution and land clearance for agriculture and urbanization) have caused many recent extinctions. The effects of climate change on habitats and life histories (e.g. breeding times) may drive many other vulnerable species to extinction.

▶ Species become extinct naturally at an estimated rate of about one species per one million species per year. This is called the background extinction rate. Since the 1500s, at least 412 vertebrate or plant species have become extinct (below). Proportionally, birds have been affected more than any other vertebrate group, although a large number of frog species are in rapid decline.

Organism	Total number of species (approx)*	Known extinctions (since ~1500 AD)*
Mammals	5487	87
Birds	9975	150
Reptiles	10,000	22
Amphibians	6700	39
Plants	300,000	114

* These numbers vastly underestimate the true numbers because so many species are undescribed.

Heinz-Josef Lücking

The dodo (skeleton left) has become synonymous with human-caused extinction ("dead as a dodo"). The dodo was endemic to the island of Mauritius in the Indian Ocean and became extinct by 1662, just 64 years after its discovery. Its extermination was so rapid that we are not even completely sure what it looked like and know very little about it.

©2018 **BIOZONE** International
ISBN: 978-1-927309-55-1
Photocopying Prohibited

3. (a) What is the total number of known extinctions over the last 500 years?_____

   (b) Why is this number probably an underestimate? _____

   _____

4. (a) There are ~10,000 living or recently extinct bird species. Assuming a background extinction rate of one species per 1,000,000 species per year, how many bird species should be becoming extinct per year? Show your working:

   _____

   _____

   _____

   (b) Since 1500 AD, 150 birds species are known to have become extinct. How many times greater is this rate of extinction than the background rate for birds (assume 1500 CE to 2015)?

   _____

   _____

   (c) Carry out the calculations above for mammals, reptiles, amphibians, and plants. Compare their current extinction rates with their background extinction rates:

   _____

   _____

   _____

   _____

   _____

5. The IUCN has established a Red List Index (RLI) for four taxonomic groups: reef forming corals, amphibians, birds, and mammals. This index focuses on the genuine status of changes. An RLI of 1.0 equates to all species qualifying as Least Concern (unlikely to become extinct in the near future). An RLI of 0 means that all species have become extinct. The figure right shows the trends in risk for the four taxonomic groups currently completed.

   (a) Which taxon is moving most rapidly towards extinction risk?

   _____

   (b) Which taxon is, on average, the most threatened?

   _____

   (c) Why would an index like this be useful and how could it help to highlight environmental issues of concern?

   _____

   _____

   _____

   _____

6. Humans are frequently not directly responsible for the extinction of a species, yet many species extinctions can be attributed to humans. Explain why these extinctions can be attributed to humans:

   _____

   _____

   _____

   _____

   _____

   _____

©2018 **BIOZONE** International
ISBN: 978-1-927309-55-1
Photocopying Prohibited

43 SNAPSHOT: Antibiotic Resistance

ENGAGE: MRSA

Methicillin-resistant *Staphylococcus aureus* (MRSA) is a strain of the bacterium *Staphylococcus aureus* that has developed resistance to beta-lactam antibiotics, a group of broad spectrum antibiotics, which includes many penicillin derivatives.

MRSA has become relatively common in hospitals, prisons, and places where there are people with impaired immune systems or open wounds. Special procedures are now used in these places to reduce the occurrence of MRSA infection, although it can now be acquired in the wider community as well.

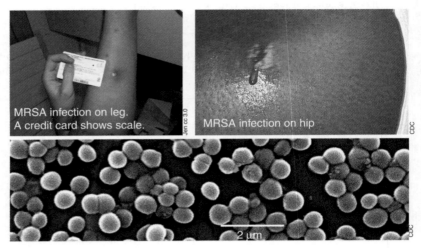

MRSA infection on leg. A credit card shows scale.

MRSA infection on hip

2 μm

1. When did you last have an infection (e.g. a cold): _____

2. Did you visit a doctor and obtain antibiotics or did you let it get better on its own? _____

3. If you took antibiotics, did you take them all? _____

4. Did they work (and how quickly)? _____

   _____

5. Do you think you might have got better just as quickly without the antibiotics, i.e. did you really need them?

   _____

   _____

EXPLORE: Antibiotic resistance

Antibiotic resistance arises when a bacterium acquires a new gene or there is a change to an existing gene that produces an antibiotic effect (gives an ability to prevent the antibiotic's action). When a bacterial population is exposed to antibiotics, a resistant population can arise if one individual cell acquires antibiotic resistance. The appearance of antibiotic resistant populations can be modeled using a spreadsheet to visualize the change in population numbers.

	A	B	C	D	E	F
1	Time	Number of susceptible bacteria		Number of resistant bacteria		Total number
2	1	100000		1		=B2+D2
3	Antibiotic	=(B2*0.25)		=D2*1		
4	2	=B3*2		=D3*2		=B4+D4
5	Antibiotic	=(B4*0.25)		=D4*1		
6	3	=B5*2		=D5*2		=B6+D6
7	Antibiotic	=(B6*0.25)		=D6*1		
8	4	=B7*2		=D7*2		=B8+D8
9	Antibiotic	=(B8*0.25)		=D8*1		
10	5	=B9*2		=D9*2		=B10+D10
11	Antibiotic	=(B10*0.25)		=D10*1		
12	6	=B11*2		=D11*2		=B12+D12
13	Antibiotic	=(B12*0.25)		=D12*1		
14	7	=B13*2		=D13*2		=B14+D14

6. The spreadsheet above gives a formula for modeling the effect of an antibiotic that kills 75% of the bacterial population each generation, starting with a population of 100,000 susceptible bacteria and 1 resistant bacterial cell. The spreadsheet assumes all surviving bacteria reproduce once per generation, that the antibiotic is present in the environment every generation, and that all resistant bacteria are fully resistant. Replicate this spreadsheet yourself (or go **BIOZONE's Resource Hub** for a copy).

   (a) What happens to the susceptible population over time? _____

   _____

   (b) What happens to the total population of bacteria over time? _____

   _____

©2018 **BIOZONE** International
ISBN: 978-1-927309-55-1
Photocopying Prohibited

(c) In reality, there is an energetic cost to a bacterial cell in maintaining resistance to an antibiotic. What might happen to the bacterial population if the antibiotics were removed halfway through the treatment?

_____

_____

_____

(d) You can increase the complexity of the spreadsheet model by adding in energy costs or the effect of variable resistance to the antibiotic. Work in pairs or groups to explore ways to do this.

EXPLORE: New mutations can enhance fitness

In 1988, Richard Lenski and his research group began the *E. coli* Long Term Evolution Experiment. They prepared 12 populations of the bacterium *E. coli* in a growth medium in which glucose was the limiting factor for growth. Each day, 1% of each population was transferred to a flask of fresh medium, and a sample of the population was frozen and stored. The experiment has now reached more than 66,000 generations. The populations were tested at intervals for mutations, but strains were not selected for any particular trait (other than being in a low glucose medium). Fitness (contribution to the next generation) in a low glucose medium has increased in all populations compared to the ancestral *E. coli* strain. However, in three strains, a hypermutation ability also arose. This increased their fitness relative to the other strains (right).

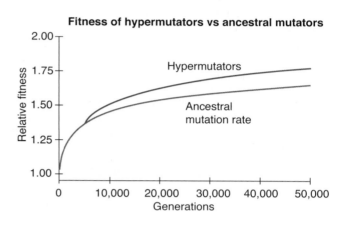
Fitness of hypermutators vs ancestral mutators

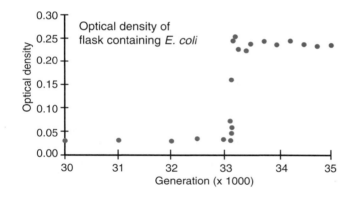
Optical density over time for strain A-3

A particularly important mutation became evident in one of the *E. coli* populations after 31,000 generations. It became able to metabolize the citrate that was a component of the growth medium. This ability gave the strain increased fitness relative to all the other populations. The mutation for citrate metabolism was noticed when the optical density (cloudiness) of the flask containing the *E. coli* suddenly increased (left), indicating an increase in the population of bacteria.

Investigations into the citrate mutation found that many other previous mutations were required before the final mutation could have an effect. Before generation 15,000, the ability to metabolize citrate was unlikely to arise. After generation 15,000, the ability to metabolize citrate became increasingly likely.

7. Why would the ability to hypermutate (produce many new mutations) increase the fitness of the strains with this trait?

_____

_____

_____

_____

8. (a) When did the mutation for citrate metabolism occur? _____

(b) What effect did it have on the fitness of the new *E coli* strain compared to the previous strain (and other strains)?

_____

(c) What would you expect to occur if the citrate mutation was mixed with the original strain? _____

_____

_____

©2018 **BIOZONE** International
ISBN: 978-1-927309-55-1
Photocopying Prohibited

EXPLORE: Antibiotic resistance

▶ Antibiotic resistance occurs when bacteria are in a selective environment created by the presence of antibiotics. Antibiotic compounds are common in the environment and were even before the invention of medical antibiotics. For example, honey, ginger, cloves, and garlic all show antibacterial properties and were extensively used before the introduction of modern antibiotics (and often still are). This use already created a selection pressure for antibiotic resistance (although less extreme than today) .

▶ Penicillin was the first commercial antibiotic produced and was introduced in large quantities in 1943, mainly to due to the need for it to be ready for the 1944 invasion of Normandy by Allied Forces. Bacterial resistance to it was already recorded by 1940. In most cases, some level of resistance to antibiotics occurs within a few years of the antibiotic being introduced.

▶ MRSA was first recorded in 1961. Outbreaks in the community began to occur from 1980. Incidence rates peaked around 2005. In the UK, a concerted effort to eliminate MRSA, including compulsory reporting and better hospital hygiene, has seen the incidence of MRSA there rapidly decline. In the United States, incidence has remained steady since around 2005. The rates are compared in the plot, below left.

Antibiotic	Discovered	Introduced	Resistance recorded
Penicillin	1940	1943	1940
Streptomycin	1944	1947	1947, 1956
Tetracycline	1948	1952	1956
Erythromycin	1952	1955	1956
Vancomycin	1956	1972	1987
Methicillin	1959	1960	1965
Gentamycin	1963	1967	1970
Linezolid	1990	2000	2001
Ceftaroline	2009	2010	2011

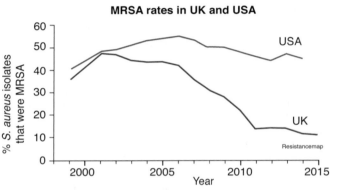

MRSA rates in UK and USA

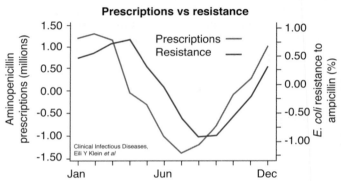

Prescriptions vs resistance

Seasonal variation in resistance

▶ Antibiotic resistance in a bacterium increases its fitness relative to susceptible bacteria in an environment that selects for antibiotic resistance (i.e. antibiotic present). There is a cost in this resistance, such as slower reproduction or a thicker cell wall and slower nutrient uptake. When the selective environment is removed, the fitness of resistant strains decreases relative to susceptible strains.

▶ This can be seen in the seasonal fluctuations of resistance (above, right). The graph shows the mean monthly prescriptions for the antibiotic aminopenicillin and the percentage resistance of *E. coli* to ampicillin (a type of aminopenicillin) over a year for the years 1999-2007. The figures are monthly means or percentage resistance relative to a mean for the entire period.

9. (a) Which antibiotic had the longest length of time before resistance was recorded? _____

   (b) When was MRSA first recorded? _____

   (c) What procedures helped reduce the incidence of MRSA in the UK? _____

   _____

10. In the plot of prescriptions vs resistance in *E. coli* (above right):

   (a) What is the approximate lag between prescriptions issued and resistance?_____

   (b) What do you think is the cause of this lag?_____

   (c) The highest resistance of *E. coli* to aminopenicillin occurs in which month in the USA? _____

   (d) Why do you think resistance increases at this time? _____

   _____

   (e) What might happen to ampicillin resistance in *E. coli* if aminopenicillin was never again prescribed after August?

   _____

   _____

©2018 **BIOZONE** International
ISBN: 978-1-927309-55-1
Photocopying Prohibited

EXPLAIN: Spreading resistance

▶ Resistant bacterial strains often occur because patients have been prescribed antibiotics unnecessarily, the dose is too low, or the patient did not finish the course of treatment. Resistant strains of bacteria can then spread and become more resistant over time, acquiring genes from other bacterial strains with different resistance properties.

How antibiotic resistance develop

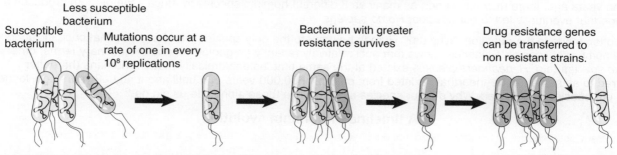

Susceptible bacterium

Less susceptible bacterium

Mutations occur at a rate of one in every 10^8 replications

Bacterium with greater resistance survives

Drug resistance genes can be transferred to non resistant strains.

Any population includes variants with unusual traits, in this case reduced sensitivity to an antibiotic. These variants arise as a result of mutations in the bacterial chromosome.

When a person takes an antibiotic, only the most susceptible bacteria will die. The more resistant cells remain and continue dividing.

If the amount of antibiotic taken is too low or not potent enough, the resistant cells survive and divide to produce a population with a higher than normal antibiotic resistance.

The antibiotic initially used against this bacterial strain will now be ineffective. The resistant cells can exchange genetic material with other bacteria, or pass on the genes for resistance to their descendants.

How resistance to antibiotics spreads and increases

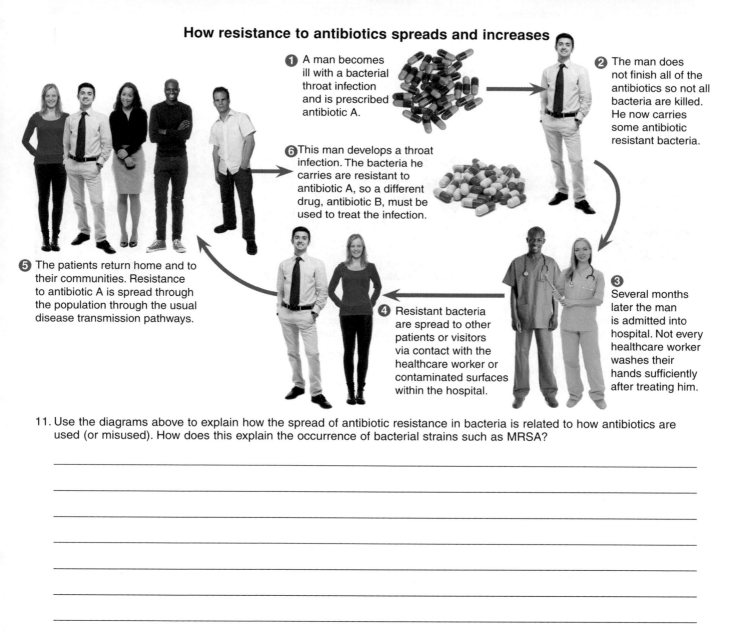

❶ A man becomes ill with a bacterial throat infection and is prescribed antibiotic A.

❷ The man does not finish all of the antibiotics so not all bacteria are killed. He now carries some antibiotic resistant bacteria.

❻ This man develops a throat infection. The bacteria he carries are resistant to antibiotic A, so a different drug, antibiotic B, must be used to treat the infection.

❺ The patients return home and to their communities. Resistance to antibiotic A is spread through the population through the usual disease transmission pathways.

❹ Resistant bacteria are spread to other patients or visitors via contact with the healthcare worker or contaminated surfaces within the hospital.

❸ Several months later the man is admitted into hospital. Not every healthcare worker washes their hands sufficiently after treating him.

11. Use the diagrams above to explain how the spread of antibiotic resistance in bacteria is related to how antibiotics are used (or misused). How does this explain the occurrence of bacterial strains such as MRSA?

_____

_____

_____

_____

_____

_____

©2018 **BIOZONE** International
ISBN: 978-1-927309-55-1
Photocopying Prohibited

44 SNAPSHOT: Human Evolution

ENGAGE: Why us?

Around 3.5 million years ago there was a rapid expansion in the number of hominins (human ancestors) in Africa. These included species of the genus *Australopithecus*, *Paranthropus*, and *Homo* (our own genus). Between 3 to 2 million years ago, there may have been as many as 8 different hominin species in Africa. One of them produced a lineage that eventually led to our species, *Homo sapiens*.

This often leads to the question: 'Why us?" Human beings are the only species left of a genus that once included a large number of species. Evidence shows that until relatively recently (in geological and evolutionary terms) other *Homo* species (e.g. *H. neanderthalensis*) existed at the same time as anatomically modern humans (hereafter referred to as humans). Neanderthals existed from possibly 600,000 years ago until about 40,000 years ago, longer than our species has existed. Why did our species survive while those alongside us did not?

A timeline of hominin evolution

Many other species of hominin existed, but our species, *Homo sapiens*, is the only one that survived to the present day.

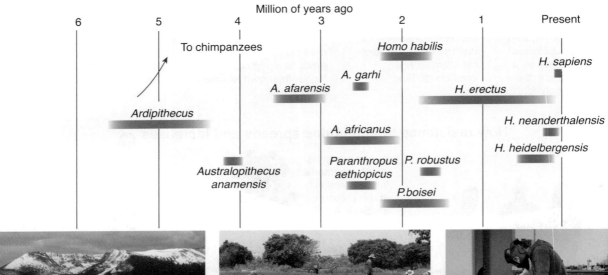

Did some structural or physiological feature allow humans to take advantage of a **change in climate** or make it that bit harder for other *Homo* species?

Was it our **cultural and social abilities** that gave us the evolutionary edge? Humans work together in social groups. Did this help us out-compete other species?

Our **large brain** gave us the ability to **make and use tools** to create ever more sophisticated objects and structures. Did this give us the edge?

1. As a class group, discuss the three hypotheses above. Which one seems most plausible? Was it all three or something else altogether? Summarize your discussion below:

_____

_____

_____

_____

EXPLORE: Group activity

The following pages present four opportunities to explore aspects of human evolution. Form groups of four. Each person in the group chooses one of the following EXPLORE aspects to research (using information on the relevant pages and any extra research they may choose to do). The four EXPLORE aspects are: climate (Q1-7), physical changes (Q8-10), molecular evidence for evolution (Q11-16), and cultural evolution (Q17-21). Once each person has become an "expert" on their particular aspect, they will report back to their group about their findings.

Each person in the group then needs to write their own summary of what they have learned about **all** four aspects and decide if there is enough evidence to attribute our success as a species to any one (or more) of the aspects presented in the EXPLORE sections (Q22-23).

LS4.C CE

©2018 **BIOZONE** International
ISBN: 978-1-927309-55-1
Photocopying Prohibited

EXPLORE: Climate change

Climate change in Africa

Earth's climate has fluctuated between warm and cool periods through its history. The fluctuations have many causes, including changes in the Earth's orbit and rotation, and changes to the Earth's surface (e.g. mountain building and continental drift). Climate change usually alters the distribution patterns of plants (vegetation) and animals. Understanding how both the climate and the distribution patterns of plants and animals have changed over time gives an insight into the selection pressures that may have influenced the evolution of our ancestors.

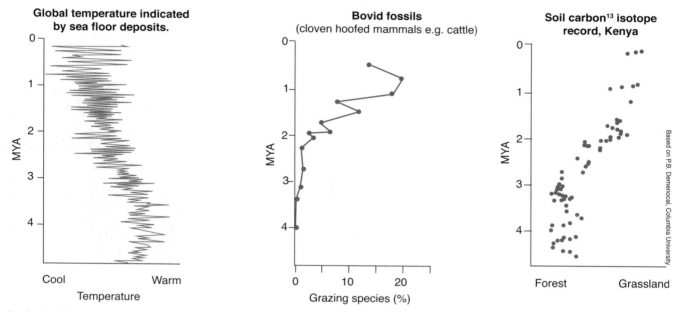

Global temperature indicated by sea floor deposits.

Cool — Warm
Temperature

Bovid fossils
(cloven hoofed mammals e.g. cattle)

Grazing species (%)

Soil carbon[13] isotope record, Kenya

Forest — Grassland

Based on P.B. Demenocal, Columbia University

Analysis of deep sea sediment and deep ice cores provides evidence about Earth's climate in the geological past (paleoclimate). From around 3 million years ago, the Earth's climate began to cool. Superimposed on this are cycles of around 23,000 years corresponding to changes in the Earth's angle of rotation. The cooling of the Earth's climate correlates with changes in the environment in Africa, especially in East Africa, which changed from being predominately forest to grasslands.

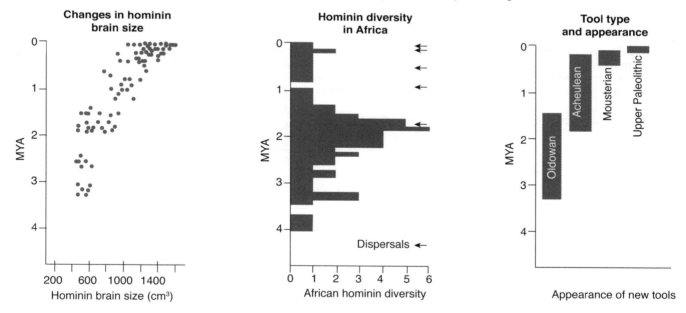

Changes in hominin brain size

Hominin brain size (cm³)

Hominin diversity in Africa

Dispersals ←

African hominin diversity

Tool type and appearance

Oldowan / Acheulean / Mousterian / Upper Paleolithic

Appearance of new tools

Important aspects of human evolution include the increase in brain volume, the development of increasingly complex tools, and dispersal out of Africa to new parts of the world. Some of these dispersals resulted in the formation of new populations and species, separate from those that remained in Africa.

2. Study the graphs above. Can you see any patterns in the data related to human evolution, dispersal, and climate?

_____

_____

_____

_____

©2018 **BIOZONE** International
ISBN: 978-1-927309-55-1
Photocopying Prohibited

 SC ESS3.D ESS2.E

Climate change in Eurasia

Modern humans evolved in Africa about 300-200,000 years ago. Although there is ongoing debate, evidence generally supports continual waves of migrations of *Homo sapiens* from as early as 180,000 years ago. Before this, perhaps 500,000 years ago, an earlier *Homo* species had migrated out of Africa to reach Eurasia where they evolved into the Neanderthals (*Homo neanderthalensis*). They were evidently very successful, but there has always been the question of what happened to them and why did modern *Homo sapiens* survive through the last glaciation whereas *H. neanderthalensis* became extinct 40,000 years ago?

Neanderthal populations and modern human migration routes through the Old World (Africa, Asia, and Europe)

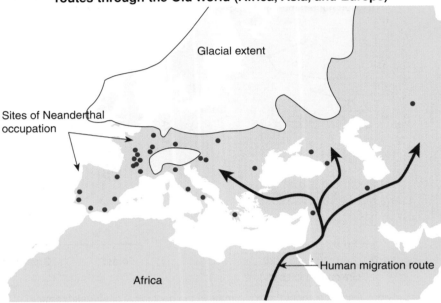

Human populations migrating out of Africa could reinforce populations affected by the glaciations in Eurasia.

▸ The map above shows the extent of the last glaciation in Eurasia, which lasted from around 115,000 to 12,000 years ago. Data from recent environmental studies suggest the onset of this glacial period in Eurasia happened extremely quickly (possibly within one or two generations).

▸ Neanderthal populations living in Eurasia would have had few large areas of refuge from the extreme climate, unlike modern humans who would have been able to "wait out" the glaciation in the relatively warmer climate of Africa.

▸ New research simulating population extinction and replacement of both Neanderthal and human populations suggests that humans may have simply outlasted Neanderthals because human populations in Eurasia could be reinforced from populations continually moving out of Africa. Neanderthals had no such separate population base, and with populations widely scattered, populations that became extinct during the glaciation were not replaced, leading to the eventual extinction of the Neanderthals.

3. Neanderthals occupied which part of the Old World? _____

4. Where did modern humans evolve? _____

5. When did modern humans evolve? _____

6. When was the last glaciation and how long did it last? _____

   _____

7. What role did human populations in Africa and the glaciers in Eurasia have on the ultimate survival of Neanderthals and modern humans?

   _____

   _____

   _____

   _____

   _____

   _____

   _____

©2018 **BIOZONE** International
ISBN: 978-1-927309-55-1
Photocopying Prohibited

EXPLORE: Physical changes in human evolution

Humans have evolved numerous characteristics that can be traced through time by studying fossils. Humans have a very large brain/skull compared to our body size and habitually walk upright. We have several skeletal modifications for this behavior. Walking upright must have had some important advantage as there has been a trend towards this since the evolution of our early ancestors, the Australopithecines.

Trends in hominin skull shape

Australopithecus afarensis	*Homo habilis*	*Homo erectus*	*Homo neanderthalensis*	*Homo sapiens*
3.9 - 2.9 mya	2.8 - 1.5 mya	1.9 mya - 600 kya	500 kya - 40 kya	200 kya - present
457 cm³	552 cm³	1016 cm³	1512 cm³	Brain volume 1335 cm³

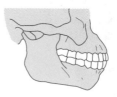

▶ The skulls above show the general trend in hominin evolution of increasing skull capacity and a trend towards a flatter, more upright face. Brain volume and skull capacity increases at the same time. The point of the spine meeting the skull (arrowed) becomes more directly under the skull over time, which helps to balance the skull on the spine when walking upright (bipedal). This reduces the muscular effort required to hold the head up.

▶ Earlier hominins had large eyebrow ridges, large jaws, cheek bones, and teeth. There is a trend to smaller teeth, especially in the molars and canines. All these changes indicate that human ancestors were eating less coarse food, requiring less chewing. Parallel to the *Homo* species were robust animals in the genus *Paranthropus*. This group had large teeth and jaws associated with a diet of coarse material such as grasses and roots (below).

Trends in hominin jaw shape

Early hominins ———————▶ **Late hominins**

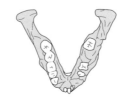

▶ Humans evolved to eat a wide range of foods. The reduction of the jaw is evidence that this food was becoming softer over time, and is probably related to a change to hunting game rather than eating coarser vegetable material. The invention of fire and cooking made food softer again and hastened this reduction in jaw size.

▶ Meat has a greater energy value than the same quantity of vegetable material. This meant *Homo* species were able to spend less time finding food and more time on cultural activities. Species such as *Paranthropus boisei* (below) spent much more time finding (and eating) food.

Australopithecus afarensis
- Relatively large canine teeth
- Relatively large jaw
- V-shaped dental arcade
- Thin tooth enamel
- Diet probably consisted of fruits with some tougher, more fibrous material

Homo erectus
- Thick jaw bones
- No chin
- Relatively large molars
- Parabolic dental arcade
- Thick tooth enamel
- Diet probably included vegetable material and a large proportion of meat

Homo sapiens
- Shortened jaw with relatively thin bones
- Chin reinforces jaw, but leaves room for tongue muscles
- Parabolic dental arcade
- Thick tooth enamel
- Small molars adapted to chewing soft/cooked food

Adaptations to a coarse diet

▶ *Paranthropus boisei* had jaws and teeth adapted to a diet of very coarse vegetation and hard seeds. Its jaws produced a massive bite force of 2161 newtons, which helped to break food up. A modern human's maximum bite force is 777 newtons.

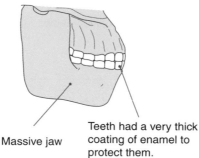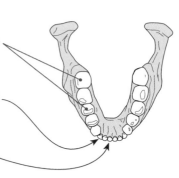

Massive jaw

Teeth had a very thick coating of enamel to protect them.

Massive molars and premolars aided effective grinding action.

Reduced size of canines permitted rotatory action, helping to grind coarse food up.

The reduced size of the incisors provides more room for molars.

©2018 **BIOZONE** International
ISBN: 978-1-927309-55-1
Photocopying Prohibited

SF ESS1.C LS4.A LS1.A

Bipedalism

Hominins stand upright and walk on two feet (they are bipedal), with modern humans being the most well adapted to this mode of locomotion. Reasons for the trend to bipedalism are varied and still debated. They include a better ability to scan for predators and food in a grassland habitat, a reduction in the surface area exposed to the midday sun, an increase in surface area exposed to the breeze for cooling, and an improved ability to use the hands to carry tools or food back to a home base.

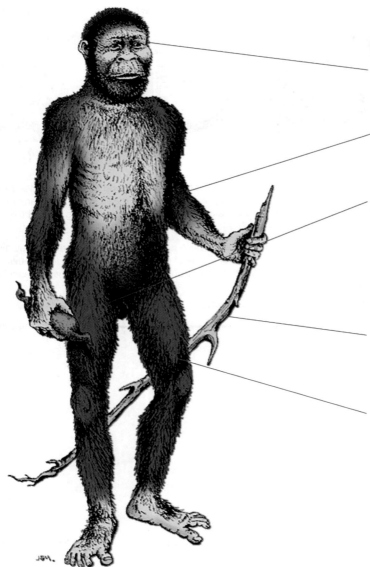

Bipedalism

Seeing over the grass
An upright posture may have helped early hominins to see predators or locate carcasses at a distance.

Carrying offspring
Walking upright enabled early hominins to carry their offspring, so the family group could move together.

Provisioning as a selection pressure
The ability to carry food while walking seems to have been important in the initial evolution of bipedalism. Females would have favored males able to provide energy-rich foods, which would improve offspring survival and increase reproductive rate. The ability to carry food from its source to a place of safety would have had a great survival advantage.

Holding tools and weapons
Tool use was probably a consequence of bipedalism, rather than a cause. Upright walking appears to have been established well before the development of hunting in early hominins.

Efficient locomotion
Once bipedalism was established, changing habitats would have provided selection pressure for greater efficiency. Being able to move across the growing savanna without expending large amounts of energy would have offered a great survival advantage.

Thermoregulation
Upright walking exposes 60% less surface area to the sun at midday and there is greater air flow across the body when it is lifted higher off the ground.

8. The trend in reduction of jaw size in hominins is a response to which evolutionary pressure? _____

_____

9. How might eating a wide variety of food provide an evolutionary advantage to a species? _____

_____

10. (a) What selection pressures are likely to have been important in the evolution of bipedalism initially?

_____

_____

(b) What environmental changes could have reinforced the advantages of bipedalism to human ancestors?

_____

_____

©2018 **BIOZONE** International
ISBN: 978-1-927309-55-1
Photocopying Prohibited

EXPLORE: Molecular evidence of human evolution

It was once thought that humans moving out of Africa simply replaced all other *Homo* species in Europe and Asia. However recent evidence from studies of the DNA of various *Homo* species and populations in Europe and Asia show that humans migrating out of Africa did indeed interbreed with those populations.

The Denisova cave finds

▶ In 2008, archeologists discovered a fragment of finger bone in the Denisova cave in Siberia. The bone fragment belonged to a juvenile female (named X-woman).

▶ Artefacts, such as a bracelet, were found at the same level as the finger bone.

▶ In 2010, a molar tooth was found at a different level to the finger bone, indicating it belonged to a different individual. A toe bone found in 2011 was at the same level as the tooth.

▶ The molar found in the Denisova cave has unique characteristics, which are not present in the molars of Neanderthals or modern humans.

▶ Carbon dating estimates the age of the artefacts and bone fragment at 40,000 years.

▶ Fossil DNA degrades quite rapidly with increasing temperature and in acidic soil conditions. The cool temperatures within the Denisova cave preserved the DNA in the fossil fragments. The fossils contained very low levels of DNA contamination from other organisms.

The Denisova cave, in the Altai mountains, Siberia, Russia

Using genome analysis to classify the Denisova cave fossils

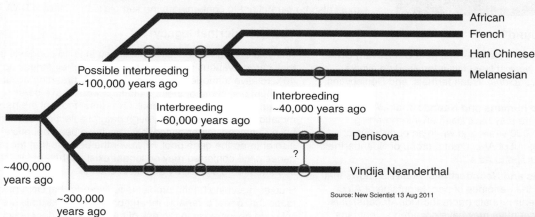

Source: New Scientist 13 Aug 2011

Nuclear DNA analysis suggests the Denisova fossils belong to a previously unknown hominin species that existed at the same time as modern humans and Neanderthals, but was genetically distinct from them (above). The fossils are called the **Denisovans**, because they have not yet been formally classified.

Nuclear genome analysis suggests the Denisovans were a sister group to the Neanderthals. They probably shared a more recent common ancestor with Neanderthals (~300,000 years ago) than with present day humans (~400,000 years ago).

The Denisovan's interbred with the ancestors of the present day Melanesians (right), and possibly with the Neanderthals, but not the ancestors of other present day populations, such as the Han Chinese. Melanesian DNA includes between 4% and 6% Denisovan DNA.

A Melanesian woman

11. How many interbreeding episodes do there appear to have been between humans and Neanderthals? _____

12. How do scientists know that the Denisovan individuals found are from separate species to humans and Neanderthals?

_____

13. (a) What modern human lineage appears to have interbred with the Denisovans: _____

(b) What percentage of DNA does this lineage appear to share with the Denisovans? _____

©2018 **BIOZONE** International
ISBN: 978-1-927309-55-1
Photocopying Prohibited

  P LS4.A

Humans interbred with Neanderthals

Neanderthals appeared about 400,000 years ago and disappeared 40,000 years ago. They lived in Europe and parts of western and central Asia. Neanderthals are the closest relative to modern humans, so there was considerable interest in mapping the Neanderthal genome. By comparing the genomes of Neanderthals and present-day humans, it may be possible to identify Neanderthal genes that have been retained in humans.

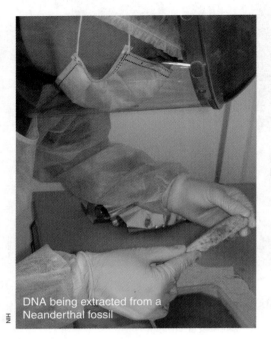

DNA being extracted from a Neanderthal fossil

Front view
Neanderthal skull   Image: Bone clones

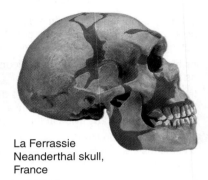

La Ferrassie
Neanderthal skull,
France

Difficulties in analyzing Neanderthal DNA

▶ The DNA is often degraded to small fragments smaller than 200 base pairs long. This makes it difficult to obtain sequence overlaps (which are critical for assembly of the genome).

▶ The DNA is often of poor quality because it has been chemically modified and degraded by the environment.

▶ Samples are often contaminated with the DNA of other organisms. Between 95-99% of the DNA obtained from the Neanderthal fossils analyzed was from microbes that colonized the bone after the Neanderthal died. Researchers must also be careful not to contaminate the sample with their own DNA.

What has been found?

The continuing analysis of Neanderthal (and Denisovan) DNA has found that there were at least five interbreeding events between humans (*H. sapiens*), Neanderthals, and Denisovans.

1. **Between archaic humans and Neanderthals**. Analysis in 2016 shows there may have been an interbreeding event around 100,000 years ago when an early wave of humans migrating out of Africa met a group of Neanderthals migrating from Europe to Asia.

2. **Between humans and Neanderthals**. Analysis shows that between 1-4% of the genomes of people outside of Africa is derived from Neanderthals (more than for Africans). It is thought these encounters may have occurred as humans migrated out of Africa ~60,000 years ago (or even earlier) and met Neanderthal populations already in the Middle East.

3. Some human populations that migrated east across Eurasia interbred with the Denisovans. Evidence of this in found in Melanesian DNA (see previous page).

4. Denisovans also interbred with Neanderthals, probably about 50,000 years ago (see previous page).

5. Denisovans interbred with an unknown group of hominins, possibly an offshoot of *H. erectus*, about 100,000 years ago.

The Neanderthal legacy

Analysis of Neanderthal DNA published in 2016 suggests that they carried various mutations that reduced their fitness up to 40% relative to modern humans. When interbreeding occurred with humans, some of these mutations would have been passed to the human gene pool. Over time, most of the harmful mutations were discarded through natural selection, but some have remained. Other genes that may have been beneficial also entered the gene pool. However the benefits that these genes once conferred may no longer exist as the human lifestyle becomes more sedentary and diets change.

Studies matching health problems to Neanderthal DNA have found that genetic variants inherited from Neanderthals are linked to an increase in the risk of heart attacks, depression, skin disorders, and nicotine addiction. However, the Neanderthal DNA may not necessarily be causing the health problem. It might just be associated with human DNA that is.

Some genes that were possibly inherited from Neanderthals or Denisovans have provided benefits. Tibetans appear to have inherited Denisovan genes that enabled high altitude adaptation. Humans may also have inherited genes associated with immunity to new diseases found outside of Africa, but already encountered by Neanderthals.

14. What percentage of Neanderthal DNA is present in modern humans? _____

15. In which group of modern humans is Neanderthal DNA mostly found and why? _____

_____

16. Describe some possible positive and negative effects of Neanderthal DNA in modern humans: _____

_____

_____

_____

_____

©2018 **BIOZONE** International
ISBN: 978-1-927309-55-1
Photocopying Prohibited

EXPLORE: Human cultural evolution

Humans have extremely complex social behaviors. Our ability to communicate information to large groups over many generations has allowed our species to dominate the Earth and change the environment to suit our needs. Was it this ability that gave us the evolutionary edge over other species of *Homo*?

Stone tools

▶ The **Paleolithic** (Old Stone Age) is a period of early cultural development spanning the emergence of the first stone tools about 3.3 mya in eastern Africa, until the development of sophisticated tool kits in the **Mesolithic** (Middle Stone Age) about 10,000 ya. These tool cultures are known mostly by their stone implements. While other materials, such as wood, were probably also used, they did not preserve well.

▶ The oldest stone tools have been dated to 3.3 million years ago and appear to have been made by *Australopithecus*, the ancestor the *Homo* genus. The tools are little more than chipped flakes of rock, but represented a huge technological leap forward. The flakes could be use for cutting or chopping, or even breaking open bones to obtain the marrow.

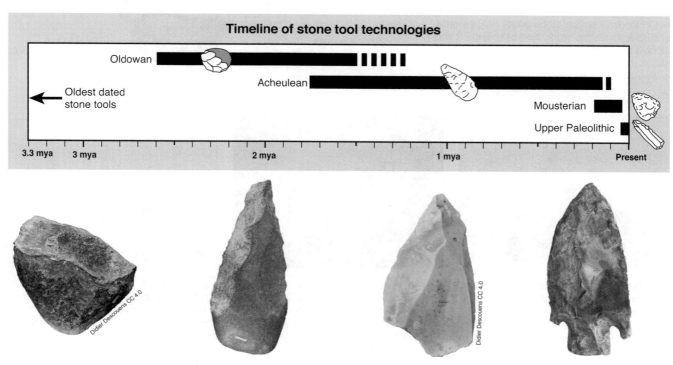

Oldowan tools are the earliest tools associated with *Homo*. They were probably made by *H. habilis* and included hammers, discs, and primitive bifaced blades.

Acheulean tools were made by Homo erectus and early *Homo sapiens*. They were bifaced hand axes that suited a wide range of uses. The tool was so successful it has been found throughout the ancient world.

Mousterian tools are associated with Neanderthals in Europe and Modern humans in Northern Africa. They are an advance on earlier Acheulean tools. Mousterian tools showed a wide range of shapes for specialized uses.

Tools from the Upper Paleolithic show advanced features and include various types of materials, including the use of bone and ivory. Tools produced include blades for cutting, spear tips and needles.

17. Describe the general trends in the design of the stone tool from Oldowan to Upper Paleolithic cultures:

_____

_____

_____

_____

_____

18. Why would a range of sophisticated tools give the species using them an advantage over other species?

_____

_____

_____

_____

©2018 **BIOZONE** International
ISBN: 978-1-927309-55-1
Photocopying Prohibited

Communication

The human brain is very large for a primate of our size, but this may not be as important as its internal organization. The most important specialization of the human brain is the capacity for language. This is a result of the development of Wernicke's and Broca's areas. Specific differences associated with the left and right hemispheres of the brain are associated with these specializations.

Wernicke's and Broca's areas

Wernicke's area is part of the brain in the cerebral cortex that is linked to speech. It is involved in the comprehension of written words and spoken language. Its function becomes apparent if it is damaged. A person will then be able to speak fluently, but produce meaningless phrases.

Broca's area is also found in the cerebral cortex but in the frontal lobe of the brain. Broca's area is involved with the production of speech. It was discovered by Pierre Paul Broca when studying two patients who had lost the ability to speak after sustaining head injuries.

Frontal lobe

Wernicke's area
The area concerned with understanding spoken words.

Broca's area
Controls the muscles of the lips, jaw, tongue, soft palate, and vocal cords during speech.

Cerebellum (largely motor control)

Evolution of communication

Didier Descouens CC SA 4.0

Australopithecus with brain endocast

Boneclones

Neanderthal hyoid bone

Communication is probably the most important aspect of human cultural evolution. Humans communicate to request help, inform others, and share attitudes as a way of social bonding. The ability to accurately convey ideas to others, both in the present and future, has allowed humans to work together, often over multiple generations, to create buildings and technology, or more efficiently gather food.

We do not know exactly when *Homo* began to use language, but comparing the features associated with language in

modern humans with fossils of ancient hominins can give us an idea. Cranial endocasts of *Homo habilis* show that the brain appears to have development in both Broca's and Wernicke's areas. However, to produce the range of sounds used by modern humans in verbal communication, the larynx and the hyoid bone must also show certain features. The hyoid bone is located in the throat and anchors the tongue muscles. The structure of the hyoid bone and its position in *Homo erectus* suggests that *H. erectus* was not able to make the same

range of sounds we can. However the structure and positioning of the hyoid bone in Neanderthals suggests they probably could make a similar range of sounds to modern humans. This suggestion is supported by computer generated models based on a Neanderthal hyoid fossil. The skull shape of Neanderthals shows an enlarged occipital lobe at the rear of the brain. The occipital lobe is involved with visual processing. It could also reflect a larger cerebellum, which is involved in the coordination of movement and spatial information.

19. What is the function of:

(a) Broca's area: _____

_____

(b) Wernicke's area: _____

_____

20. *Homo erectus* probably couldn't communicate verbally in the same way as *Homo sapiens*, but Neanderthals probably could. Explain why:

_____

_____

21. What might have been a consequence of the Neanderthal brain having a greater emphasis on spatial awareness and coordination than the human brain?

_____

_____

©2018 **BIOZONE** International
ISBN: 978-1-927309-55-1
Photocopying Prohibited

EXPLAIN: Where did we come from?

22. Return to your group and share the information you found about the aspect of human evolution you investigated. Summarize the information you have gathered as a group in the spaces below:

(a) Climate change: _____

_____

_____

_____

(b) Physical changes in human evolution: _____

_____

_____

_____

(c) Molecular evidence of human evolution: _____

_____

_____

_____

(d) Human cultural evolution: _____

_____

_____

_____

23. Decide if there is enough evidence to attribute our success as a species to any one (or more) of the four aspects presented in this activity. Justify your decisions with evidence. You may attach extra material to this page:

_____

_____

_____

_____

_____

_____

_____

_____

_____

_____

_____

_____

_____

_____

_____

_____

_____

45 Continental Drift

ENGAGE: Modeling continental drift

On the opposite page are three identical sets of four puzzle pieces representing three different time periods from the four continent world of Square World.

1. (a) Cut out the three sets of "continents", A, B, C (don't get them mixed up!). Each set of four continents forms a larger square (Square World). The colored objects (red, blue, green) represent sediments found across Square World. There are three ways that Square World can be formed from the four continents. Your task is to determine what the three ways are and the order in which they "formed" (from the oldest Square World to the youngest Square World). None of the continents need to be rotated from their current orientation. Once you think you have successfully made the three complete Square Worlds, glue them into the spaces provided below:

Youngest

Middle

Oldest

©2018 **BIOZONE** International
ISBN: 978-1-927309-55-1
Photocopying Prohibited

Three identical sets of puzzle pieces (continents)

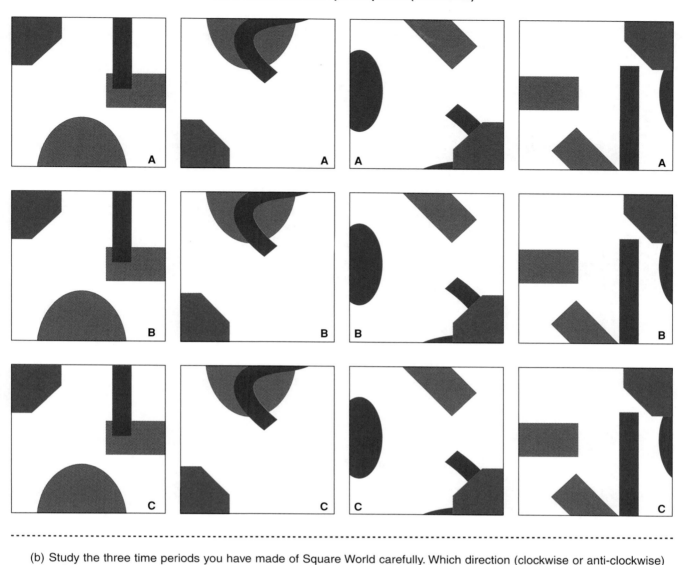

(b) Study the three time periods you have made of Square World carefully. Which direction (clockwise or anti-clockwise) is Square World "rotating"? Use evidence from the patterns in the continents to make a case for the direction of rotation in Square World.

_____

_____

_____

_____

_____

_____

_____

_____

_____

_____

_____

_____

_____

_____

_____

©2018 **BIOZONE** International
ISBN: 978-1-927309-55-1
Photocopying Prohibited

This page has been deliberately left blank

©2018 **BIOZONE** International
ISBN: 978-1-927309-55-1
Photocopying Prohibited

EXPLORE: Continental drift

▶ Continental drift (the movement of the Earth's continents relative to each other) is a measurable phenomenon; it has continued throughout Earth's history. Movements of up to 2-11 cm a year have been recorded between continents using GPS. The movements of the Earth's seven major crustal plates are driven by a geological process known as **plate tectonics**. Some continents are drifting apart while others are moving together. Many lines of evidence show that the modern continents were once joined together as 'supercontinents'. One supercontinent, Gondwana, was made up of the southern continents some 200 mya.

▶ Continental drift can also be used to explain the distribution of both living and extinct organisms. Related organisms that are now separated by oceans were once living side by side, before the continents moved apart.

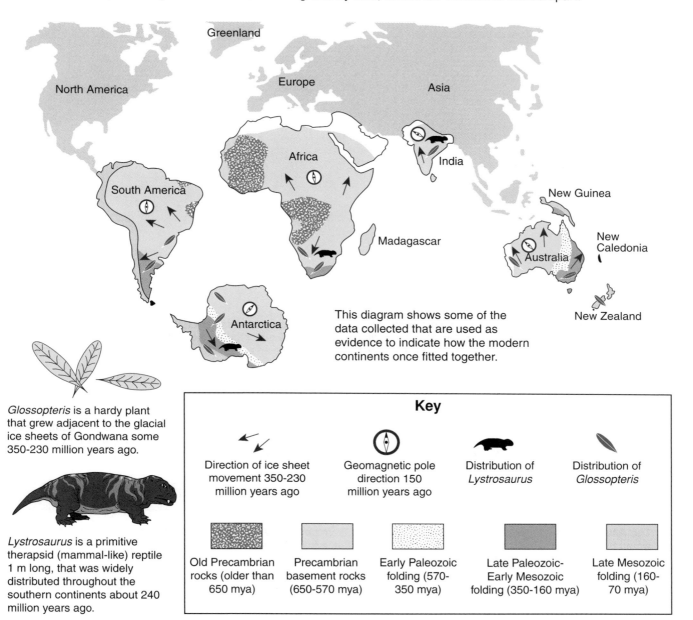

This diagram shows some of the data collected that are used as evidence to indicate how the modern continents once fitted together.

Glossopteris is a hardy plant that grew adjacent to the glacial ice sheets of Gondwana some 350-230 million years ago.

Lystrosaurus is a primitive therapsid (mammal-like) reptile 1 m long, that was widely distributed throughout the southern continents about 240 million years ago.

Key

Direction of ice sheet movement 350-230 million years ago	Geomagnetic pole direction 150 million years ago	Distribution of *Lystrosaurus*	Distribution of *Glossopteris*

Old Precambrian rocks (older than 650 mya)	Precambrian basement rocks (650-570 mya)	Early Paleozoic folding (570-350 mya)	Late Paleozoic-Early Mesozoic folding (350-160 mya)	Late Mesozoic folding (160-70 mya)

2. Name the modern landmasses (continents and large islands) that made up the supercontinent of Gondwana:

_____

_____

3. Cut out the southern continents on page 199 and arrange them to recreate the supercontinent of Gondwana. Take care to cut the shapes out close to the coastlines. When arranging them into the space showing the outline of Gondwana on page 198, take into account the following information:
   (a) The location of ancient rocks and periods of mountain folding during different geological ages.
   (b) The direction of ancient ice sheet movements.
   (c) The geomagnetic orientation of old rocks (the way that magnetic crystals are lined up in ancient rock gives an indication of the direction the magnetic pole was at the time the rock was formed).
   (d) The distribution of fossils of ancient species such as *Lystrosaurus* and *Glossopteris*.

©2018 **BIOZONE** International
ISBN: 978-1-927309-55-1
Photocopying Prohibited

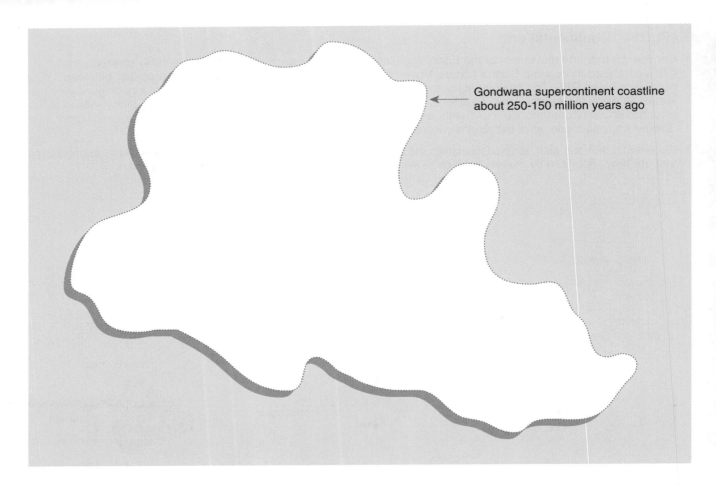

Gondwana supercontinent coastline about 250-150 million years ago

4. Once you have positioned the modern continents into the pattern of the supercontinent, mark on the diagram:
   (a) The likely position of the South Pole 350-230 million years ago (as indicated by the movement of the ice sheets).
   (b) The likely position of the geomagnetic South Pole 150 million years ago (as indicated by ancient geomagnetism).

5. State what general deduction you can make about the position of the polar regions with respect to land masses:

   _____

6. Fossils of *Lystrosaurus* are known from Antarctica, South Africa, India and Western China. With the modern continents in their present position, *Lystrosaurus* could have walked across dry land to get to China, Africa and India. However, it was not possible for it to walk to Antarctica. Use continental drift to explain the distribution of this ancient species:

   _____

   _____

7. The Atlantic Ocean is currently opening up at the rate of 2 cm per year. At this rate in the past, calculate how long it would have taken to reach its current extent, with the distance from Africa to South America being 2300 km (assume the rate of spreading has been constant):

   _____

   _____

8. Explain how continental drift provides evidence to support evolutionary mechanisms: _____

   _____

   _____

   _____

   _____

   _____

   _____

   _____

   _____

©2018 **BIOZONE** International
ISBN: 978-1-927309-55-1
Photocopying Prohibited

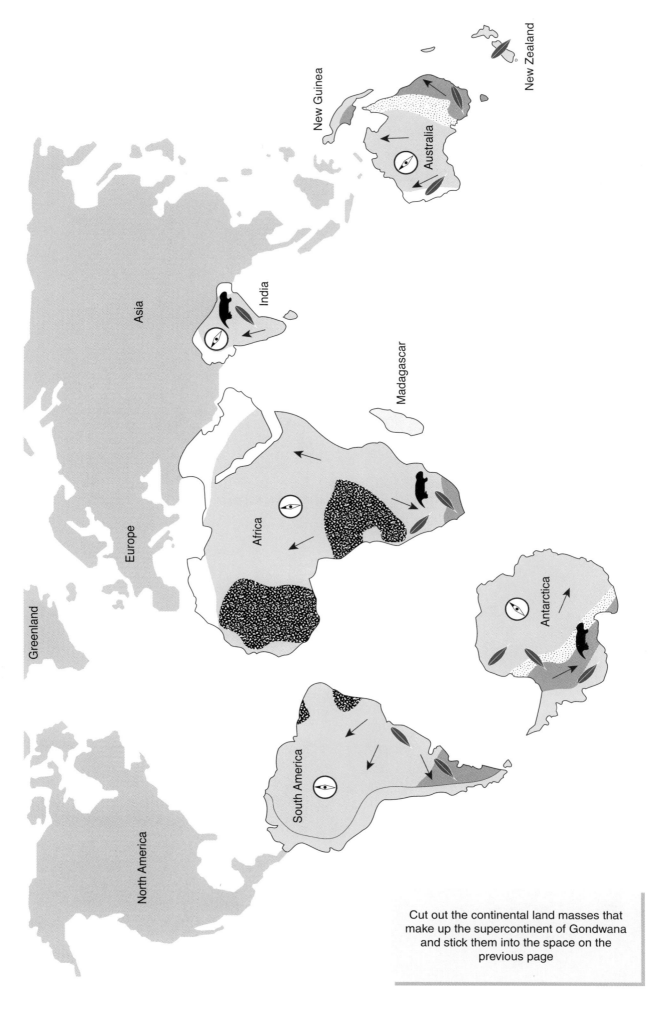

Cut out the continental land masses that make up the supercontinent of Gondwana and stick them into the space on the previous page

©2018 **BIOZONE** International
ISBN: 978-1-927309-55-1
Photocopying Prohibited

This page has been deliberately left blank

©2018 **BIOZONE** International
ISBN: 978-1-927309-55-1
Photocopying Prohibited

46 The Rise of the Tyrants Revisited

Tyrannosaurus

Tarbosaurus

At the start of this chapter you were shown a map that showed the distribution of members of the group of dinosaurs Tyrannosauroidea. The largest (and one of the last) members of this group was *Tyrannosaurus rex*, which lived in what is now North America. Fossils of the first members of the group were found in England and Siberia in rocks 100 million years older than those in which fossils of *Tyrannosaurus rex* are found. Using the knowledge and information you have gained from this chapter, you should now be able to explain the distribution of the Tyrannosauroidea fossils and the diversity of the group.

1. Use your knowledge to present an argument for the current fossil distribution of the Tyrannosauroidea, including how the group might have evolved and how their fossils arrived at their current locations.

©2018 **BIOZONE** International
ISBN: 978-1-927309-55-1
Photocopying Prohibited

47 Summative Assessment

1. Identify the type of weathering occurring in the following photos.

(a) _____   (b) _____   (c) _____

(d) Explain how the geological feature in the image below was probably formed:

_____
_____
_____
_____
_____
_____
_____
_____
_____
_____
_____

_____
_____
_____
_____

2. (a) The diagram below shows the current shape of a short part of a river's course. Draw over the top of the diagram what you might expect the river to look like at a time in the future:

(b) Explain how erosion can cause this to occur: _____

_____
_____
_____
_____

      ESS1.C ESS2.C LS4.A LS4.B LS4.C ETS1.B P CE SF

The Sacramento–San Joaquin Delta (image below) is an extensive tidal estuary, receiving the entire run-off of the Central Valley. An inverted delta (meaning its narrow end is at the seafront), the delta supplies much of the water for central and southern California, delivered through the Central Valley Project (CVP) and the State Water Project (SWP).

Most of the delta has been claimed for agriculture but land subsidence and reduced freshwater inflows have resulted in salt water intrusion. In some areas the land has subsided to greater than 5 meters below sea level. Aging levees are now also prone to failure, increasing the risks of inundation of agricultural land.

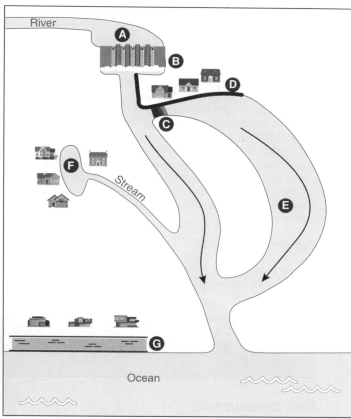

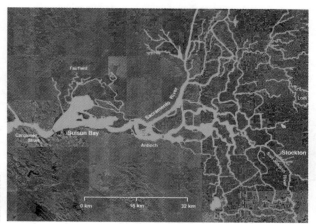

3. The diagram above right shows a managed river system depicting the typical infrastructure employed to manage water flow and prevent flood damage and erosion.

(a) Study the definitions below and match each one first with its identifying structure from the word list below and then its corresponding letter on the diagram.

Structure word list: weir, seawall, reservoir, dam, levee, detention basin, flood bypass

_____ Shoreline wall or embankment to hold back waves.   _____

_____ Embankment (often earthen) alongside a river, used to contain or control water flows.   _____

_____ Pond to capture storm water runoff.   _____

_____ Land area to convey floodwaters away from an area, reducing flow in the main river.   _____

_____ Lake created by water impounded behind a dam.   _____

_____ Barrier to hold back water and regulate flows.   _____

_____ Low wall built to alter the speed or direction of water as it flows over.   _____

(b) Describe the possible hazards faced by the houses shown at point D on the map and explain the role of structure D in reducing these hazards:

_____

_____

_____

_____

_____

_____

_____

_____

©2018 **BIOZONE** International
ISBN: 978-1-927309-55-1
Photocopying Prohibited

Levee failures and salt water intrusion are an increasing problem in the Sacramento-San Joaquin Delta. Pre-development (top panel below), the delta was a rich and diverse tidal marsh, exposed to regular flood events. The extensive post-development levee system (lower panel below) allowed farmers to drain and reclaim almost 2000 km^2) of the Delta.

The peat, exposed to air, has oxidized and the land has subsided so much that currently most of the Delta is below sea level by 3 m or more. Land subsidence endangers the levees, which are now aging, and this can trigger levee failures and flooding (inundation) such as the one right and the failure of the Upper Jones Tract levee in 2004 in which Woodward Island was flooded with 190 million cubic meters of water.

Usually, there is enough freshwater stored within the many reservoirs in the system to maintain fresh water flows and prevent salt water intrusion. However, during prolonged droughts (e.g. 2012-2017) inflows are reduced and groundwater pumping (to supply needs) further increases subsidence and the risk of levee failure.

The Sacramento-San Joaquin Delta in flood (2009). King Edward Island to the right is inundated.

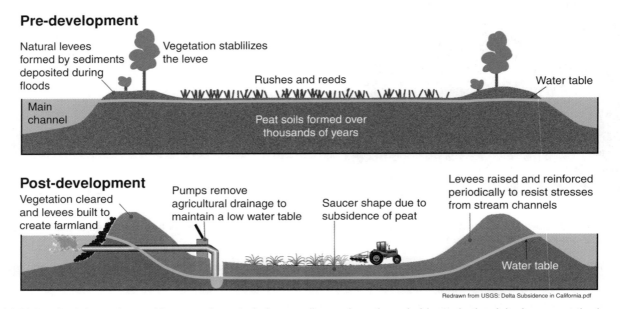

Redrawn from USGS: Delta Subsidence in California.pdf

4. (a) Using the information and images above to help you, discuss how the subsidence in the delta increases the hazards to the agriculture in the area:

_____

_____

_____

_____

_____

_____

_____

_____

_____

_____

_____

_____

_____

_____

_____

(b) Devise a possible plan to help reduce the impact of these hazards. Your plan might include management and/or engineering options and can include text and annotated drawings. Attach your plan to this page. If you need further information, visit **BIOZONE's Resource Hub:**

©2018 **BIOZONE** International
ISBN: 978-1-927309-55-1
Photocopying Prohibited

5. Wind is an agent of erosion. How much it erodes soil depends on the moisture content of the soil. The data below shows the effect of wind velocity on the erosion of soil:

Wind velocity (m/s)	Soils lost (kg/m²/minute)	
	2.67% moisture	5.20% moisture
2	0.2	0.1
4	0.8	0.2
6	1.3	0.35
8	2.2	0.6
10	3.6	0.8

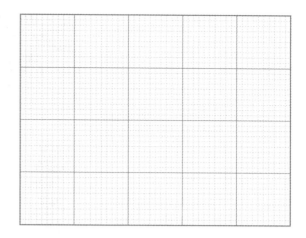

(a) Graph the data on the grid provided:

(b) What is the effect of soil moisture of erosion? _____

_____

(c) What is the effect of wind velocity on erosion? _____

_____

6. Consider the diagram below showing the evolution of a hypothetical fish species A over time.

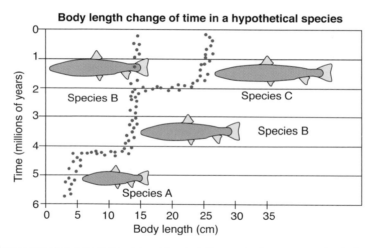

Body length change of time in a hypothetical species

(a) Does the appearance of new species over time resemble phyletic gradualism or punctuated equilibrium? Explain:

_____

_____

_____

(b) Suppose the species shown in the diagram have a generation time of 12 years. The time between the disappearance of species A and the appearance of species B is estimated to be 150,000 years (a short time in geological terms). Over this time, the mean body length from species A to species B increased by a little more than 10 centimeters. What might you expect to see in the fossil record for the evolutionary history of these two fish species ?

_____

_____

_____

(c) How many generations passed between the disappearance of species A and the appearance of species B?

_____

_____

©2018 **BIOZONE** International
ISBN: 978-1-927309-55-1
Photocopying Prohibited

You saw earlier in this chapter how geographical isolation as a result of geologic changes resulted in adaptive radiation of cichlid fish species in the rift lakes of East Africa, which is the center of cichlid diversity. The radiation originated in Lake Tanganyika, where seven lineages diversified to occupy all available freshwater fish niches. The radiations in Lakes Victoria and Malawi began with a single Tanganyikan lineage and diversified in a similar way to occupy the available niches. Within each lake, equivalent ecotypes occupy the same niche and show similar morphologies, coloration, and reproductive strategies. Both species pictured right are browsers of benthic algae.

Similarly, in the diverse family of rove beetles, the phenomenon of social parasitism has evolved independently in at least 12 geographically isolated lineages. These beetles mimic different species of army ants in body shape, behavior, and pheromone chemistry, tricking the ants into accepting the beetles into the colony, where they then consume the ant young.

Tropheus sp. L. Tanganyika

Pseudotropheus sp. L. Malawi

Eciton burchellii, army ant

Rove beetle mimic

Rove beetles: Isolated images show the usual morphology of a generalized free living species (left) next to army ant mimics. Photo shows a rove beetle mimic and its host ant.

7. (a) Use the information above and what you have learned during this chapter to explain adaptive radiation and the occurrence of similar life histories, morphologies, and behaviors in geographically separated taxa:

_____

_____

_____

_____

_____

_____

_____

_____

_____

_____

_____

_____

Innovation and extinction

The diversification of the cichlids in the East African rift lakes was associated with the evolution of particularly strong pharyngeal jaws, which allow diverse feeding ecologies but also prevent the jaws opening very wide. In Lake Victoria, introduction of the Nile perch in the 1950s and increasing enrichment of the lake have been associated with the extinction of more than 200 species (about 40%) of the lake's endemic cichlids (species found nowhere else). The Nile perch has a wider gape and can feed more rapidly than the cichlids, which were quickly outcompeted.

Haplochromis nyererei

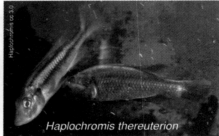
Haplochromis thereuterion

Many of the surviving cichlid species in L. Victoria, including those above, have retreated to refugia (small areas of habitat) or adapted to the human-induced changes in the lake. Adaptations include changes in foraging behavior and feeding apparatus, and morphological changes to allow better vision and increased oxygen uptake in turbid water.

(b) "A natural or human-induced change in the environment can lead to the expansion of some species, the emergence of new species, or the extinction of species". Use the information provided above and on **BIOZONE's Resource Hub** to evaluate the evidence for this statement. Produce a written synopsis suitable for oral presentation (attach here).

©2018 **BIOZONE** International
ISBN: 978-1-927309-55-1
Photocopying Prohibited

In 2004, a fossil of an unknown vertebrate was discovered in northern Canada and subsequently called *Tiktaalik roseae*. The *Tiktaalik* fossil was quite well preserved and many interesting features could be identified. These are shown on the photograph of the fossil below.

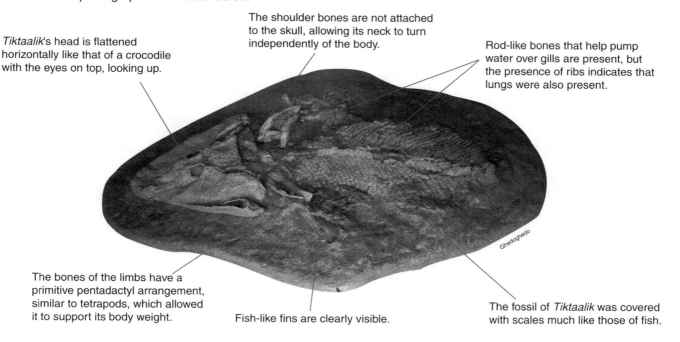

The shoulder bones are not attached to the skull, allowing its neck to turn independently of the body.

Tiktaalik's head is flattened horizontally like that of a crocodile with the eyes on top, looking up.

Rod-like bones that help pump water over gills are present, but the presence of ribs indicates that lungs were also present.

The bones of the limbs have a primitive pentadactyl arrangement, similar to tetrapods, which allowed it to support its body weight.

Fish-like fins are clearly visible.

The fossil of *Tiktaalik* was covered with scales much like those of fish.

8. Use the information above to place *Tiktaalik* on the time line of vertebrate evolution. Discuss the evidence for your decision.

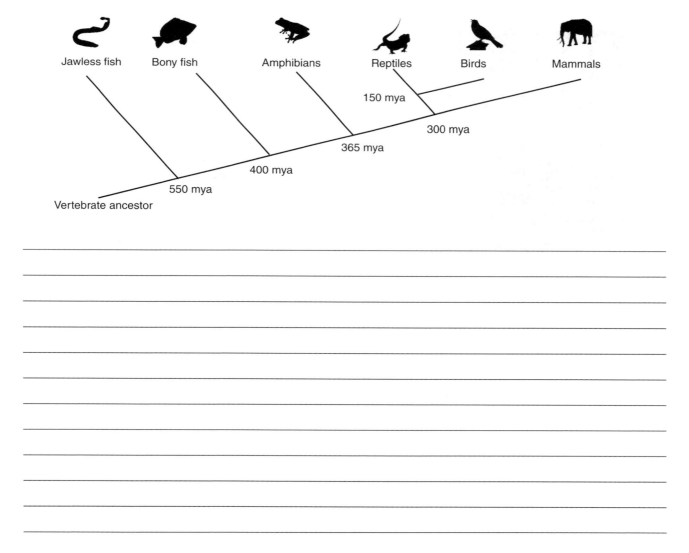

Jawless fish   Bony fish   Amphibians   Reptiles   Birds   Mammals

150 mya

300 mya

365 mya

400 mya

550 mya

Vertebrate ancestor

_____

_____

_____

_____

_____

_____

_____

_____

_____

_____

_____

_____

_____

©2018 **BIOZONE** International
ISBN: 978-1-927309-55-1
Photocopying Prohibited

9. Drosophilidae (fruit flies) are a group of small flies found almost everywhere in the world. Two genera, *Drosophila* and *Scaptomyza* are found in the Hawaiian islands and between them there are more than 800 species present on a land area of just 16,500 km². It is one of the densest concentrations of related species found anywhere. The flies range from 1.5 mm to 20 mm in length and display a startling range of wing forms and patterns, body shapes and colors, and head and leg shapes. Genetic analyses show that they are all related to a single species that may have arrived on the islands around 8 million years ago. Older species appear on the older islands and more recent species appear as one moves from the oldest to the newest islands.

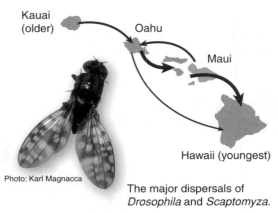

Photo: Karl Magnacca

Kauai (older)
Oahu
Maui
Hawaii (youngest)

The major dispersals of *Drosophila* and *Scaptomyza*.

(a) What evolutionary pattern is shown by the Hawaiian fruit flies:

_____

(b) Suggest why so many fruit fly species are present in Hawaii: _____

_____

(c) Describe the relationship between the age of the islands and the age of the fly species: _____

_____

(d) Account for this relationship: _____

_____

10. Cycads have a scattered distribution, being found in mainly tropical parts of Australia, southern Africa, Malaysia and the Americas. These places are separated by large areas of ocean. The seeds of most cycads sink in water and disperse only short distances on land (they don't tend to be dispersed by flying animals). The cycad fossil record dates back to around 300-280 million years ago.

A modern cycad

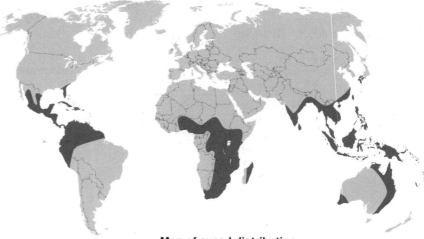

Map of cycad distribution

Present an argument (set of reasons) for the distribution of modern cycads: _____

_____

_____

_____

_____

_____

_____

_____

_____

©2018 **BIOZONE** International
ISBN: 978-1-927309-55-1
Photocopying Prohibited

Instructional Segment 4

Inheritance of Traits

Anchoring Phenomenon

Pale and interesting: Albinism is widespread throughout the animal kingdom.

48 60

How are characteristics of one generation passed to the next?

☐ 1 Describe the accumulation of experimental evidence through the 20th century that led to our understanding about the role of DNA and chromosomes in coding for inherited traits. What questions did they ask about inheritance and how did they go about answering them? How did the accumulation of evidence lead to Watson and Crick building their DNA model in 1953? Find out more about their model to understand its significance to our understanding of inheritance.

49

☐ 2 Make and use a model that incorporates some of the features of Watson and Crick's 1953 model. How does the molecule you have built carry information? What happens if you change a component of the model?

50

☐ 3 The discovery of DNA's structure paved the way for modern genetics, including the continuing study of how changes in DNA (mutations) affect the appearance (phenotype) of organisms. Explain how the experiments of Mendel with peas and Thomas Hunt Morgan with fruit fly mutations provided early evidence for the role of chromosomes as the carriers of genetic material. Outline the evidence from experiments with *Neurospora* that mutations have direct effects on the functioning of metabolic pathways.

51 52

☐ 4 Use information from phenotype studies to explore the effect of heritable single gene mutations on phenotype and the occurrence of specific diseases (e.g. polycystic kidney disease and sickle cell disease). Obtain information about how genotype matching is used to match patients for organ and tissue transplants. What genes are important in tissue typing and why?

52

What allows traits to be transmitted from parents to offspring?

☐ 5 Recall from the previous chapter that phenotypic variation in populations provides the raw material for natural selection. Provide evidence of this variation from your own observations and explain how variation arises through mutation and sexual reproduction. Use a physical model to visualize and provide evidence for how sexual reproduction produces variation in populations.

53 54 55

☐ 6 Recall how mutations may cause a change in phenotype. Provide examples as evidence that some of these mutations may be viable but result in a genetic disease. These may be mutations to autosomal chromosomes (e.g. sickle cell disease) or sex chromosomes (e.g. Turner syndrome). Explain how mutations may be beneficial in some circumstances or in some environments.

54 55

☐ 7 Use Punnett squares as a model to show how variation can arise from the mating of two biological parents. Analyze the quantity and proportion of possible outcomes to explain the variation we see in the offspring of genetic crosses (including in the inheritance of genetic diseases). You can also use interactive computer simulations to predict the outcomes of crosses based on certain parental genotypes.

56 61

☐ 8 Use pedigrees as another model to look at patterns of inheritance across generations. Evaluate possible genetic combinations and predict the chances of traits appearing in certain individual offspring.

57 61

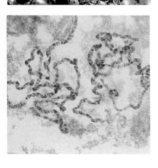

☐ 9 Describe examples to show how the environment can affect expression of the phenotype. Analyze the frequency or distribution of traits in a real population to construct an argument for the effect of environment on phenotype.

58

How does variation affect a population under selective pressure?

☐ 10 Draw connections between the variation in populations and how selection acts on the phenotype of the individuals within those populations. Analyze data from real populations to explain how the frequencies of particular traits can change from generation to generation and how populations adapt and evolve. Recall how some mutations may have a selective advantage in some situations. How does the frequency of these mutations globally provide evidence for their origin and spread (CCR5Δ32) or for their selective advantage in certain environments (HbS)?

55 59 61

48 Pale and Interesting

ANCHORING PHENOMENON: Albinism is widespread throughout the animal kingdom

Albinism is an inherited genetic disease resulting in the absence of pigmentation or coloration. The condition is widespread throughout the animal kingdom, but more common in birds, reptiles and amphibians than it is in mammals. Affected individuals have a very characteristic appearance, white or extremely pale skin (and hair if present). In mammals this is accompanied by red or pink eye color. Individuals with the affliction are subjected to selection pressures as a result of a physical appearance differing from the norm.

Muntuwandi CC 3.0

1. From your own observations and utilizing your prior knowledge, suggest how common albinism is within a population:

   _____

   _____

   _____

   _____

2. (a) Albinism is an inherited condition. What type of inheritance pattern do you think it shows? _____

   (b) Explain your answer: _____

   _____

3. (a) What environmental conditions might negatively affect an albino individual? _____

   _____

   _____

   (b) How could these conditions influence an individual's survival? _____

   _____

   _____

   _____

   _____

©2018 **BIOZONE** International
ISBN: 978-1-927309-55-1
Photocopying Prohibited

49 Experiments Showed DNA Carries the Code

ENGAGE: Experiments with a bacterium showed that DNA is the material of inheritance

By the 1900s, scientists knew that chromosomes were passed on from generation to generation. Scientists also knew that chromosomes contained DNA and proteins but they didn't know which of these was responsible for the inheritance of traits. It was only after a series of experiments, each building upon earlier findings, that this question was answered.

The first two crucial experiments involved strains of the bacterium *Streptococcus pneumoniae* (right), which can cause pneumonia and numerous other illnesses. The S strain has a smooth capsule, which makes it pathogenic (disease-causing). The R (rough) strain has no capsule and is harmless.

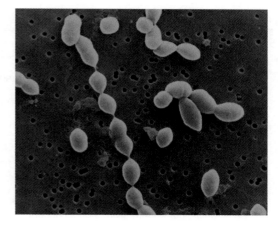

EXPLORE: Experiments with *Streptococcus*

Griffith (1928)

▶ Griffith used mice as a 'model organism' to determine how the disease pneumonia was contracted from the bacterium *S. pneumoniae*.

▶ He injected cultures of the harmless strain (R), disease-causing strain (S), and heat-killed S strain into living mice (1-3).

▶ He then injected mice with a mix of heat-killed S cells and harmless R cells (4). The result for this experiment was not what he had expected.

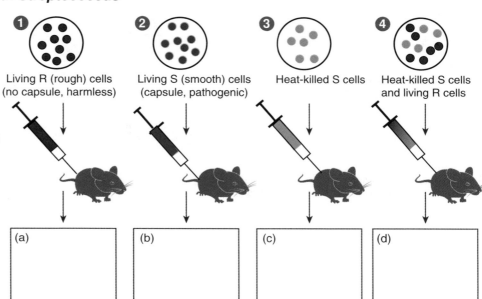

1 Living R (rough) cells (no capsule, harmless) **2** Living S (smooth) cells (capsule, pathogenic) **3** Heat-killed S cells **4** Heat-killed S cells and living R cells

(a) (b) (c) (d)

1. In the boxes (a)-(d) predict the outcome of each treatment (mouse lives or dies and bacterial strain present in the blood).

Avery-MacLeod-McCarty (1944)

▶ In Griffith's experiment, the mouse injected with heat killed S cells and living R cells actually died, but all the others lived. Was your prediction correct? Griffith concluded that the living R cells had been transformed into pathogenic cells by a heritable substance from the dead S cells.

▶ What was the unknown transformation factor in Griffith's experiment? Avery and his coworkers designed an experiment to determine if the unknown heritable factor was RNA, DNA, or protein.

▶ They broke open the heat-killed pathogenic cells and treated samples with agents that inactivated either protein, DNA, or RNA. They then tested the samples for their ability to transform harmless bacteria. The results are presented right.

2. Circle the samples on the right where transformation occurred.

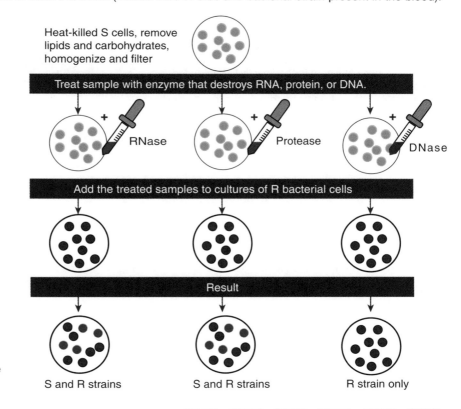

Heat-killed S cells, remove lipids and carbohydrates, homogenize and filter

Treat sample with enzyme that destroys RNA, protein, or DNA.

+ RNase   + Protease   + DNase

Add the treated samples to cultures of R bacterial cells

Result

S and R strains   S and R strains   R strain only

©2018 **BIOZONE** International
ISBN: 978-1-927309-55-1
Photocopying Prohibited

CE LS3.A LS1.A

3. Describe the role of the following in Avery's experiment:

    (a) RNase: _____

    (b) Protease: _____

    (c) DNase: _____

4. Which molecule did Avery identify as being responsible for the transformation? _____

5. How did Avery's experiment build on Griffith's findings? _____

    _____

    _____

Hershey and Chase (1952)

▸ Despite the findings of Avery and his colleagues, the scientific community were slow to accept the role of DNA as the carrier of the code. The approaches they used were not fashionable and some scientists criticized the results, saying the procedures they used led to protein contamination. At the time, protein was still favored as the carrier of the code because nucleic acids were not believed to have any biological activity and their structure was not defined. Despite the importance of the work, Avery and his colleagues were overlooked for a Nobel Prize.

▸ The work of Hershey and Chase followed the work of Avery and his colleagues and was instrumental in the acceptance of DNA as the hereditary material. Hershey and Chase worked on viruses called **phages**, which infect bacteria. Phages are composed of only DNA and protein. When they infect, they inject their DNA into the bacterial cell, leaving their protein coat stuck to the outside.

▸ Hershey and Chase used two batches of phage. **Batch 1** phage were grown with radioactively labeled sulfur, which was incorporated into the phage protein coat.
Batch 2 phage were grown with radioactively labeled phosphorus, which was incorporated into the phage DNA.

▸ The phage were mixed with bacteria, which they then infected. When the bacterial cells were separated from the phage coats, Hershey and Chase looked at where the radioactivity had ended up (right). Hershey and Chase showed conclusively that DNA is the only material transferred from phage to bacteria when bacteria are infected.

Batch 1 phage
Radioactive protein capsule (^{35}S)

Protein

Batch 2 phage
Radioactive DNA (^{32}P)

DNA

Homogenization (blending) separates phage outside the bacteria from the cells and their contents. After centrifugation (spinning down), the cells and their DNA form a pellet. Viral protein coats are left in the supernatant (liquid).

Centrifuge

The radioactivity is in the liquid supernatant

The radioactivity is in the pellet

6. (a) How did the Hershey-Chase experiment provide evidence that nucleic acids, not protein are the hereditary material?

    _____

    _____

    (b) How would the results of the experiment have differed if proteins carried the genetic information? _____

    _____

    (c) Why do you think the Hershey-Chase experiment was so successful in convincing the scientific community at the time that DNA was the material that carried the genetic code? Why weren't Avery's experiments equally successful?

    _____

    _____

    _____

    _____

©2018 **BIOZONE** International
ISBN: 978-1-927309-55-1
Photocopying Prohibited

EXPLAIN: How did the work of many scientists contribute to the discovery of DNA's structure?

DNA is easily extracted and isolated from cells. This was first done in 1869, but it took the work of many scientists working in different areas many years to determine DNA's structure. In particular, four scientists, Watson, Crick, Franklin, and Wilkins are now recognized as having made significant contributions in determining the structure of DNA. Once the structure of DNA was known, scientists could determine how it was replicated and how it could pass information from one generation to the next.

Discovering the structure of DNA ... a story of collaboration and friction

Although Watson and Crick are often credited with discovering DNA's structure, the contributions of many scientists were important. This includes not only the contributions from scientists at the time, but also from earlier researchers whose findings contributed to the body of existing knowledge.

Personal conflicts and internal politics probably prevented DNA's structure being determined earlier. Professional friction between Rosalind Franklin and Maurice Wilkins meant that they worked independently of each other. Watson and Crick analyzed some of Franklin's results, notably 'photo 51', without her knowledge or consent and Watson himself recalls that he tended to dismiss her. Photo 51 was crucial to Watson and Crick's model because it showed that DNA was a double helix. Only later did he acknowledge her considerable contribution.

Franklin was conservative by nature and opposed to prematurely building theoretical models until there was enough data to guide the model building. However, when she saw Watson and Crick's model, she readily accepted it. Despite her contribution, Franklin did not receive the Nobel prize, which cannot be awarded posthumously.

IMAGE: A. Barrington-Brown, © Gonville and Caius College, Cambridge / Coloured by Science Photo Library

James Watson (left) and Francis Crick (right) in 1953 with their DNA model.

Date	Researcher	Experiment, discovery, or event
1944	Oswald Avery, Colin MacLeod, Maclyn McCarty	An experimental demonstration that DNA is the substance that causes bacterial transformation. The aim of the work was to characterize the agent responsible for the transformation phenomenon first described in Griffith's experiment of 1928.
Late 1940s	Linus Pauling	Determined by X-ray crystallography that proteins have a helical structure.
1949	Erwin Chargaff	Chargaff's rules: DNA contains equal proportions of bases A and T and G and C. This hinted at DNA rather than protein being the genetic material.
1951	Rosalind Franklin, Maurice Wilkins	Both studied the structure of DNA using X-ray crystallography. They both worked at King's College but did not get on well.
1951	James Watson, Francis Crick	Build their first DNA model, a three stranded helix, with phosphate groups to the inside and bases to the outside. Franklin points out that their model is incorrect and not consistent with the data.
1952	Alfred Hershey, Martha Chase	Showed conclusively that DNA is the only material transferred from phage to bacteria when bacteria are infected.
1952	Rosalind Franklin	Produces "photo 51", showing DNA is a helix. She was working on a less hydrated form of DNA and did not return to the photo again until 1953.
30 Jan. 1953	Watson, Crick, and Wilkins	Wilkins shows Watson and Crick "photo 51" without Franklin's approval or knowledge. It provided the structural information they needed to finalize their model, completed on 7 March 1953.
16 Apr. 1958	Rosalind Franklin	Franklin dies at age 37 of ovarian cancer. She was never nominated for a Nobel Prize.
1962	Watson, Crick, and Wilkins	Win the Nobel Prize in Physiology or Medicine. Franklin was not acknowledged.

7. As a class, work together to produce a timeline of the events leading to the discovery of DNA's structure. In small groups, choose one event or researcher and explain its/their significance. You can use the information presented in this activity, as well as any information provided on **BIOZONE's Resource Hub**. Take a photograph of your completed timeline and attach it to this page. How much was the discovery of DNA's structure a collaboration between scientists working in related fields? Were Watson and Crick wrong to use information they did not have permission to use? How much did their less cautious approach accelerate the determination of DNA's structure? Attach any extra notes or comments to this page.

©2018 **BIOZONE** International
ISBN: 978-1-927309-55-1
Photocopying Prohibited

50 Modeling the Structure of DNA

ENGAGE: Models can be used to show DNA structure

Models of DNA (deoxyribonucleic acid) range from simple models (right) to highly sophisticated models constructed with the latest technology. The model of DNA built by Watson and Crick in 1953 was inspired but it was also based on information from a number of sources that together enabled Watson and Crick to visualize what the molecule would look like. Rosalind Franklin's Photo 51 was crucial to this story. Watson and Crick's model was fundamental to our understanding of inheritance but it would be another 8 years before researchers determined exactly how the code determined the order of amino acids.

A model needs evidence: Photo 51

The X-ray diffraction patterns of the famous "photo 51" (recreated in the illustration right) provide measurements of different parts of the molecule and the position of different groups of atoms.

The X pattern indicates a helix, but Watson and Crick realized that the apparent gaps in the X (labeled **A**) were due to the repeating pattern of a double helix. The diamond shapes (in blue) indicate the helix is continuous and of constant dimensions and that the sugar-phosphate backbone is on the outside of the helix. The distance between the dark horizontal bands allows the calculation of the length of one full turn of the helix.

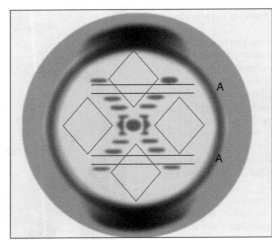

A model can provide different levels of information

The models below show how DNA can be represented in models of different complexity. Each one provides more information about the molecule than the one before it in the sequence.

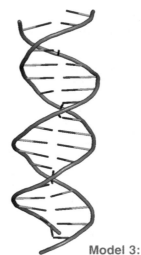

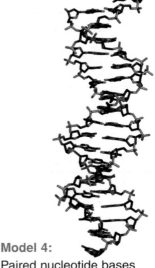

Model 1:
DNA has a double helix structure consisting of two strands.

Model 2:
Rungs between the two strands join them together.

Model 3:
The rungs consist of nitrogen-containing nucleotide bases, of which there are four types: cytosine, guanine, adenine, or thymine.

Model 4:
Paired nucleotide bases attach to a DNA backbone of alternating deoxyribose sugar molecules joined to a phosphate via phosphodiester bonds.

1. (a) What do all of the DNA models above have in common? _____

   (b) What makes some of the models better than others? _____

   _____

   _____

2. Describe one limitation of Watson and Crick's DNA model: _____

   _____

LS1.A LS3.A CE   ©2018 **BIOZONE** International
ISBN: 978-1-927309-55-1
Photocopying Prohibited

EXPLORE: Building physical models helps to understand the structure of DNA

The following exercise will help you understand the structure of DNA and learn the base pairing rule for DNA.

The way the nucleotide bases pair up between strands is very specific. The chemistry and shape of each base means they can only bond with one other DNA nucleotide. Use the information in the table below if you need help remembering the base pairing rule while you are constructing your DNA molecules.

DID YOU KNOW?

Chargaff's rules

Before Watson and Crick described the structure of DNA, an Austrian chemist called Chargaff analyzed the base composition of DNA from a number of organisms. He found that the base composition varies between species but that within a species the percentage of A and T bases are equal and the percentage of G and C bases are equal. Validation of Chargaff's rules was the basis of Watson and Crick's base pairs in the DNA double helix model.

DNA base pairing rule			
Adenine	always pairs with	**Thymine**	A ←→ T
Thymine	always pairs with	**Adenine**	T ←→ A
Cytosine	always pairs with	**Guanine**	C ←→ G
Guanine	always pairs with	**Cytosine**	G ←→ C

3. Cut out each of the nucleotides on page 217 by cutting along the columns and rows (see arrows indicating two such cutting points). Although drawn as geometric shapes, these symbols represent chemical structures.

4. Place one of each of the four kinds of nucleotide on their correct spaces below:

> Place a cut-out symbol for **thymine** here

Thymine

> Place a cut-out symbol for **cytosine** here

Cytosine

> Place a cut-out symbol for **adenine** here

Adenine

> Place a cut-out symbol for **guanine** here

Guanine

5. Identify and label the following features on the adenine nucleotide above: **phosphate, sugar, base, hydrogen bonds**.

6. Create one strand of the DNA molecule by placing the 9 correct 'cut out' nucleotides in the labeled spaces on the following page (DNA molecule). Make sure these are the right way up (with the P on the left) and are aligned with the left hand edge of each box. Begin with thymine and end with guanine.

7. (a) Now create the complementary strand of DNA by using the base pairing rule above.

(b) The nucleotides have to be arranged upside down. What does this tell you about the DNA molecule?

_____

8. Once you have checked that the arrangement is correct, glue, paste, or tape these nucleotides in place.

9. Predict what you think would happen to the DNA structure if there was a mistake in the DNA sequence and mismatched base pairing occurred (e.g. A paired with C). You can use your model to visualize this if you want:

_____

_____

10. In what way is this simple model deficient as a representation of DNA? _____

_____

©2018 **BIOZONE** International
ISBN: 978-1-927309-55-1
Photocopying Prohibited

DNA molecule

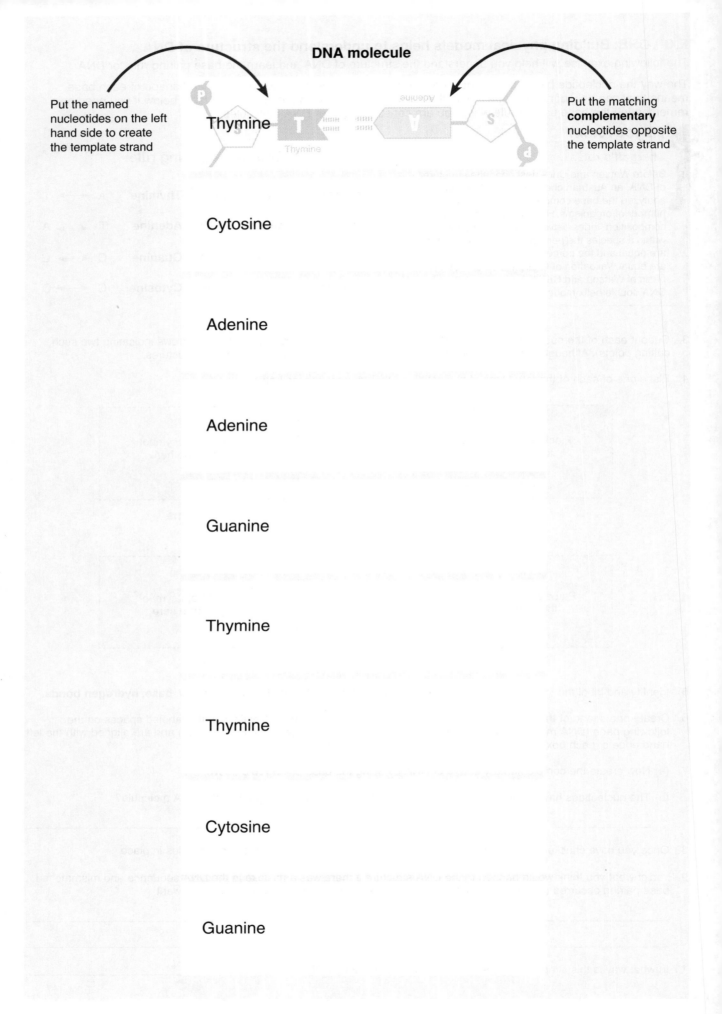

Put the named nucleotides on the left hand side to create the template strand

Thymine

Put the matching **complementary** nucleotides opposite the template strand

Cytosine

Adenine

Adenine

Guanine

Thymine

Thymine

Cytosine

Guanine

©2018 **BIOZONE** International
ISBN: 978-1-927309-55-1
Photocopying Prohibited

Nucleotides

Tear out this page and separate each of the 24 nucleotides
by cutting along the columns and rows (see arrows indicating the cutting points).

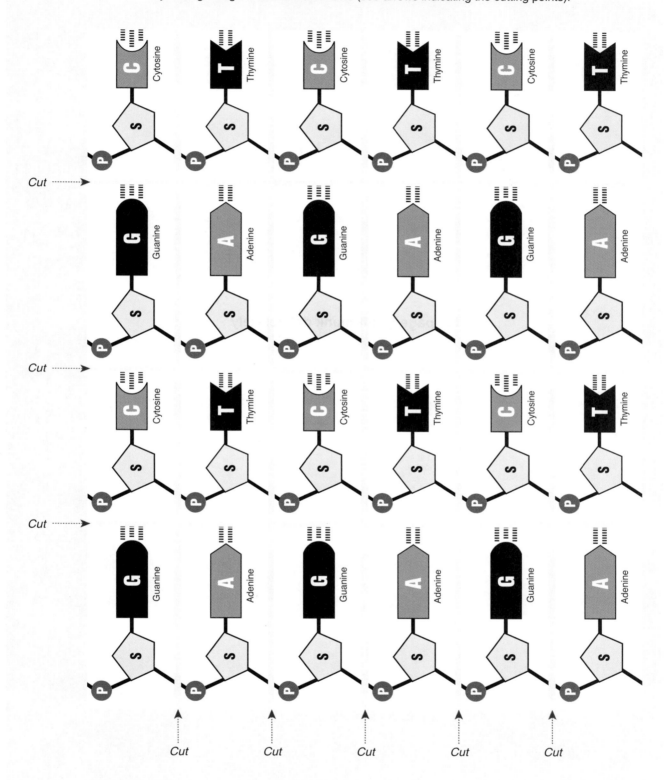

©2018 **BIOZONE** International
ISBN: 978-1-927309-55-1
Photocopying Prohibited

This page is left blank deliberately

©2018 **BIOZONE** International
ISBN: 978-1-927309-55-1
Photocopying Prohibited

51 Genome Studies

ENGAGE: Thomas Hunt Morgan's fruit fly breeding experiments

Red eye
White eye

Thomas Hunt Morgan

▸ In the early 1900s, Thomas Hunt Morgan was carrying out methodical, carefully documented breeding experiments on the common fruit fly, *Drosophila melanogaster*. He was trying to find out if a large-scale mutation in them would result in a new species.

▸ During his work he noticed that one fruit fly had white eyes instead of the usual red. He mated this individual with a red eyed individual to see what the offspring would look like. The offspring all had red eyes. However, when the second generation individuals were mated together the white eye phenotype returned. Not only that, but only males had white eyes. This was the first of many *Drosophila* mutations documented.

▸ Before beginning his experiments, Morgan (right) was critical of Mendel's theory of inheritance and mistrusted chromosomal theory (proposed independently by Boveri and Sutton in 1902). Nor did he believe that Darwin's concept of natural selection could account for the emergence of new species. However, after carefully carrying out experiments on thousands of fruit flies, he set aside his own skepticism and was forced to reconsider all these ideas. Morgan's work was instrumental in progressing the field of modern genetics. He was the first person to successfully map the position of genes on chromosomes, which provided evidence for the role of chromosomes as the carriers of genetic material.

1. (a) Thomas Hunt Morgan initially set out trying to discover if a new fruit fly species would result from a large scale mutation. What did he actually discover?

_____

_____

   (b) Scientific inquiry is a dynamic process. It should build on the findings of others and should not be approached with any bias. How did Morgan's work demonstrate this principle?

_____

_____

_____

EXPLORE: What are traits?

Traits are particular variants of phenotypic (observed physical) characters, e.g. blue eye color. Phenotypic characters may be controlled by one gene or many genes and can show continuous variation, e.g. the continuum of skin and eye colors in human populations, or discontinuous variation, e.g. discrete flower colors in pea plants.

2. For the phenotypic characters pictured above, identify the two traits evident in each photograph:

   (a) Eye color: _____

   (b) Flower color:_____

   (c) Skin color: _____

©2018 **BIOZONE** International
ISBN: 978-1-927309-55-1
Photocopying Prohibited

CE LS3.B LS3.A

Gregor Mendel, a 19th century Austrian monk, used pea plants to study inheritance. Using several phenotypic characteristics he was able to show that their traits were inherited in predictable ways.

One of Mendel's experiments involved the seeds of pea plants. He noticed that pea plant seeds exhibited two phenotypes, round and wrinkled (below left).

Mendel noticed that traits present in a parental generation (e.g. wrinkled seeds) could be absent in the next. However when plants from the first generation were bred together, the wrinkled seed trait reappeared in the second generation.

From this Mendel determined that even though the physical characteristics (traits) may be absent from a particular generation the genetic material (genes) for that trait must still be passed to the offspring.

Gregor Mendel. (1822-1884)

Wrinkled   Round

Parents:
Plants producing wrinkled peas and round peas were bred (crossed)

F₁ generation:
The plants in this first generation (F$_1$) produced only round peas. F$_1$ plants were then bred together.

F₂ generation:
The offspring of this F$_1$ x F$_1$ cross produced some round and some wrinkled peas. The wrinkled trait reappeared.

EXPLORE: Genes are located on chromosomes

Humans have long realized that certain phenotypic characteristics (traits) are passed on from one generation to the next. Mendel's work showed that sometimes these characteristics do not show in every generation, but will show again in later generations, as seen in the pea example (above). This inheritance pattern was confirmed by the work of Thomas Hunt Morgan. He not only confirmed Mendel's findings, but provided a mechanism to explain his observations.

Cross 1: A cross between the mutant (white eye) male and a red-eyed female produced only red-eyed offspring.

Cross 2: Morgan crossed the offspring from the first test. White eyed mutants appeared, consistent with Mendel's observations. However only males had white eyes.

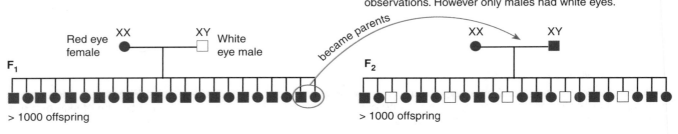

Red eye female XX   XY White eye male

became parents

F₁

> 1000 offspring

F₂

> 1000 offspring

Morgan knew that fruit flies have two sex chromosomes. Males have a XY configuration, females have a XX configuration. Because the white-eyed trait was seen only in males of the second generation, he concluded that white-eye gene must be carried on one of the sex chromosomes (it is a sex-linked trait). If the 'white-eye' gene was on the Y chromosome all the males would have white eyes, so Morgan concluded it must be on the X chromosome.

These (and subsequent) experiments carried out by Morgan and other scientists established the chromosomal theory of heredity. That is, the genes for visible characteristics (phenotypes) are located on chromosomes and passed on when an individual inherits that chromosome. He later showed that genes occupy specific regions on a chromosome and developed the first chromosome maps.

3. (a) Why did Morgan think that the 'white-eye' mutation was on the X chromosome and not the Y chromosome?

_____

_____

(b) How did Morgan's results support those of Mendel? _____

_____

_____

_____

©2018 **BIOZONE** International
ISBN: 978-1-927309-55-1
Photocopying Prohibited

EXPLAIN: What can mutations tell us about gene function?

One way to determine the role of a gene is to alter (mutate) it and see the effect on the organism. Mutations often alter cellular processes. By altering a gene and observing the effect, geneticists gain insight into the gene's function.

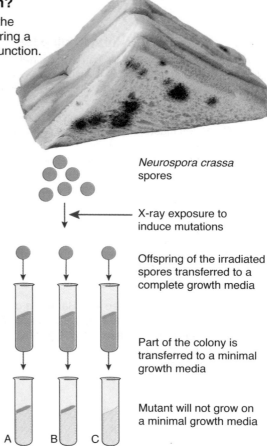

Two American scientists, George Wells Beadle and Edward Lawrie Tatum, are credited with the discovery that genes code for proteins.

Beadle and Tatum used X-rays to induce mutations in the bread mold *Neurospora crassa* (right). They showed that these mutations caused changes in the specific enzymes involved in metabolic pathways. This led to the proposal that there was a direct link between genes and enzyme-catalyzed reactions. They called it the "one gene, one enzyme" hypothesis. Although this is now considered too simplistic, their work was ground-breaking.

Neurospora crassa spores

X-ray exposure to induce mutations

Offspring of the irradiated spores transferred to a complete growth media

Part of the colony is transferred to a minimal growth media

Mutant will not grow on a minimal growth media

A   B   C

▶ *N.crassa* grows on a minimal growth media containing only a few simple sugars, inorganic salts, and the vitamin biotin. It can make all of the components it needs from these few molecules.

▶ Beadle and Tatum proposed that if they mutated a gene required to make one of the enzymes used in metabolism, then the mutant strain wouldn't grow on the minimal media. The first part of their method is outlined (right).

▶ When culture #299 did not grow in the minimal media they knew it contained a mutation. They transferred it to media containing either vitamins or amino acids and found it did not grow with amino acid supplements, but did grow with vitamin supplements. Therefore, culture #299 could not make one of the vitamins it needed for growth.

▶ After testing a range of vitamins they found that the culture would only grow if vitamin B6 was present, indicating one of the enzymes in B6 synthesis was non-functional. From this they concluded that gene mutations can affect metabolic pathways.

Today the entire DNA sequences (**genomes**) of many organisms are known and stored in large databases. Scientists can also target a particular gene sequence and, using a variety of techniques, cause it to mutate. When they come across a gene sequence of interest they use powerful computing tools to search the databases for similar sequences. In this way they can predict what the function of a gene may be and predict what the effect of the mutation may be.

4. (a) Identify the tube containing the mutant strain of *N.crassa* from the diagram above: _____

(b) Explain why the absence of growth in minimal growth media indicated the presence of a mutation: _____

_____

_____

_____

_____

5. (a) What evidence is there from the information above that Beadle and Tatum's experiments were very time consuming?

_____

(b) Briefly explain how modern technology allows today's scientists to discover the effect of a gene mutation much faster than in Beadle and Tatum's era:

_____

_____

_____

_____

_____

©2018 **BIOZONE** International
ISBN: 978-1-927309-55-1
Photocopying Prohibited

52 Modern Genetics

ENGAGE: Polycystic kidney disease

At age 32, Samuel began having abdominal and back pain and noticing blood in his urine. His doctor ran some tests and discovered that Samuel was suffering from a disorder called polycystic kidney disease (PKD).

▸ PKD causes clusters of cysts (fluid filled sacs) to form on the kidneys and has a number of negative effects on health. If left untreated, the cysts can cause serious kidney damage over time and the kidneys may lose function altogether. There is currently no cure for PKD, but there are a number of treatments available to slow down the rate of cyst formation and manage the symptoms.

▸ Samuel's doctor told him that PKD is an inherited disorder, so the disease can be passed on from parents to their children. PKD can arise if a person has a mutation in one of three genes, but the most common form of PKD occurs if there is a mutation to either of the genes called PKD-1 or PKD-2.

▸ The doctor recommended that Samuel's children undergo genetic testing for PKD mutations because people often do not show any symptoms until they are in the 30s or 40s. After talking to a genetic counselor, Samuel and his wife decided to have the children tested. The results showed that the youngest child had a mutation, but the eldest did not.

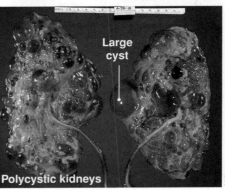

Large cyst

Polycystic kidneys

1. Using your prior knowledge of inheritance, suggest why one of Samuel's children carried the PKD mutation, but the other did not:

It is possible for the older child to not have it because there is only a 50% chance of inheriting the disorder.

2. Why do you think it was important that Samuel's children were tested for gene mutations associated with PKD?

I think it is important so they can look out for it as they get older. They will be able to treat it in it's early stages.

3. Samuel and his wife consulted a genetic counselor before having their children genetically tested for PKD mutations. Go to BIOZONE's Resource Hub and find out more about genetic counseling. What are the advantages and disadvantages of genetic counseling? Summarize your findings here (you can use more paper if you need):

From what I learned in class, the counceling is there to make sure they want to know if their children has it. For example, in the video we saw in class, we saw that the women testing for Huntington disease also had to get counceling as well.

LS3.A ETS1.B CE

©2018 BIOZONE International
ISBN: 978-1-927309-55-1
Photocopying Prohibited

EXPLORE: Characteristics can be recessive or dominant

Thomas Hunt Morgan was the first to determine that genes are located on chromosomes. We now know that in sexually reproducing organisms, chromosomes are generally found in pairs. Each parent contributes one chromosome to the pair. The pairs are called **homologues** or **homologous pairs**. Each homologue carries an identical assortment of genes, but the version of the gene (the **allele**) from each parent may differ. This is the basis for the different traits we see expressed in individuals. This diagram below shows the position of three different genes on the same chromosome that control three different traits (R, S, and C). A photographic example of the principle in each case is shown. Unlike the hypothetical example shown, guinea pigs (cavies) have 64 chromosomes in 32 pairs.

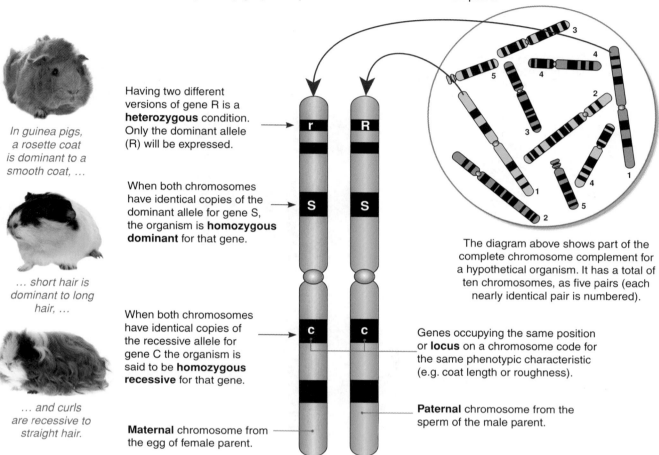

In guinea pigs, a rosette coat is dominant to a smooth coat, ...

Having two different versions of gene R is a **heterozygous** condition. Only the dominant allele (R) will be expressed.

When both chromosomes have identical copies of the dominant allele for gene S, the organism is **homozygous dominant** for that gene.

... short hair is dominant to long hair, ...

When both chromosomes have identical copies of the recessive allele for gene C the organism is said to be **homozygous recessive** for that gene.

... and curls are recessive to straight hair.

Maternal chromosome from the egg of female parent.

The diagram above shows part of the complete chromosome complement for a hypothetical organism. It has a total of ten chromosomes, as five pairs (each nearly identical pair is numbered).

Genes occupying the same position or **locus** on a chromosome code for the same phenotypic characteristic (e.g. coat length or roughness).

Paternal chromosome from the sperm of the male parent.

Studying dominance in guinea pigs

Coat color in guinea pigs is controlled by two alleles for black and white color. The diagram on the right shows what offspring are produced when two heterozygous parents are crossed.

Heterozygous black   Heterozygous black

X

4. Based on the results, state:

(a) The dominant allele: _____black_____

(b) The recessive allele: _____white_____

5. Use the guinea pig results to construct definitions for dominant and recessive alleles:

The dominant alleles are stronger and are more likely to show physically

Homozygous black   Heterozygous black   Heterozygous black   Homozygous white

6. Predict the color of the offspring from a cross between two white guinea pigs. Explain your answer: I think the offspring will all be white, because you need two recessive alleles for it to show physically.

©2018 **BIOZONE** International
ISBN: 978-1-927309-55-1
Photocopying Prohibited

EXPLAIN: Inheritance of autosomal dominant PKD

▶ Gene mutations are the major cause of PKD and, because there are three different genes involved, PKD can take on several different forms. The most common form is called autosomal dominant PKD (ADPKD) and is caused by a mutation to either the PKD1 or PKD2 gene.

▶ Autosomal means the gene is located on a non-sex chromosome.

▶ A less common form of the disease is inherited only when a person has two copies of the mutated PKDH1 gene. It is called autosomal recessive PKD (ARPKD).

▶ The inheritance pattern for ADPKD is described below.

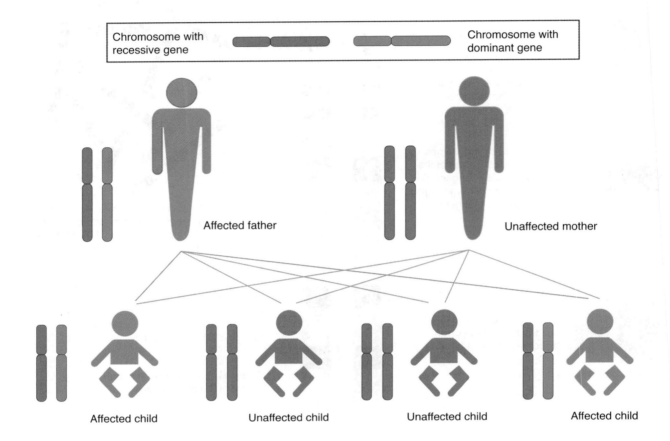

7. (a) Explain what autosomal dominant means: _____

_____

_____

(b) What is the probability that the offspring of an affected parent and unaffected parent will have ARPKD?

_____

8. (a) Does sex affect the inheritance of PKD?_____

(b) Explain your answer: _____

_____

9. How would the inheritance of an autosomal dominant disorder differ from that of an autosomal recessive disorder?

_____

_____

_____

_____

©2018 **BIOZONE** International
ISBN: 978-1-927309-55-1
Photocopying Prohibited

EXPLORE: Kidney transplantation

The PKD1 gene is located on chromosome 16. It codes for a membrane-spanning protein called polycystin-1. Signal molecules bind to parts of the protein outside the kidney cell membrane and trigger a cascade of events inside the kidney cell to regulate many aspects of cell function, including growth and development.

More than 250 mutations affect PKD1, many of which produce a small, non-functioning version of the gene. The mutations disrupt the normal signaling role of polycystin-1. As a result, the cells grow and divide abnormally and form cysts.

Symptoms of PKD include high blood pressure, blood in urine, pain, and kidney stones. Sometimes, the kidneys fail and patients must be placed on dialysis (which removes wastes from the blood) until a suitable donor for a kidney transplant can be found.

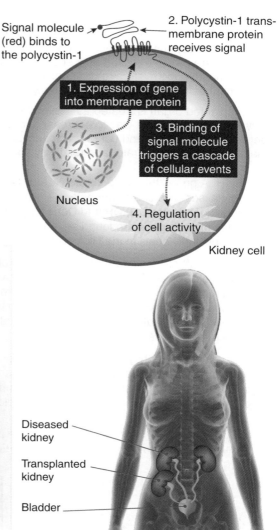

Signal molecule (red) binds to the polycystin-1

2. Polycystin-1 trans-membrane protein receives signal

1. Expression of gene into membrane protein

3. Binding of signal molecule triggers a cascade of cellular events

Nucleus

4. Regulation of cell activity

Kidney cell

- Healthy individuals can live normally with one fully functioning kidney. Therefore, many transplanted kidneys come from living donors. Kidneys are also taken from an organ donor who has just died.
- When a transplant is performed, the damaged kidney is left in place and the new kidney transplanted into the lower abdomen (right). Provided recipients comply with medical requirements (e.g. correct diet and medication) over 85% of kidney transplants are successful.

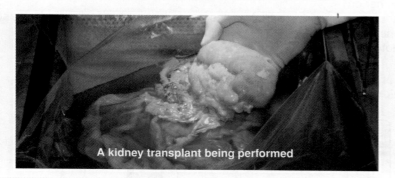

A kidney transplant being performed

Diseased kidney

Transplanted kidney

Bladder

In the US, 115,000 people are on a wait list to receive organ transplants. Kidney transplants are the most common form of organ transplantation, with 16,000 kidney transplants performed in the US in 2017. People requiring an organ donation go on to a national database (waiting list). The waiting list is managed by a non-profit organization called the United Network for Organ Sharing (UNOS). Their job is to match organs to recipients to ensure the best possible outcome. Many factors determine a person's priority level on the waiting list. When an organ becomes available, a complex computer program generates a list of potential recipients ranked according to specific criteria.

10. Work in pairs or small groups to research the criteria that UNOS uses to match organs to suitable recipients. You could use the organ donor website and other information via **BIOZONE's Resource Hub** to begin your research. Summarize your results here and present your findings as an oral or written report, as a poster, or as a slide-show presentation.

©2018 **BIOZONE** International
ISBN: 978-1-927309-55-1
Photocopying Prohibited

EXPLAIN: How does donor matching improve the success of organ transplants?

One of the most important tests performed before an organ transplant is **HLA testing**. HLA (human leukocyte antigens) are proteins (called antigens) found on the surface of all body cells except red blood cells. These antigens enable the immune system to distinguish its own cells from foreign ("non-self") cells. Cells displaying "foreign" HLA antigens (i.e. from another person) are rejected (attacked and destroyed) by the immune system.

A transplanted organ is treated as foreign by the recipient's immune system unless the HLA antigens match. This is why the most important way to decrease the risk of rejection is by HLA typing. This involves matching HLA profiles between the recipient and donor as closely as possible. Now, with advanced gene sequencing techniques, HLA matching is rapid, accurate, and relatively inexpensive.

▸ The HLA genes that code for the HLA proteins fall into two classes. The class I genes are A, B, and C. The class II genes are DR, DP and DQ. Many different alleles of each gene are found so there is huge variation amongst individuals in the HLA proteins.

▸ Each parent carries three class-I and three class-II HLA alleles on each of the two copies of chromosome 6, making twelve HLA alleles (and their corresponding proteins) in each cell. Six are inherited from the mother and six from the father (below left). The six alleles from each parent are termed a haplotype.

▸ HLA alleles are tightly linked (very close together on the chromosome) as shown on the chromosome below center. Therefore, these alleles tend to be inherited together (crossing over is rare between haplotypes).

▸ In the inheritance of HLA alleles, the haplotypes are inherited according to Mendelian rules.

▸ A child inherits one haplotype from each parent so for all children sharing the same parents, there will only be four possible haplotype combinations (as shown on the diagram below right).

▸ In HLA typing, 6-10 HLA proteins are matched. The most commonly tested for transplantation are HLA-A, B, and C, and HLA-DP, DQ, and DR.

▸ Some haplotypes occur with greater frequency in some human populations than others.

▸ HLA alleles are codominant (equally dominant) so their proteins are fully and equally expressed.

Body cell showing cell surface HLA antigens

This cell has two haplotypes producing two sets of cell surface antigens:
Maternal: A1-B1-C1-DP1-DQ1-DR1
Paternal: A2-B2-C2-DP2-DQ2-DR2

Chromosome 6 and location of HLA genes

Human chromosome 6 showing the positions of the HLA alleles. Notice how close they are. Crossing over is unlikely between haplotypes.

Inheritance of HLA haplotypes

Class-I HLA: A, B, C

Class-II HLA: DR, DP, DQ

In the inheritance of HLA alleles, there is a 25% chance two siblings will have a complete match.

11. (a) Use the diagram of chromosome 6 above to explain why HLA alleles tend to be inherited together as a unit:

_____

_____

_____

(b) How does this help to explain their pattern of inheritance?_____

_____

_____

_____

©2018 **BIOZONE** International
ISBN: 978-1-927309-55-1
Photocopying Prohibited

(c) Use the inheritance diagram opposite (far right) to determine the probability that any two siblings with the same parents will have a one haplotype match (share one haplotype). Explain your reasoning:

_____

_____

12. Predict the relative success of a transplant where the donor and recipient have 3 HLA proteins in common compared to when the donor and recipient have 6 HLA proteins in common:

_____

_____

13. (a) Explain why matching donor and recipient HLA profiles can be very difficult: _____

_____

_____

_____

(b) Explain why you are more likely to find a successful match within your immediate relatives (e.g. full siblings, half siblings, parents) than in the wider public (e.g. amongst your friends).

_____

_____

_____

14. (a) Study the graph (right) and explain the relationship between the number of HLA mismatches and graft survival over time:

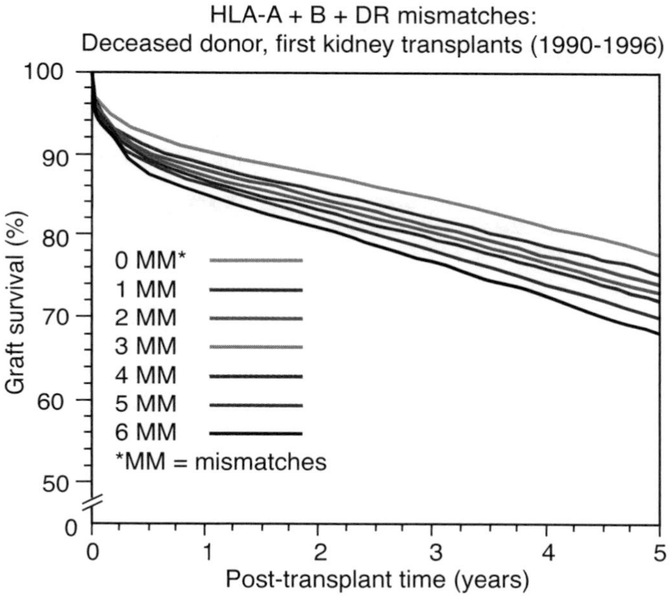

_____

_____

_____

_____

_____

HLA-A + B + DR mismatches:
Deceased donor, first kidney transplants (1990-1996)

Graft survival (%) vs Post-transplant time (years)

0 MM*
1 MM
2 MM
3 MM
4 MM
5 MM
6 MM
*MM = mismatches

(b) Explain why the recipient's body will reject the transplanted organ if there are too many HLA mismatches:

_____

_____

_____

_____

_____

15. Explain how advances in technology have helped to overcome some of the problems associated with the transplantation of donated organs. If you wish to explore this engineering connection further, visit **BIOZONE's Resource Hub**:

_____

_____

_____

_____

©2018 **BIOZONE** International
ISBN: 978-1-927309-55-1
Photocopying Prohibited

53 Variation

ENGAGE: Individuals within a population vary

Land snails: *Cepaea vindobonensis*

Mallard ducklings

Icelandic horses

In the previous chapter you noticed that your classmates showed variation for many characteristics such as height, weight, eye color, and skin tone, and explored how variation is the basis for natural selection. In this activity we are going to look at the basis of variation and explain why individuals of a species often have a similar appearance but are rarely ever identical. This diversity of physical characteristics (phenotypes) in a population or species is called variation. Part of this phenotypic variation is the result of variation in the genotypes of individuals (their genetic makeup) and part is the result of environmental influences, such as nutrition.

1. Would you expect foot length to vary within a population? _____

2. The data in the table below shows foot length for 20 adults.

   (a) In the space, construct a tally chart for the data.

   (b) Plot the data as a histogram on the grid below.

Adult foot length (mm)			
265	272	257	315
300	320	250	250
215	330	240	270
252	270	265	350
315	300	290	310

Tally chart

3. (a) Were you correct in your prediction about foot length? _____

   (b) What factors do you think could contribute to the natural variation observed in populations?

   _____

   _____

   _____

EXPLORE: How does variation arise?

Variation is important for species survival in a changing environment. Individuals with characteristics that benefit them in the conditions at the time will survive, whereas those with characteristics less suited to those conditions are less likely to survive or reproduce successfully. Variation arises as a result of a number of different factors (below).

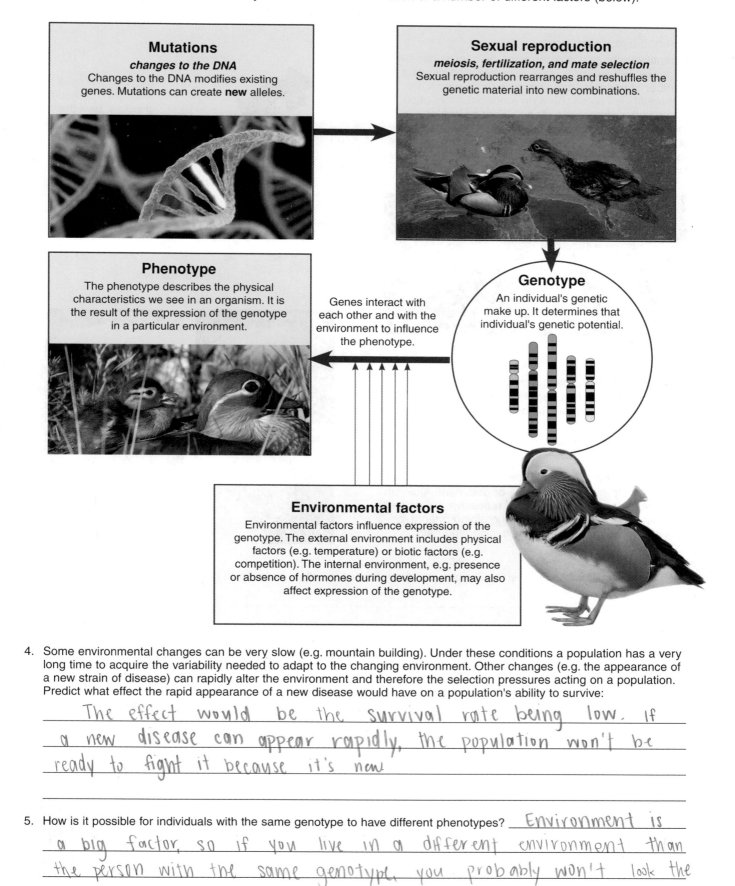

Mutations

changes to the DNA
Changes to the DNA modifies existing genes. Mutations can create **new** alleles.

Sexual reproduction

meiosis, fertilization, and mate selection
Sexual reproduction rearranges and reshuffles the genetic material into new combinations.

Phenotype

The phenotype describes the physical characteristics we see in an organism. It is the result of the expression of the genotype in a particular environment.

Genes interact with each other and with the environment to influence the phenotype.

Genotype

An individual's genetic make up. It determines that individual's genetic potential.

Environmental factors

Environmental factors influence expression of the genotype. The external environment includes physical factors (e.g. temperature) or biotic factors (e.g. competition). The internal environment, e.g. presence or absence of hormones during development, may also affect expression of the genotype.

4. Some environmental changes can be very slow (e.g. mountain building). Under these conditions a population has a very long time to acquire the variability needed to adapt to the changing environment. Other changes (e.g. the appearance of a new strain of disease) can rapidly alter the environment and therefore the selection pressures acting on a population. Predict what effect the rapid appearance of a new disease would have on a population's ability to survive:

The effect would be the survival rate being low. If a new disease can appear rapidly, the population won't be ready to fight it because it's new

5. How is it possible for individuals with the same genotype to have different phenotypes? *Environment is a big factor, so if you live in a different environment than the person with the same genotype, you probably won't look the same.*

©2018 **BIOZONE** International
ISBN: 978-1-927309-55-1
Photocopying Prohibited

EXPLORE: Examples of genetic variation

As we know individuals show particular variants of phenotypic characters called traits, e.g. blue eye color.

▶ Traits that show continuous variation are called quantitative traits (they can be measured or quantified).

▶ Traits that show discontinuous variation are called qualitative traits (they fall into discrete categories).

Quantitative traits

Quantitative traits are determined by a large number of genes. For example, skin color has a continuous number of variants from very pale to very dark. Individuals fall somewhere on a normal distribution curve of the phenotypic range. Other examples include height in humans for any given age group, length of leaves in plants, grain yield in corn, growth in pigs, and milk production in cattle. Most quantitative traits are also influenced by environmental factors.

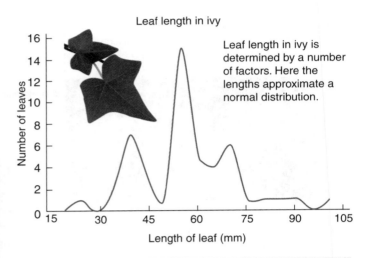

Leaf length in ivy

Leaf length in ivy is determined by a number of factors. Here the lengths approximate a normal distribution.

Grain yield in corn   Growth in piglets

Qualitative traits

Qualitative traits are determined by a single gene with a very limited number of variants present in the population. For example, blood type (ABO) in humans has four discontinuous traits A, B, AB or O. Individuals fall into separate categories. Comb shape in poultry (right) is a qualitative trait and birds have one of four phenotypes depending on which combination of four alleles they inherit. The dash (missing allele) indicates that the allele may be recessive or dominant. Albinism is the result of the inheritance of recessive alleles for melanin production. Those with the albino phenotype lack melanin pigment in the eyes, skin, and hair.

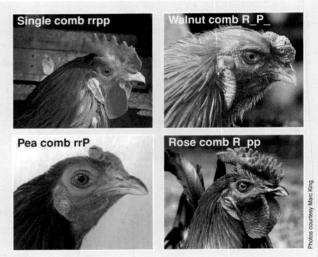

Single comb rrpp

Walnut comb R_P_

Pea comb rrP_

Rose comb R_pp

Photos courtesy Marc King

6. Explain the difference between continuous and discontinuous variation, including its genetic basis: __A continuous variation means that it is determined by a large number of genes. On the other hand, a discontinuous variation means it is determined by a single gene with a very limited number of variants.__

7. Identify each of the following phenotypic traits as continuous (quantitative) or discontinuous (qualitative):

(a) Wool production in sheep: __quantitative__

(b) Hand span in humans: __quantitative__

(c) Blood groups in humans: __qualitative__

(d) Albinism in mammals: __qualitative__

(e) Body weight in mice: __quantitative__

(f) Flower color in snapdragons: __quantitative__

(g) HLA haplotypes in humans: __qualitative__

©2018 BIOZONE International
ISBN: 978-1-927309-55-1
Photocopying Prohibited

54 Sexual Reproduction Produces Genetic Variation

ENGAGE: Why do some species show more variation than others?

The photo, below left, shows aphids feeding on a plant. You will notice that all the aphids look the same. Now study the photo of the snails (*Cepaea nemoralis*). Note the variations in the shells. No two appear to be the same!

How is it that the aphids look very similar yet there is huge variation between the snail shells? The aphids have reproduced asexually and the offspring are clones of a single parent. This explains why all the individuals look so similar (they are!). The snails are the result of sexual reproduction, in which sex cells (eggs and sperm) of the male and female combine to produce new individuals. Most sexually reproducing organisms are diploid (2N), meaning they have two full sets of chromosomes. One set comes from the mother (maternal) and one set from the father (paternal). Sexual reproduction involves a diploid organism producing haploid gametes (cells with one set of chromosomes). When gametes fuse in fertilization to produce a zygote (fertilized egg) the diploid number of chromosomes is restored.

EXPLORE: How does variation arise in sex cells?

The sex cells of the parent snails are produced by a special type of cell division called **meiosis**. During meiosis the chromosomes are duplicated and then there are two divisions so that a single copy of each chromosome ends up in each sex cell (i.e. the sex cells are haploid).

During meiosis, genes can be exchanged between homologous chromosomes by a process called crossing over. The exchange of genetic material means that the gametes will have different combinations of genes. This produces variation amongst the individuals produced when these gametes are joined in fertilization.

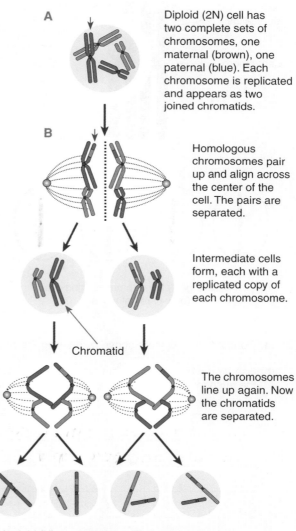

A Diploid (2N) cell has two complete sets of chromosomes, one maternal (brown), one paternal (blue). Each chromosome is replicated and appears as two joined chromatids.

B Homologous chromosomes pair up and align across the center of the cell. The pairs are separated.

Intermediate cells form, each with a replicated copy of each chromosome.

Chromatid

The chromosomes line up again. Now the chromatids are separated.

Haploid (N) gametes form. They each have one complete set of chromosomes.

1. (a) Compare the appearance of the chromosome (arrowed red) at points A and B in the diagram. What do you think has happened to cause the different appearance at B?

    crossing over has occured in
    homologous chromosome pairs

    (b) Use this information to explain why sexual reproduction produces more variation than asexual reproduction:

    In sexual reproduction, you
    get 23 chromosomes from
    both parents. In the end,
    you get 46, but there are
    different variations. In
    asexual reproduction, you get
    46 chromosomes from 1 parent,
    causing the off spring to
    be identical.

©2018 **BIOZONE** International
ISBN: 978-1-927309-55-1
Photocopying Prohibited

CE   LS3.B

- Meiosis creates genetic variation in the gametes as alleles are reshuffled into different combinations. This variation arises through two processes: crossing over and independent assortment. As a result of meiosis, siblings with the same biological parents can appear very different, although there is often a family resemblance (right).

- **Crossing over** is the mutual exchange of pieces of chromosomes (and their genes) between homologous chromosomes. Crossing over results in recombination of alleles in the gametes (below).

- **Independent assortment** is the random alignment and distribution of homologous chromosomes to the gametes, i.e the chromosomes separate and segregate independently of each other.

Shuffling the genetic material

- Chromosomes replicate during interphase, before meiosis, to produce replicated chromosomes with sister chromatids held together at the centromere (see below).

- When the replicated chromosomes are paired during the first stage of meiosis, non-sister chromatids may become entangled and segments may be exchanged in a process called **crossing over**.

- **Crossing over** results in the **recombination** of alleles (variations of the same gene) producing greater variation in the offspring than would otherwise occur.

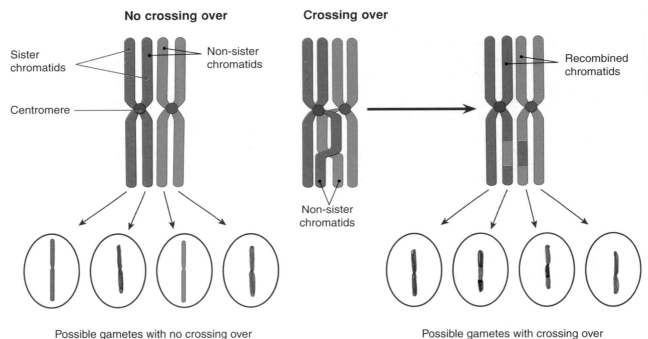

2. (a) Complete the diagram above by drawing the possible gametes in the empty circles. Two have been completed:

(b) Use your drawings to explain how crossing over and recombination can increase variation in the gametes (and hence the offspring):

Crossing over creates different variations of genes.
on homologous chromosome pairs which increases variation
in the gametes and offspring.

©2018 **BIOZONE** International
ISBN: 978-1-927309-55-1
Photocopying Prohibited

Distribution of chromosomes into sex cells is random

▸ The law of independent assortment states that the alleles for separate traits are passed independently of one another from parents to offspring. In other words, the allele a gamete receives for one gene does not influence the allele received for another gene.

▸ Independent assortment produces 2^x different possible chromosome combinations (where x is the number of chromosome pairs). Because of crossing over and recombination of alleles, the real number is much larger.

▸ For the example right, there are two chromosome pairs. The number of possible allele combinations in the gametes is $2^2 = 4$ (only two of the four possible combinations are shown).

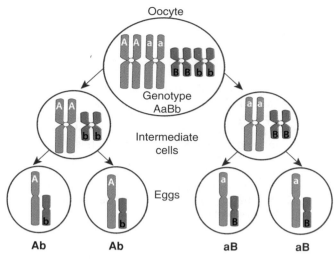

Oocyte

Genotype
AaBb

Intermediate
cells

Eggs

Ab   Ab   aB   aB

3. (a) What is independent assortment? ___It is the random alignment and distribution of homologous chromosomes to the gametes.___

(b) How does it increase variation in the gametes? ___Independent assortment increases variation in the gametes. by increasing the number of gametes___

4. (a) In the circles right, draw the two gamete combinations not shown in the diagram above:

(b) For each of the following 2N chromosome numbers, calculate the number of possible allele combinations in the gametes:

   i. 8 chromosomes: ___4___

   ii. 24 chromosomes: ___12___

   iii. 64 chromosomes: ___32___

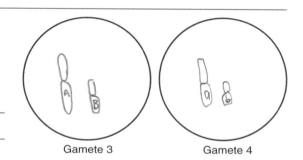

Gamete 3   Gamete 4

5. Homologous chromosomes are shown below. Possible crossover points are marked with numbered arrows.

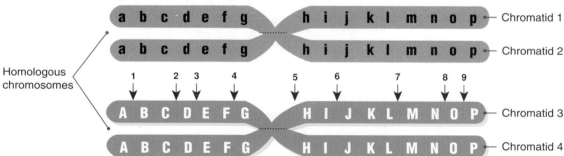

(a) Draw the gene sequence for the four chromatids (above) after crossing over has occurred at crossover point 2:

(b) Which genes have been exchanged between the homologous chromosomes?

   ___2 & 3___

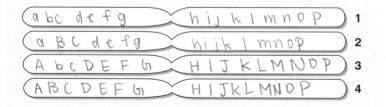

©2018 **BIOZONE** International
ISBN: 978-1-927309-55-1
Photocopying Prohibited

EXPLORE: Use a model to understand how meiosis produces variation

Modeling meiosis using ice-block sticks can help to understand how meiosis introduces variation into a population.

Background

Each of your somatic cells contain 46 chromosomes, 23 **maternal chromosomes** and 23 **paternal chromosomes**. Therefore, you have 23 homologous (same) pairs. For simplicity, the number of chromosomes studied in this exercise has been reduced to four (two homologous pairs). To study the effect of crossing over on genetic variability, you will look at the inheritance of two of your own traits: the ability to **tongue roll** and **handedness**. This activity will take 25-45 minutes.

Chromosome number	Phenotype	Genotype
10	Tongue roller	TT, Tt
10	Non-tongue roller	tt
2	Right handed	RR, Rr
2	Left handed	rr

Record your phenotype and genotype for each trait in the table (right). If you have a dominant trait, you will not know if you are heterozygous or homozygous for that trait, so you can choose either genotype for this activity.

Trait	Phenotype	Genotype
Handedness		
Tongue rolling		

BEFORE YOU START THE SIMULATION:

Partner up with a classmate. Your gametes will combine with theirs (fertilization) at the end of the activity to produce a child. Decide who will be the female, and who will be the male. You will need to work with this person again at step 6.

i. Collect four ice-blocks sticks. These represent four chromosomes. Color two sticks blue or mark them with a P. These are the paternal chromosomes. The plain sticks are the maternal chromosomes. Write your initial on each of the four sticks. Label each chromosome with their chromosome number (below). Label four sticky dots with the alleles for each of your phenotypic traits, and stick each onto the appropriate chromosome. For example, if you are heterozygous for tongue rolling, the sticky dots with have the alleles T and t, and they will be placed on chromosome 10. If you are left handed, the alleles will be r and r and be placed on chromosome 2 (below).

ii. Randomly drop the chromosomes onto a table. This represents a cell in either the testes or ovaries. **Duplicate** your chromosomes (to simulate DNA replication) by adding four more identical ice-block sticks to the table (below). This represents **interphase**.

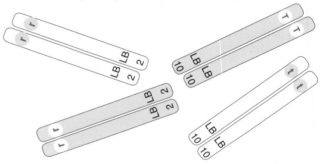

iii. Simulate the first stage of meiosis by lining the duplicated chromosome pair with their homologous pair (below). For each chromosome number, you will have four sticks touching side-by-side (A). At this stage **crossing over** occurs. Simulate this by swapping sticky dots from adjoining homologues (B).

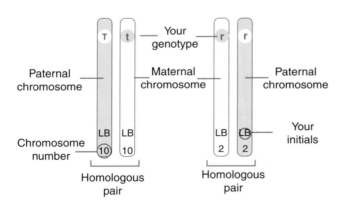

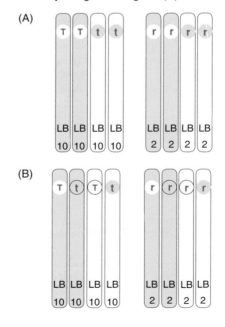

©2018 **BIOZONE** International
ISBN: 978-1-927309-55-1
Photocopying Prohibited

iv. Randomly align the homologous chromosome pairs to simulate alignment across the cell's center (equator) (as occurs in the next phase of meiosis). Simulate the separation of the chromosome pairs. For each group of four sticks, two are pulled to each pole (end) of the cell.

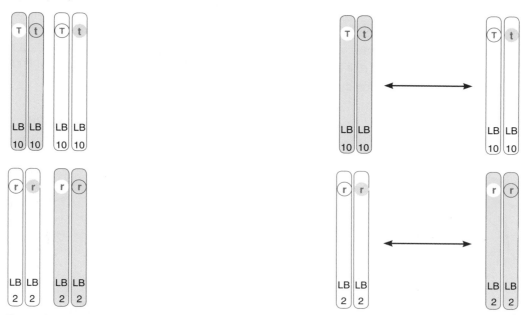

v. Two intermediate cells are formed. If you have been random in the previous step, each intermediate cell will have half the diploid chromosome number (it will be haploid) and it will contain a mixture of maternal and paternal chromosomes. This is the end of the first division of meiosis. Your cells now need to undergo the second division. Repeat steps iii and iv but this time there is no crossing over and you are now separating replicated chromosomes, not homologues. At the end of the process each intermediate cell will have produced two haploid gametes (below).

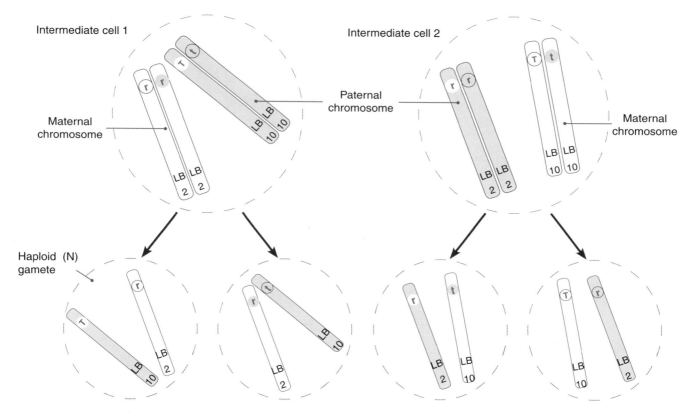

vi. Pair up with the partner you chose at the beginning of the exercise to carry out **fertilization**. Randomly select one sperm and one egg cell. The unsuccessful gametes can be removed from the table. Combine the chromosomes of the successful gametes. You have created a child! Fill in the following chart to describe your child's genotype and phenotype for tongue rolling and handedness.

Trait	Phenotype	Genotype
Handedness		
Tongue rolling		

©2018 **BIOZONE** International
ISBN: 978-1-927309-55-1
Photocopying Prohibited

EXPLAIN: Genetic variability is reduced by linked genes

▶ Linked genes are genes found on the same chromosome. Linked genes tend to be inherited together (recall the HLA alleles) and so fewer genetic combinations of their alleles are possible. The closer genes are to each other on the chromosome, the more 'tightly' they are linked. Crossing over is rare between tightly linked genes.

▶ Linkage is indicated in genetic crosses when a greater proportion of the offspring from a cross are of the parental type (than would be expected if the alleles were on separate chromosomes and assorting independently).

▶ Linkage reduces the genetic variation that can be produced in the offspring.

Inheritance of linked genes

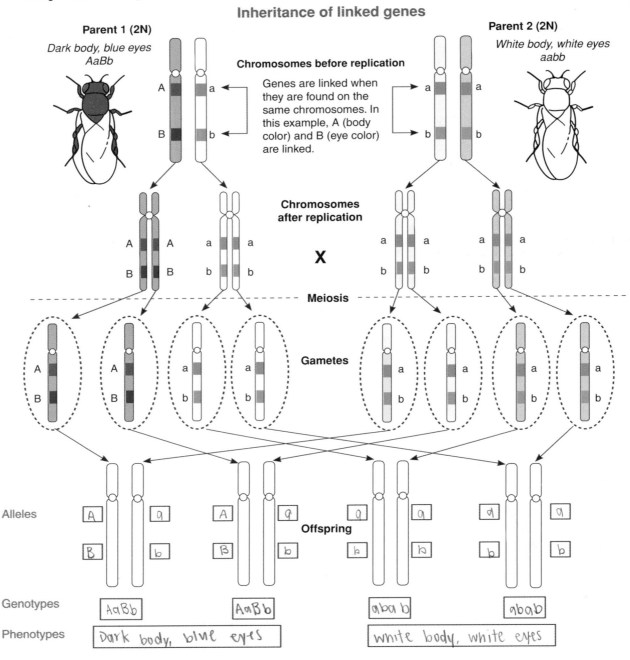

6. (a) In the boxes above, write the alleles, the genotypes, and the phenotypes of the offspring:

   (b) How many different genotypes/phenotypes are possible for the cross pictured? ___2 different types.___

   (c) Are they the same as the parental genotypes/phenotypes? YES / NO (circle correct answer).

7. Explain why linkage reduces genetic variability: ___Linkage reduces genetic variability because it prevents crossing over.___

©2018 BIOZONE International
ISBN: 978-1-927309-55-1
Photocopying Prohibited

EXPLAIN: How do abnormal chromosome numbers arise?

When homologous chromosomes or sister chromatids fail to separate properly during meiosis it results in daughter cells with abnormal chromosome numbers. Chromosome numbers can be too high or too low. This is called **non-disjunction**, and is shown below.

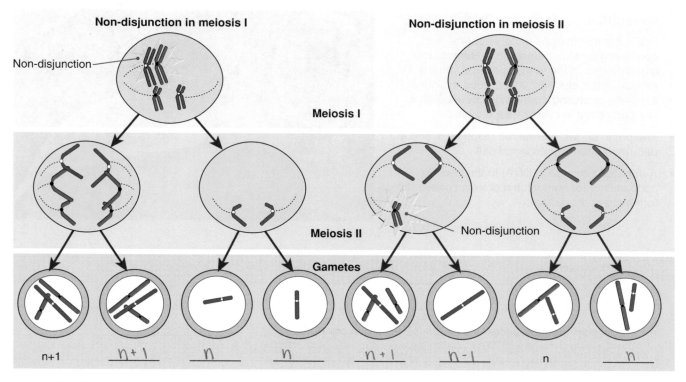

8. (a) The chromosome numbers for two gametes in the diagram above have been described. Using the same notation, complete the descriptions for the remaining gametes:

   (b) Suggest why non-disjunction in meiosis I causes more abnormal gametes than when non-disjunction occurs in meiosis II:

   There aren't enough chromosomes in the last set of cells.

Turner Syndrome is caused by non-disjunction

▸ Turner Syndrome is a chromosomal condition affecting females. it occurs as a result of non-disjunction and is not (usually) an inherited disease.

▸ Females have two sex chromosomes (XX), but a female with Turner Syndrome will have only one normal X chromosome. The other X chromosome is either missing (right) or structurally altered (e.g. missing parts).

▸ A gene called *SHOX* involved in controlling skeletal development is located on sex (X and Y) chromosomes. The loss of one copy of this gene affects bone development in women with Turner Syndrome. Developmental issues include a short stature (most common) and abnormal development of arms and legs. Other health issues (e.g. infertility and kidney and heart problems) are also common.

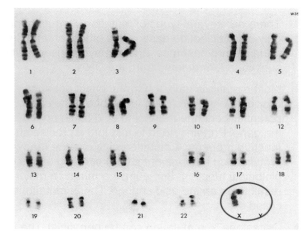

9. Using the same notation as in 8(a), describe the gamete for an individual with Turner Syndrome: n-1

10. Why do you think normal skeletal development is affected in a person with Turner Syndrome? Because they don't have the genes to code for skeletal development

©2018 **BIOZONE** International
ISBN: 978-1-927309-55-1
Photocopying Prohibited

55 Mutations Produce Variation

ENGAGE: Sickle cell red blood cells

▸ The photo on the right shows human red blood cells (RBCs).

▸ RBCs contain millions of copies of the iron-containing protein, hemoglobin. Hemoglobin binds oxygen in the lung capillaries (where oxygen levels are high) and carries it around the body, releasing it where it is needed as a key participant in cellular respiration.

▸ The photo shows normal red blood cells and a deformed (sickled) red blood cell.

▸ A mutation (change in DNA) in the genetic code coding for hemoglobin protein causes the formation of the sickle cell.

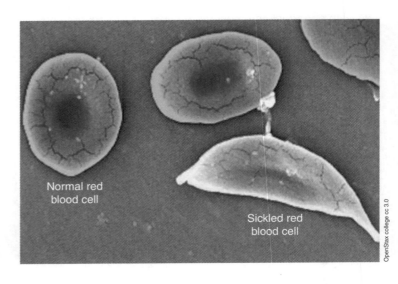

Normal red blood cell

Sickled red blood cell

OpenStax college cc 3.0

1. (a) Describe the differences in shape between the normal red blood cell and the sickled red blood cell: _____

_____

_____

(b) Predict some consequences of a person having sickle cell RBCs: _____

_____

_____

_____

EXPLORE: The effect of mutations

▸ Mutations are changes in the DNA sequence and occur through errors in DNA copying or as a result of agents (such as UV light) that can damage the DNA. Changes to the DNA modifies existing genes and can create variation in the form of new alleles. Mutations are the source of all new genetic variation.

▸ There are several types of mutation. Some change only one nucleotide base, while others change large parts of chromosomes. Bases may be inserted into, substituted, or deleted from the DNA.

▸ Most mutations have a harmful effect because changes to the DNA sequence of a gene can potentially change the amino acid chain encoded by the gene. Proteins need to fold into a precise shape to function properly. A mutation may change the way the protein folds and prevent it from carrying out its usual biological function. Sometimes a mutation may not result in an amino acid change. These mutations are called silent.

▸ Occasionally, a mutation can be beneficial. The mutation may produce a new or more efficient protein that improves the survival of the organism. Beneficial mutations are common in bacteria and viruses. In viruses (e.g. *Influenzavirus* top right) genes coding for the glycoprotein spikes (arrowed) are constantly mutating, producing new strains that avoid detection by the host's immune system. In bacteria, such as MRSA (below right), mutations that provide resistance to antibiotics are favored in an environment where antibiotics are present.

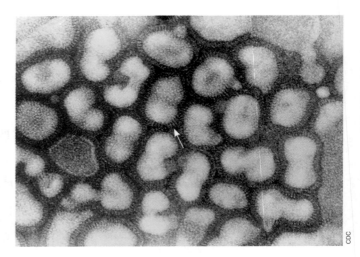

CDC

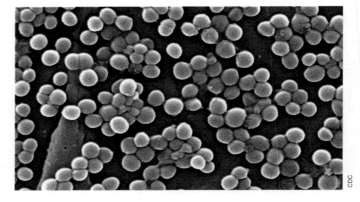

CDC

  LS3.B CE

©2018 **BIOZONE** International
ISBN: 978-1-927309-55-1
Photocopying Prohibited

2. How might a mutation cause a beneficial effect on protein function? _____

_____

_____

_____

3. How might a mutation have a harmful effect on protein function? _____

_____

_____

EXPLORE: How do mutations alter phenotype?

▶ The effect of mutation on protein structure can be illustrated by exploring a mutation to the hemoglobin protein (HbA).

▶ Hemoglobin is 146 amino acids long. A single base substitution mutation in the DNA sequence coding for hemoglobin changes one amino acid and produces an abnormal hemoglobin protein called hemoglobin S (HbS).

▶ The presence of HbS alters the shape of the red blood cells carrying it, producing a deformed sickle (crescent) appearance. The mutation is described below.

| Normal DNA sequence for β-chain of hemogloblin | GTG | CAC | CTG | ACT | CCT | GAG | GAG |
| | CAC | GTG | GAC | TGA | GGA | CTC | CTC |

| Normal mRNA sequence | GUG | CAC | CUG | ACU | CCU | GAG | GAG |

| Normal amino acid sequence | Val | His | Leu | Thr | Pro | Glu | Glu |

Normal red blood cell phenotype

| Mutated DNA sequence for β-chain of hemoglobin. | GTG | CAC | CTG | ACT | CCT | GTG | GAG |
| | CAC | GTG | GAC | TGA | GGA | CAC | CTC |

| Mutated mRNA sequence | GUG | CAC | CUG | ACU | CCU | GUG | GAG |

| Mutated amino acid sequence | Val | His | Leu | Thr | Pro | Val | Glu |

Mutated red blood cell phenotype

4. (a) In your own words describe what change has occurred in the mutated DNA sequence: _____

_____

_____

(b) Explain how this mutation affects the amino acid sequence: _____

_____

_____

©2018 **BIOZONE** International
ISBN: 978-1-927309-55-1
Photocopying Prohibited

EXPLAIN: The physiological changes caused by sickle cell disease

Mutations cause many diseases, including sickle cell disease. Sickle cell disease (formerly sickle cell anemia) is an inherited disease affecting the production of the oxygen-transporting hemoglobin protein. In the sickle cell mutation, the single mistake in the DNA code results in valine being added to the amino acid sequence instead of glutamic acid.

Why does changing one amino acid affect the hemoglobin protein?

▸ Each amino acid has a unique side chain with specific chemical properties. The way these amino acid side chains interact with each other causes a protein to fold up in a very specific way.

▸ A mutation resulting in the change of an amino acid within a sequence can disrupt the way a protein folds and this may alter the protein's shape and function.

▸ This can be illustrated by the HbS mutation. In normal hemoglobin, glutamic acid (Glu) is found at position 6. Glutamic acid is a 'water-loving' (hydrophilic) amino acid with a negatively charged side group. In the mutated sequence, valine (Val) is found in position 6. Valine is a 'water-hating' (hydrophobic) amino acid with no charged side group. It bonds with other hydrophobic amino acids and, as a result, the hemoglobin takes up a different shape.

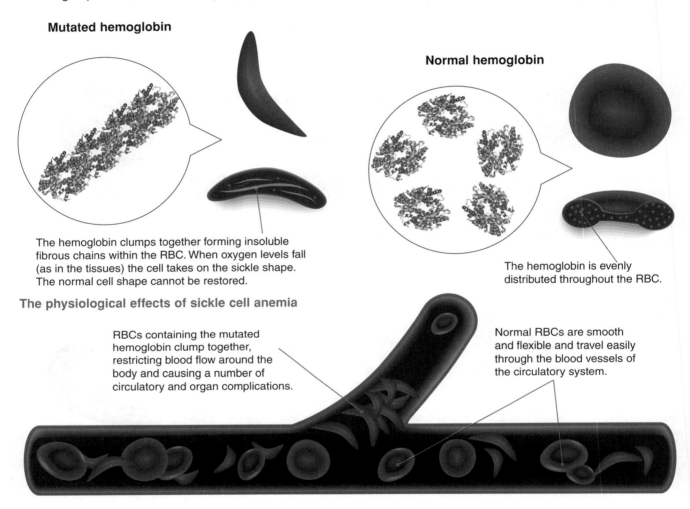

Mutated hemoglobin

Normal hemoglobin

The hemoglobin clumps together forming insoluble fibrous chains within the RBC. When oxygen levels fall (as in the tissues) the cell takes on the sickle shape. The normal cell shape cannot be restored.

The hemoglobin is evenly distributed throughout the RBC.

The physiological effects of sickle cell anemia

RBCs containing the mutated hemoglobin clump together, restricting blood flow around the body and causing a number of circulatory and organ complications.

Normal RBCs are smooth and flexible and travel easily through the blood vessels of the circulatory system.

▸ The inflexible sickle-shaped RBCs stick together in clumps. They can form clots in the capillaries and so reduce blood flow to vital organs.

▸ Sickled RBCs have a life span of 10-20 days, so there is a shortage of RBCs and hemoglobin. This causes anemia, one of the symptoms of sickle cell disease.

▸ Blood hemoglobin levels in a person with sickle cell are 6-8 g per 100 mL. Reduced hemoglobin means reduced capacity to carry oxygen around the body so a person with sickle cell disease often feels very tired.

▸ Sickle cell is associated with pain and an increased risk of stroke, high blood pressure, and organ damage.

▸ The flexible, smooth RBCs can easily flow through the capillaries. They do not clump together so blood flow is not restricted.

▸ Normal RBCs have lifespan of 120 days. There are always sufficient RBCs available to transport oxygen.

▸ Hemoglobin levels are a measure of RBCs. Normal blood hemoglobin levels are >11 g per 100 mL. The RBCs have sufficient capacity to transport oxygen from the lungs to the body's tissues.

©2018 **BIOZONE** International
ISBN: 978-1-927309-55-1
Photocopying Prohibited

5. How does the sickle cell mutation affect the phenotype of the RBC? _____

_____

_____

6. The hemoglobin mutation resulting in sickle cell disease shows how change occurs on many different organizational levels. Briefly outline how the DNA, protein, cellular, and organism level are affected:

_____

_____

_____

_____

_____

_____

_____

_____

_____

_____

Why is the mutation still present in the population?

▶ If an individual inherits one mutated (HbS) gene and one normal (HbA) gene they are said to be carriers. The alleles are codominant (equally expressed) so they have both normal and sickle cell RBCs. They produce enough normal hemoglobin to function normally under most conditions and generally do not show any signs of sickle cell disease.

▶ The sickle cell mutation (HbS) can be lethal if two copies of the allele are inherited (HbSS), but if one copy is inherited (HbAS) it can provide protection against malaria. This is because the malarial parasite cannot infect the deformed RBCs. A high frequency of the mutation is present in the population where malaria is naturally present in the population all the time (endemic).

▶ A less well known mutation (HbC) to the same gene, discovered in populations in Burkina Faso, Africa, results in a 29% reduction in the likelihood of contracting malaria if the person has one copy of the mutated gene and a 93% reduction if the person has two copies. In addition, the anemia that person suffers as a result of the mutation is much less pronounced than in the HbS mutation.

Malaria parasites in blood

Malaria parasites

RBC

Graph of sickle cell allele variation vs mortality in children aged 2-16 months in Western Kenya

— HbAS - sickle cell trait
— HbAA - normal phenotype
— HbSS - sickle cell disease

7. The CDC has carried out a study to determine if malaria mortality rates vary with sickle cell allele type. The results of their study in Western Kenya are presented right.

(a) Describe the results obtained in the CDC study:

_____

_____

_____

(b) Explain why, despite its many negative effects, the sickle cell mutation is still present in the human population today:

_____

_____

_____

_____

©2018 **BIOZONE** International
ISBN: 978-1-927309-55-1
Photocopying Prohibited

EXPLAIN: How does the CCR5 mutation provide protection from HIV?

Before large scale global travel became common, human populations were largely isolated, and populations showed regional variations in particular alleles and allele frequencies. Some of this allelic diversity affects our immune system and causes some of the variation in immunity that is seen in humans. This diversity may provide natural resistance to some diseases in some people. An example of this is the CCR5 mutation described below.

What is HIV?

HIV/AIDS is an incurable disease caused by the human immunodeficiency virus (HIV). It has killed more than 35 million people.

HIV infects the T helper cells of the immune system. It uses the cells to replicate itself in large numbers, then the newly formed viral particles exit the cell to infect more T helper cells.

Many T helper cells are destroyed in the process of HIV replication. T helper cells are a crucial component of the body's immune system, so when their levels become too low, the immune system can no longer fight off infections and a person often dies as a result.

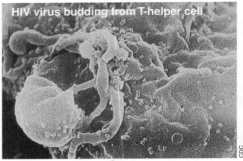
HIV virus budding from T-helper cell

Genetic diversity and HIV-1 resistance

The HIV-1 virus enters T helper cells by docking with the CCR5 membrane receptor (right) encoded by the CCR5 gene. People with a mutation in the gene (called CCR5Δ32) have resistance to HIV-1. The CCR5Δ32 mutation produces a premature stop codon in the mRNA. This shortens the membrane receptor protein produced so it does not protrude outside of the cell and the HIV cannot bind and enter the cell.

CCR5 receptor (yellow) in the membrane (gray)

The CCR5Δ32 mutation is found in populations of people of European descent. In some areas of northern Europe, up to 18% of the population have the mutation. The mutation is virtually absent in Asian, Middle Eastern, and American Indian populations.

▶ People with the mutation in one allele produce T-cells with a reduced number of CCR5 receptors. HIV-1 infects these cells slowly, taking 2-3 years longer than normal to progress to AIDS.

▶ People with mutations in both alleles produce T-cells with no CCR5 receptors. HIV-1 is effectively unable to infect these cells.

Distribution of the CCR5Δ32 mutation

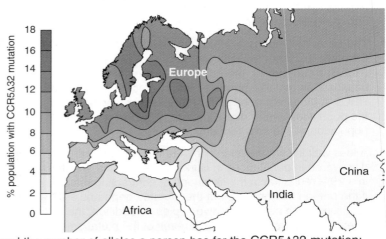

8. Explain the relationship between HIV resistance and the number of alleles a person has for the CCR5Δ32 mutation:

_____

_____

_____

_____

9. Based on the distribution of the CCR5Δ32 mutation above:

(a) Identify the most likely origin of the mutation: _____

(b) Describe the evidence to support your answer: _____

_____

_____

_____

_____

©2018 **BIOZONE** International
ISBN: 978-1-927309-55-1
Photocopying Prohibited

56 Mendelian Genetics

ENGAGE: Can you roll your tongue?

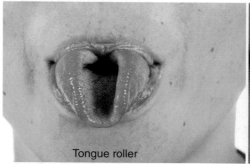
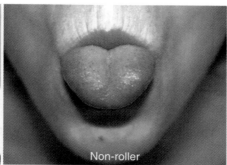

Tongue roller | Non-roller

Are you able to roll your tongue, or are you a non-roller and no matter how hard you try your tongue will not roll up to form a tube? The ability to roll your tongue is often used as an example of simple Mendelian inheritance.

In the 1940s, geneticist Alfred Sturtevant proposed that tongue rolling was inherited according to simple Mendelian rules. Under this scenario, the allele for tongue rolling is dominant, so if you have at least one dominant allele (T) you will be a tongue roller. Non-rollers have two copies of the recessive allele (tt).

1. (a) Can you roll your tongue? _____

   (b) State whether each of your parents is able to roll their tongue: _____

   _____

   (c) Using the T and t notation for the tongue rolling alleles, decide what your genotype could be and record it here:

   _____

   (d) When you compare your phenotype with that of your parents, does it make sense? Explain why or why not:

   _____

   _____

   _____

You may have found in your own investigation above that both of your parents are non-rollers but you (or one of your siblings) can actually roll your tongue. How can this be?

Since Sturtevant's work in the 1940s we have learned that the genetics of tongue rolling is not as simple as first thought. A study by Komai in 1951 obtained the results presented in the table on the right.

2. (a) Calculate the percentage of rollers from each cross and enter these values on the table.

   (b) What percentage of R offspring were produced by the NR x NR crosses?

Parents	R offspring	NR offspring	% R
R x R	928	104	90%
R x NR	468	217	68%
NR x NR	48	92	34%

R = roller, NR = non-roller.

   _____

   (c) Does this result suggest tongue rolling is inherited in a classic Mendelian inheritance pattern? Explain your answer:

   _____

   _____

So it's not that simple! Further studies involving identical twins showed that in some cases one of the pair could tongue roll while the other could not. Given the (nearly) identical nature of their DNA, we would expect both to have the same phenotype for tongue rolling. We now know that tongue rolling is not determined solely by genetics and that there are probably environmental factors in play also. You may still see many references using tongue rolling as an example of Mendelian inheritance, but keep in mind that it is not that simple.

©2018 **BIOZONE** International
ISBN: 978-1-927309-55-1
Photocopying Prohibited

SPQ LS3.B

EXPLORE: ACHOO Syndrome

▶ Have you ever walked outside into bright light and started to sneeze? If so, you may have ACHOO Syndrome. Although it sounds like a fictitious condition, ACHOO Syndrome is real and affects 18-35% of Americans.

▶ Other names for ACHOO Syndrome include sun sneezing or photic sneeze reflex.

▶ When affected people move from dim light into bright light, they sneeze a number of times in quick succession. The number of sneezes is usually 2-3, but some people can sneeze as many as 40 times.

▶ Why would sunlight make you sneeze? Although not well understood, current research suggests that when a specific nerve responsible for sensation and movement in the face is stimulated by bright light, it indirectly stimulates a sneeze reflex.

ACHOO Syndrome is inherited

Studies have shown that a single mutation (C replaces T) on chromosome 2 is responsible for ACHOO Syndrome. The mutation increases the likelihood of photic sneezing by 1.3 times and exhibits a Mendelian inheritance pattern. It is inherited in an autosomal dominant manner.

3. Conduct an experiment with your classmates. Have everyone walk out from being in a relatively dim room into bright sunlight. Have each student record if they sneezed or not when exposed to the sunlight.

   (a) What percentage of the class are photic sneezers? _____

   _____

   (b) How do your class results compare with the percentage of people affected nationally? _____

   _____

   _____

   (c) Predict what may happen if people close their eyes before stepping into the bright sunlight: _____

   _____

   _____

4. Now repeat the experiment as described above, but make sure everyone has their eyes closed when they step into the bright sunlight. Take care not to walk into anything when your eyes are closed (have someone guide you out).

   (a) What percentage of the class sneezed? _____

   (b) Do your class results differ from the first experiment? _____

   _____

   _____

   (c) Suggest why you might expect a difference in the results: _____

   _____

   _____

   _____

5. ACHOO Syndrome is inherited in an autosomal dominant pattern.

   (a) Assuming only one parent has ACHOO, what are the chances of any of their children having ACHOO?

   _____

   (b) What proportion of the children would have ACHOO if both parents were sufferers? _____

   _____

   (c) If you do not have ACHOO syndrome, can you pass on the trait to your children?_____

©2018 **BIOZONE** International
ISBN: 978-1-927309-55-1
Photocopying Prohibited

EXPLORE: Mendel's pea experiments

Some of the best known experiments in phenotypes are the experiments carried out by Gregor Mendel on pea plants. Mendel carried out breeding experiments to study seven phenotypic characters of the pea plant.

As described earlier, during one of the experiments (below) he noticed how traits expressed in one generation disappeared in the next, but then reappeared in the generation after that.

In his experiments Mendel used true breeding plants. When self-crossed (self-fertilized), true breeding plants are homozygous for the gene in question and therefore produce offspring with the same phenotypes as the parents. In the experiment described below, Mendel crossed plants true breeding for round seeds with plants true breeding for wrinkled seeds.

Peas show noticeably different traits for pod shape and color, seed shape and color, flower color, height, and position of the flowers on the stem.

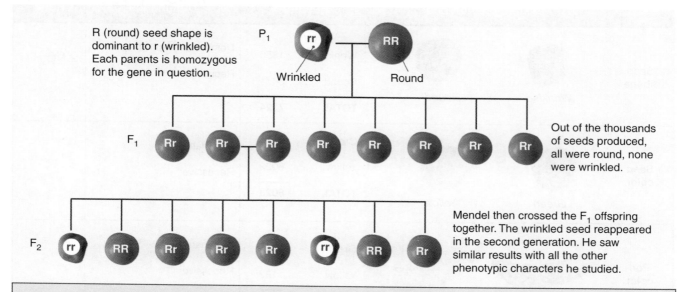

R (round) seed shape is dominant to r (wrinkled). Each parents is homozygous for the gene in question.

Out of the thousands of seeds produced, all were round, none were wrinkled.

Mendel then crossed the F₁ offspring together. The wrinkled seed reappeared in the second generation. He saw similar results with all the other phenotypic characters he studied.

How can these results be explained? Mendel was able to explain his observations in the following way:

▸ Traits are determined by a unit, which passes unchanged from parent to offspring (we now know that these units are genes).

▸ Each individual inherits one unit (gene) for each trait from each parent (each individual has two units).

▸ Traits may not physically appear in an individual, but the units (genes) for them can still be passed to its offspring.

6. In Mendel's pea experiments investigating wrinkled versus round peas above:

(a) What was the ratio of smooth seeds to wrinkled seeds in the F₂ generation? ___3 : 1___

(b) Why did the wrinkled seed trait not appear in the F₁ generation? __The wrinkled gene is__ __recessive, all the seeds in F₁ are homozygous.__

(c) Why did the wrinkled seed appear in the F₂ generation? __In crossing 2 homozygous__ __seeds, there is a 25% chance that their offspring__ __is wrinkled (rr)__

7. Another of Mendel's experiments involved crossing true breeding parents with green seeds and yellow seeds together. Yellow seed is dominant over green seed. Based on Mendel's experiment on seed shape above:

(a) What seed color(s) would you expect to observe in the F₁ offspring? __yellow__

(b) What seed color(s) would you expect to observe in the F₂ offspring? __yellow & green__

(c) You can test your prediction by drawing a diagram in the box right to illustrate the results of the experiment.

©2018 **BIOZONE** International
ISBN: 978-1-927309-55-1
Photocopying Prohibited

8. Mendel examined seven phenotypic traits (table right). Some of his results from crossing plants heterozygous for the gene in question are shown below. The numbers in the results column represent how many offspring had those particular traits.

(a) Study the results for each of the six experiments below. Determine which of the two phenotypes is dominant, and which is the recessive. Place your answers in the spaces in the dominance column in the table below.

(b) Calculate the ratio of dominant phenotypes to recessive phenotypes (to two decimal places). The first one has been done for you (5474 ÷ 1850 = 2.96). Place your answers in the spaces provided in the table below:

Phenotypic characters of the pea plant:
- Flower color (violet or white)
- Pod color (green or yellow)
- Height (tall or short)
- Position of the flowers on the stem (axial or terminal)
- Pod shape (inflated or constricted)
- Seed shape (round of wrinkled)
- Seed color (yellow or green)

Trait	Possible phenotypes		Results		Dominance	Ratio
Seed shape	*Wrinkled*	*Round*	Wrinkled Round **TOTAL**	1850 5474 **7324**	Dominant: Round Recessive: Wrinkled	2.96:1
Seed color	*Green*	*Yellow*	Green Yellow **TOTAL**	2001 6022 **8023**	Dominant: Yellow Recessive: Green	3:1
Pod color	*Green*	*Yellow*	Green Yellow **TOTAL**	428 152 **580**	Dominant: green Recessive: yellow	2.82:1
Flower position	*Axial*	*Terminal*	Axial Terminal **TOTAL**	651 207 **858**	Dominant: axial Recessive: terminal	3.14:1
Pod shape	*Constricted*	*Inflated*	Constricted Inflated **TOTAL**	299 882 **1181**	Dominant: inflated Recessive: constricted	2.95:1
Stem length	*Tall*	*Dwarf*	Tall Dwarf **TOTAL**	787 277 **1064**	Dominant: Tall Recessive: dwarf	2.84:1

9. Mendel's experiments identified that two heterozygous parents should produce offspring in the ratio of 3 of the dominant phenotype to 1 of the recessive phenotype.

(a) Which three of Mendel's experiments provided ratios closest to the theoretical 3:1 ratio? Pod shape, seed shape, seed color

(b) Suggest why these results deviated less from the theoretical ratio than the others: The bigger the sample size, the closer the ratio.

©2018 **BIOZONE** International
ISBN: 978-1-927309-55-1
Photocopying Prohibited

EXPLORE: Monohybrid cross

Monohybrid crosses are used to show single gene inheritance patterns between two individuals. Monohybrid crosses can be used to determine allele dominance. A simple square grid called a **Punnett square** is used to determine all of the possible outcomes of a genetic cross.

The diagram on the right shows how to draw a Punnett square for a monohybrid cross for a trait defined by the alleles A and a. Both parents have the allele combination Aa. Crosses can also be drawn using circles and arrows or lines to illustrate segregation of the gametes and fertilization as below.

Note: The dominant allele is always written first, no matter what the parent.

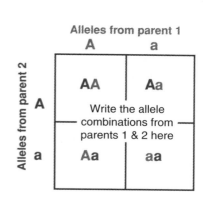

Alleles from parent 1

	A	a
A	AA	Aa
a	Aa	aa

Write the allele combinations from parents 1 & 2 here

Alleles from parent 2

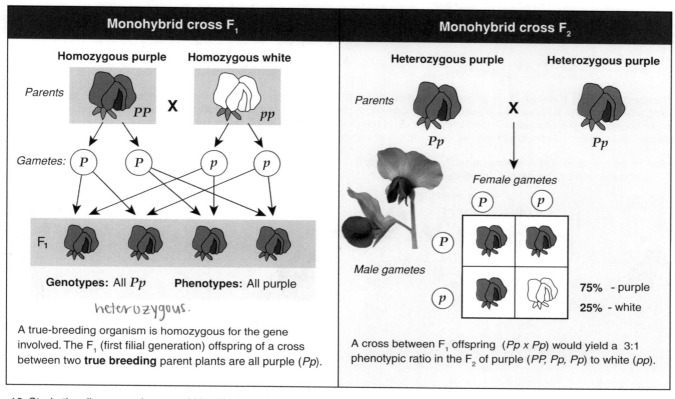

Monohybrid cross F₁

Homozygous purple X **Homozygous white**

Parents — PP X pp

Gametes: P P p p

F₁

Genotypes: All Pp **Phenotypes:** All purple

heterozygous.

A true-breeding organism is homozygous for the gene involved. The F₁ (first filial generation) offspring of a cross between two **true breeding** parent plants are all purple (Pp).

Monohybrid cross F₂

Heterozygous purple X **Heterozygous purple**

Parents — Pp X Pp

Female gametes P p

Male gametes P p

75% - purple
25% - white

A cross between F₁ offspring (Pp x Pp) would yield a 3:1 phenotypic ratio in the F₂ of purple (PP, Pp, Pp) to white (pp).

10. Study the diagrams above and identify the dominant allele and the recessive allele. Explain your reasoning: _____

dominant is purple
recessive is white.

11. Complete the crosses below:

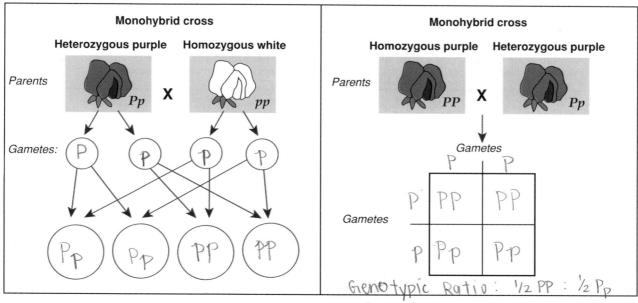

Monohybrid cross

Heterozygous purple X **Homozygous white**

Parents — Pp X pp

Gametes: P P P P

Pp Pp PP PP

Monohybrid cross

Homozygous purple X **Heterozygous purple**

Parents — PP X Pp

Gametes

	P	P
P	PP	PP
P	Pp	Pp

Gametes

Genotypic Ratio: ½ PP : ½ Pp
Phenotypic ratio: 100% purple

©2018 **BIOZONE** International
ISBN: 978-1-927309-55-1
Photocopying Prohibited

EXPLORE: The test cross

▶ It is not always possible to determine an organism's genotype by its appearance because gene expression is complicated by patterns of dominance and by gene interactions. The **test cross** is a special type of back cross used to determine the genotype of an organism with the dominant phenotype for a particular trait.

▶ The principle is simple. The individual with the unknown genotype is bred with a homozygous recessive individual for the trait(s) of interest. The homozygous recessive can produce only one type of allele (recessive), so the phenotypes of the offspring will reveal the genotype of the unknown parent (below). The test cross can be used to determine the genotype of single genes or multiple genes.

The common fruit fly (*Drosophila melanogaster*) is often used to illustrate basic principles of inheritance because it has several easily identified phenotypes, which act as genetic markers. Once such phenotype is body color. Wild type (normal) *Drosophila* have yellow-brown bodies. The allele for yellow-brown body color (E) is dominant. The allele for an ebony colored body (e) is recessive. The test crosses below show the possible outcomes for an individual with homozygous and heterozygous alleles for ebony body color.

A. A homozygous recessive female (ee) with an ebony body is crossed with a homozyogous dominant male (EE).

B. A homozygous recessive female (ee) with an ebony body is crossed with a heterozygous male (Ee).

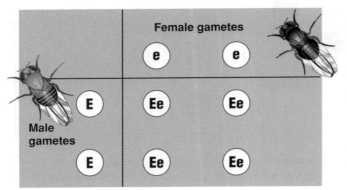

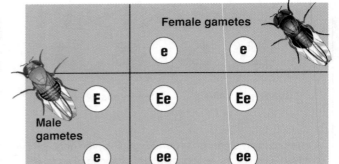

Cross A:
(a) Genotype frequency: 100% Ee
(b) Phenotype frequency: 100% yellow-brown

Cross B:
(a) Genotype frequency: 50% Ee, 50% ee
(b) Phenotype frequency: 50% yellow-brown, 50% ebony

DID YOU KNOW?

In crosses involving *Drosophila*, the wild-type alleles are dominant but are given an upper case symbol of the mutant phenotype. This is because there are many alternative mutant phenotypes to the wild type, e.g. vestigial wings and curled wings are both mutant phenotypes, whereas the wild type is straight wing. This can happen in other crosses too, e.g. guinea pigs, if a letter is already used for another character.

12. In *Drosophila*, the allele for brown eyes (b) is recessive, while the red eye allele (B) is dominant. Describe (using text or diagrams) how you would carry out a test cross to determine the genotype of a male with red eyes:

13. Of the test cross offspring, 50% have red eyes and 50% have brown eyes.

What is the genotype of the male *Drosophila*?_____

©2018 **BIOZONE** International
ISBN: 978-1-927309-55-1
Photocopying Prohibited

EXPLAIN: Sickle cell inheritance patterns

A monohybrid cross studies the inheritance pattern of one gene. The offspring of these crosses occur in predictable ratios. In the examples below you will use monohybrid crosses to determine the genotype and phenotype outcomes for sickle cell disease in humans. Recall that this condition is inherited in an autosomal recessive pattern (although the alleles are codominant). Two copies of the recessive allele must be inherited for an individual to have sickle cell anemia. An individual with only one copy is a carrier for sickle cell trait (they have both normal and sickled RBCs). The inheritance pattern for two parents heterozygous for the sickle cell mutation is shown below.

14. Write the phenotype for each individual on the diagram below:

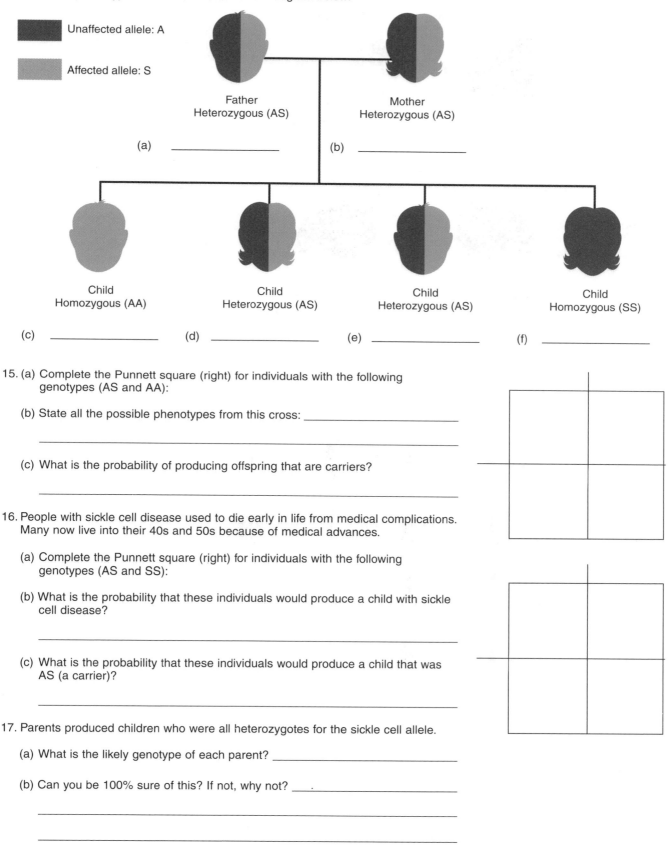

Unaffected allele: A

Affected allele: S

Father
Heterozygous (AS)

Mother
Heterozygous (AS)

(a) _____

(b) _____

Child
Homozygous (AA)

Child
Heterozygous (AS)

Child
Heterozygous (AS)

Child
Homozygous (SS)

(c) _____   (d) _____   (e) _____   (f) _____

15. (a) Complete the Punnett square (right) for individuals with the following genotypes (AS and AA):

(b) State all the possible phenotypes from this cross: _____

_____

(c) What is the probability of producing offspring that are carriers?

_____

16. People with sickle cell disease used to die early in life from medical complications. Many now live into their 40s and 50s because of medical advances.

(a) Complete the Punnett square (right) for individuals with the following genotypes (AS and SS):

(b) What is the probability that these individuals would produce a child with sickle cell disease?

_____

(c) What is the probability that these individuals would produce a child that was AS (a carrier)?

_____

17. Parents produced children who were all heterozygotes for the sickle cell allele.

(a) What is the likely genotype of each parent? _____

(b) Can you be 100% sure of this? If not, why not? _____

_____

_____

©2018 **BIOZONE** International
ISBN: 978-1-927309-55-1
Photocopying Prohibited

EXPLORE: Dihybrid cross

A dihybrid cross studies the inheritance pattern of two genes. In crosses involving unlinked autosomal genes, the offspring occur in predictable ratios.

▶ There are four types of gamete produced in a cross involving two genes, where the genes are carried on separate chromosomes and are sorted independently of each other during meiosis.

▶ The two genes in the example below are on separate chromosomes and control two unrelated characteristics, hair color and coat length. Black (B) and short (L) are dominant to white (b) and long (l).

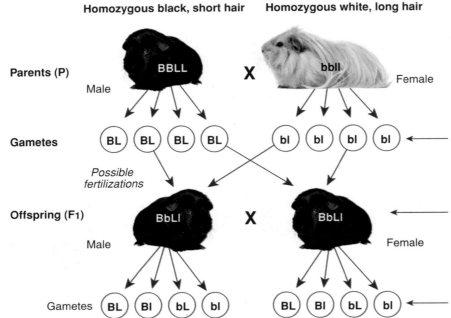

Homozygous black, short hair   **Homozygous white, long hair**

Parents (P) — Male — BBLL X bbll — Female

Gametes — BL BL BL BL   bl bl bl bl

Possible fertilizations

Offspring (F1) — Male — BbLl X BbLl — Female

Gametes — BL Bl bL bl   BL Bl bL bl

Parents: The notation P is used for a cross between true breeding (homozygous) parents.

Gametes: Each parent produces only one type of gamete. This is because each parent is homozygous for both traits.

F1 offspring: There is one kind of gamete from each parent, therefore only one kind of offspring produced in the first generation. The notation F1 is used to denote the heterozygous offspring of a cross between true breeding parents.

F2 offspring: The F1 were mated with each other. Each individual from the F1 is able to produce four different kinds of gamete.

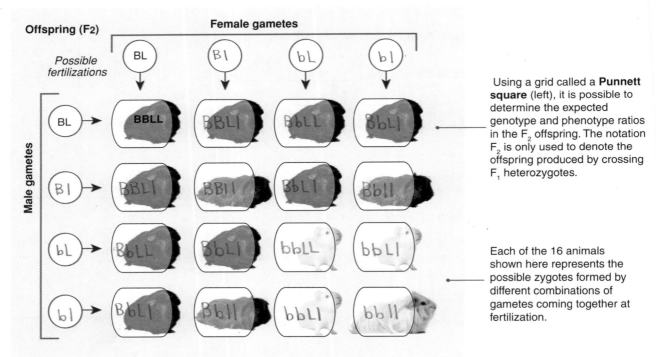

Offspring (F2) — Female gametes

Possible fertilizations

Male gametes

	BL	Bl	bL	bl
BL	BBLL	BBLl	BbLL	BbLl
Bl	BBLl	BBll	BbLl	Bbll
bL	BbLL	BbLl	bbLL	bbLl
bl	BbLl	Bbll	bbLl	bbll

Using a grid called a **Punnett square** (left), it is possible to determine the expected genotype and phenotype ratios in the F2 offspring. The notation F2 is only used to denote the offspring produced by crossing F1 heterozygotes.

Each of the 16 animals shown here represents the possible zygotes formed by different combinations of gametes coming together at fertilization.

18. (a) Fill in the gametes and complete the Punnett square above.

(b) Identify the number of each phenotype in the offspring, together with all the genotypes producing the phenotype in each case. Write them below:

Black, short: BBLL, BbLl, BBLl, BbLl   White, short: bbLL -1, bbLl -2
                1   2   2   4

Black, Long: BBll -1   White, long: bbll -1
              Bbll -2

First
Outside
Inner
Last

©2018 **BIOZONE** International
ISBN: 978-1-927309-55-1
Photocopying Prohibited

19. In guinea pigs, rough coat **R** is dominant over smooth coat **r** and black coat **B** is dominant over white **b**. The genes are not linked. A homozygous rough black animal was crossed with a homozygous smooth white:

(a) State the genotype of the F₁: _____RrBb_____

(b) State the phenotype of the F₁: _Heterozygous rough black_

(c) Use the Punnett square to show the outcome of a cross between the F₁ (the F₂):

	RB	Rb	rB	rb
RB	RRBB	RRBb	RrBB	RrBb
Rb	RRBb	RRbb	RrBb	Rrbb
rB	RrBB	RrBb	rrBB	rrBb
rb	RrBb	Rrbb	rrBb	rrbb

(d) Using ratios, state the phenotypes of the F₂ generation:

_____9 rough / black :_____
_____3 rough / white : 3 smooth /_____
_____black : 1 smooth / white_____

(e) Use the next Punnett square to show the outcome of a back cross of the F₁ to the rough, black parent:

(f) Using ratios, state the phenotype of the offspring of this back cross:

RrBb
RRBB

_____1 rough black : 0 smooth : 0 smooth white :_____
_____0 rough white._____

	RB	RB	RB	RB
RB	RRBB	RRBB	RRBB	RRBB
Rb	RRBb	RRBb	RRBb	RRBb
rB	RrBB	RrBB	RrBB	RrBB
rb	RrBb	RrBb	RrBb	RrBb

(g) A rough black guinea pig was crossed with a rough white guinea pig produced the following offspring: 3 rough black, 2 rough white, and 1 smooth white. What are genotypes of the parents? Explain your reasoning:

_____Rough Black : RrBb ; Rough_____
_____white Rrbb._____
_____Because there is a smooth_____
_____white guinea pig, both parents_____
_____need to carry a recessive_____
_____gene._____

20. In humans, two genes affecting the appearance of the hands are the gene for thumb hyperextension (curving) and the gene for mid-digit hair. The allele for curved thumb, **H**, is dominant to the allele for straight thumb, **h**. The allele for mid digit hair, **M**, is dominant to that for an absence of hair, **m**.

(a) Give all the genotypes of individuals who are able to curve their thumbs, but have no mid-digit hair:

_____HHmm   or   Hhmm_____

(b) Complete the Punnett square to show the possible genotypes from a cross between two individuals heterozygous for both alleles: HhMm

	HM	hM	Hm	hm
HM	HHMM	HhMM	HHMm	HhMm
hM	HhMM	hhMM	HhMm	hhMm
Hm	HHMm	HhMm	HHmm	Hhmm
hm	HhMm	hhMm	Hhmm	hhmm

(c) State the phenotype ratios of the F₁ progeny:

_____9 curved thumb with mid-digit hair:_____
_____3 straight thumb with mid-digit hair:_____
_____3 curved thumb without mid-digit hair:_____
_____1 straight thumb without mid-digit hair_____

(d) What is the probability that one of the offspring would have mid-digit hair (show your working)?

_____75% because 12 of 16_____
_____is 75._____

©2018 **BIOZONE** International
ISBN: 978-1-927309-55-1
Photocopying Prohibited

EXPLORE: Probability

Many events cannot be predicted with absolute certainty, however we can determine how likely it is that an event will happen. This calculated likelihood of an event occurring is called **probability**. The probability of an event ranges from 0 to 1. The sum of all probabilities equals 1.

In biology, probability is used to calculate the statistical significance of a difference between means or the probability of an event occurring, e.g. getting an offspring with a certain genotype and phenotype in a genetic cross.

$$\text{Probability of an event happening} = \frac{\text{Number of ways it can happen}}{\text{Total number of outcomes}}$$

▶ Tossing a coin and predicting whether it will land heads (H) up or tails (T) up is a good example to illustrate probability.

▶ There are two possible outcomes; the coin will either land heads up or tails up, and only one outcome can occur at a time. Therefore the probability of a coin landing heads up is 1/2. The likelihood of a coin landing tails up is also 1/2.

▶ Remember probability is just an indication of how likely something will happen. Even though we predict that heads and tails will come up 50 times each if we toss a coin 100 times, it might not be exactly that.

21. Calculate the probability that a 6 will occur when you roll a single dice (die):

_____

The rules for calculating probability

Probability rules are used when we want to predict the likelihood of two events occurring together or when we want to determine the chances of one outcome over another. The rules are useful when we want to determine the probably of certain outcomes in genetic crosses, especially when large numbers of alleles are involved. The probability rule used depends on the situation.

PRODUCT RULE for independent events
For independent events, A & B, the probability (P) of them both occurring (A&B) = P(A) X P(B)
Example: If you roll two dice at the same time, what is the probability of rolling two sixes?
Solution: The probability of getting six on two dice at once is 1/6 x 1/6 = 1/36.

SUM RULE for mutually exclusive events
For mutually exclusive events, A & B, the probability (P) that one will occur (A or B) = P(A) + P(B)
Example: A single die is rolled. What are the chances of rolling a 2 **or** a 6?
Solution: P(A or B) = P(A) + P(B). 1/6 + 1/6 = 2/6 (1/3). There is a 1/3 chance that a 2 or 6 will be rolled.

22. In a cross Aa x Aa, use the sum rule to determine the probability of the offspring having a dominant phenotype:

_____

23. Use the product rule to determine the probability of a first and second child born to the same parents both being boys?

_____

24. In a cross of rabbits both heterozygous for genes for coat color and length (BbLl x BbLl), determine the probability of the offspring being BbLl. HINT: Calculate probabilities for Bb and Ll separately and then use the product rule. Test your calculation using the Punnett square (right).

_____

25. In a cross of two individuals with various alleles of four unlinked genes: AaBbCCdd x AabbCcDd, explain how you would calculate the probability of getting offspring with the dominant phenotype for all four traits?

_____

_____

_____

©2018 **BIOZONE** International
ISBN: 978-1-927309-55-1
Photocopying Prohibited

EVALUATE: Testing the outcome of genetic crosses against predicted ratios

The chi-squared test for goodness of fit (χ^2) can be used for testing the outcome of dihybrid crosses against an expected (predicted) Mendelian ratio.

Using χ^2 in Mendelian genetics

▸ In genetic crosses, certain ratios of offspring can be predicted based on the known genotypes of the parents The chi-squared test is a statistical test to determine how well observed offspring numbers match (or fit) expected numbers. Raw counts should be used and a large sample size is required for the test to be valid.

▸ In a chi-squared test, the null hypothesis predicts the ratio of offspring of different phenotypes is the same as the expected Mendelian ratio for the cross, assuming independent assortment of alleles (no linkage, i.e. the genes involved are on different chromosomes).

▸ Significant departures from the predicted Mendelian ratio indicate linkage (the genes are on the same chromosome) of the alleles in question.

▸ In a *Drosophila* genetics experiment, two individuals were crossed (the details of the cross are not relevant here). The predicted Mendelian ratios for the offspring of this cross were 1:1:1:1 for each of the four following phenotypes: gray body-long wing, gray body-vestigial wing, ebony body-long wing, ebony body-vestigial wing.

▸ The observed results of the cross were not exactly as predicted. The following numbers for each phenotype were observed in the offspring of the cross:

Gray body, vestigial wing	Gray body, long wing	Ebony body, long wing	Ebony body, vestigial wing
88	**98**	**102**	**112**

Table 1: Critical values of χ^2 at different levels of probability. By convention, the critical probability for rejecting the null hypothesis (H_0) is 5%. If the test statistic is less than the tabulated critical value for $P = 0.05$ we cannot reject H_0 and the result is not significant. If the statistic is greater than the tabulated value for $P = 0.05$ we reject (H_0) in favor of the alternative hypothesis.

Degrees of freedom	Level of probability (P)					
	0.50	0.20	0.10	0.05	0.02	0.01
1	0.455	1.64	2.71	3.84	5.41	6.64
2	1.386	3.22	4.61	5.99	7.82	9.21
3	2.366	4.64	6.25	7.82	9.84	11.35
4	3.357	5.99	7.78	9.49	11.67	13.28
5	4.351	7.29	9.24	11.07	13.39	15.09
	Do not reject H_0			Reject H_0		

Steps in performing a χ^2 test

1 Enter the observed value (O).
Enter the values of the offspring into the table (below) in the appropriate category (column 1).

2 Calculate the expected value (E).
In this case the expected ratio is 1:1:1:1. Therefore the number of offspring in each category should be the same (i.e. total offspring/ no. categories). 400 / 4 = 100 (column 2).

3 Calculate O-E and $(O-E)^2$
The difference between the observed and expected values is calculated as a measure of the deviation from a predicted result. Since some deviations are negative, they are all squared to give positive values (columns 3 and 4).

4 Calculate χ^2
For each category, calculate $(O - E)^2 / E$. Then sum these values to produce the χ^2 value (column 5).

$$\chi^2 = \sum \frac{(O - E)^2}{E}$$

5 Calculate degrees of freedom
The probability that any particular χ^2 value could be exceeded by chance depends on the number of degrees of freedom. This is simply one less than the total number of categories (this is the number that could vary independently without affecting the last value) In this case 4 - 1 = 3.

6 Use χ^2 table
On the χ^2 table with 3 degrees of freedom, the calculated χ^2 value corresponds to a probability between 0.2 and 0.5. By chance alone a χ^2 value of **2.96** will happen 20% to 50% of the time. The probability of 0.0 to 0.5 is higher than 0.05 (i.e 5% of the time) and therefore the null hypothesis cannot be rejected. We have no reason to believe the observed values differ significantly from the expected values.

	1	2	3	4	5
Category	O	E	O-E	$(O\_E)^2$	$(O\_E)^2/E$
GB, LW	98	100	-2	4	0.04
GB, VW	88	100	-12	144	1.44
EB, LW	102	100	2	4	0.04
EB, VW	112	100	12	144	1.44
				χ^2	2.96

©2018 **BIOZONE** International
ISBN: 978-1-927309-55-1
Photocopying Prohibited

26. Students carried out a pea plant breeding experiment in which they crossed two plants heterozygous for seed shape and color. The predicted Mendelian ratios for the offspring were **9:3:3:1** for each of the four following phenotypes: round-yellow seed, round-green seed, wrinkled-yellow seed, wrinkled-green seed.

The observed results of the cross were not exactly as predicted. The numbers of offspring with each phenotype are provided below:

Observed results of the pea plant cross			
Round-yellow seed	441	Wrinkled-yellow seed	143
Round-green seed	159	Wrinkled-green seed	57

(a) State your null hypothesis for this investigation (H_0)

_____

_____

(b) State the alternative hypothesis (H_A): _____

_____

Use the chi-squared test to determine if the differences between the observed and expected phenotypic ratios are significant. Use the table of critical values for χ^2 at different P values on the previous page.

(c) Enter the observed and expected values (number of individuals) and complete the table to calculate the χ^2 value.

Category	O	E	O – E	$(O – E)^2$	$\frac{(O – E)^2}{E}$
Round-yellow seed					
Round-green seed					
Wrinkled-yellow seed					
Wrinkled-green seed					
					Σ

(d) Calculate the χ^2 value using the equation $\chi^2 = \sum \frac{(O - E)^2}{E}$ (right hand column of the table):

(e) Calculate the degrees of freedom: _____

(f) Using the χ^2 table, state the P value corresponding to your calculated χ^2 value: _____

(g) State your decision (circle one): reject H_0 / do not reject H_0

27. In another experiment, a group of students bred two corn plants together. One corn plant was known to have grown from a kernel that was colorless (c) and did not have a waxy endosperm (w). The other corn plant was grown from a seed that was colored (C) but with a waxy endosperm (W). When the corn ear was mature the students removed it and counted the different phenotypes in the corn kernels.

Observed results of corn kernels			
Colored - waxy	201	Colorless - waxy	86
Colored - not waxy	85	Colorless - not waxy	210

From the observed results the students argued two points:
(1) The plant with the dominant phenotype must have been heterozygous for both traits.
(2) The genes for kernel color and endosperm waxiness must be linked (on the same chromosome).

(a) Defend the students' first argument: _____

_____

(b) On a separate sheet, use a chi-squared test to provide evidence for or against the students' second argument:

©2018 **BIOZONE** International
ISBN: 978-1-927309-55-1
Photocopying Prohibited

Elaborate: Computer simulations can be used to predict inheritance patterns

Computer simulations can be used to simulate the inheritance of particular genes. These provide a way to investigate inheritance even when you do not have access to the live organisms themselves, and they allow you to do a large number of crosses in a very short time.

Drosophila sciencecourseware.org

This virtual genetics lab provides a virtual lab environment where you can order fruit flies, mate them, and then sort and analyze the offspring. Along the way, you make and test a hypothesis. An online note book lets you record your results and you can even take a quiz.

To access *sciencecourseware.org*, visit their home page: http://www.sciencecourseware.org/vcise/drosophila/ and follow the instructions provided. Alternatively, access the site through **BIOZONE's Resource Hub.**

EXAMPLE: A wild type female is crossed with a male with three mutations to the antennae, eyes, and body color. The offspring produced from the cross are sorted in to groups and the phenotypes are given in the table below.

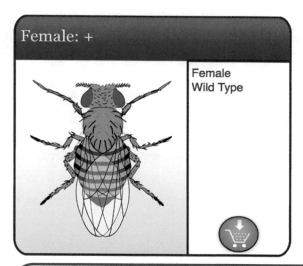

Female: +

Female
Wild Type

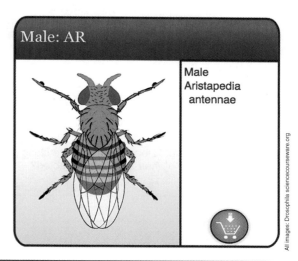

Male: AR

Male
Aristapedia
antennae

All images: Drosophila sciencecourseware.org

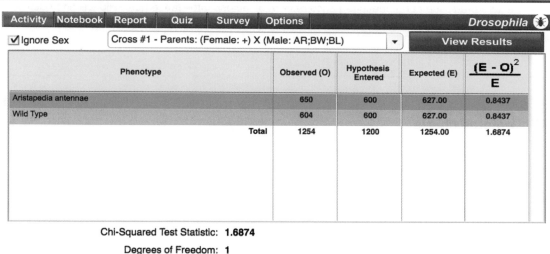

| Activity | Notebook | Report | Quiz | Survey | Options | | | *Drosophila* ✲ |

☑ Ignore Sex   Cross #1 - Parents: (Female: +) X (Male: AR;BW;BL) ▼   **View Results**

Phenotype	Observed (O)	Hypothesis Entered	Expected (E)	$\dfrac{(E - O)^2}{E}$
Aristapedia antennae	650	600	627.00	0.8437
Wild Type	604	600	627.00	0.8437
Total	1254	1200	1254.00	1.6874

Chi-Squared Test Statistic: **1.6874**

Degrees of Freedom: **1**

Level of Significance: **0.1939**

Enter New Hypothesis

Print Analysis   **Add to Notebook**   **Return to Menu**

Chi Square Analysis of Data and Your Hypothesis

The Chi Square analysis test the statistical difference between the expected and observed results, and uses the degrees of freedom to obtain the level of significance. If it is <0.05, it suggests that your hypothesis is unlikely. Click "Add to Notebook" to record the analysis, or "Enter New Hypothesis" to enter a different hypothesis.

Once the mating has been completed you can propose a hypothesis and then test the significance of the results by carrying out a Chi-squared test. The hypothesis for the breeding experiment above was that there would be equal numbers of each phenotype. The results for the cross show a variation from this prediction. Using the degrees of freedom table from page 252 we can see that the tabulated critical value at P= 0.05 and one degree of freedom is 3.84. The chi-squared result was 1.6874. This is lower than 3.84 so we do not reject the null hypothesis (the departure of the observed results from the expected are not significant).

©2018 **BIOZONE** International
ISBN: 978-1-927309-55-1
Photocopying Prohibited

57 Pedigree Analysis

ENGAGE: Lactose intolerance

Do you know someone who can't eat dairy products because they make them feel unwell? These people react badly to lactose, the sugar found in milk products. The condition is called lactose intolerance, and people with it can suffer a range of symptoms including abdominal bloating and cramps, flatulence, diarrhea, nausea, stomach rumbling, and vomiting.

The condition arises because the gene that produces the lactase enzyme is turned off. Lactase breaks lactose into single sugar units so that they can be absorbed. In a person with lactose intolerance, the undigested lactose passes into the large intestine (bowel) where bacterial fermentation produces large amounts of gas and acid. The unabsorbed sugars and metabolic products cause water to flow into the bowels resulting in diarrhea.

Lactose intolerance is inherited and its prevalence varies between regions. Less than 10% of the population in Northern Europe are lactose intolerant, but in East Asia it can be as high as 90%.

1. Talk to your classmates. Does anyone in the class have lactose intolerance? If so, are any of their other family members affected? How many of their family are affected?

_____

_____

_____

2. If possible, make a prediction about the inheritance pattern of lactose intolerance here. You will find out if you are right later on in this activity.

_____

_____

EXPLORE: Pedigree charts

Pedigree charts are a way of showing inheritance patterns over a number of generations. They are often used to study the inheritance of genetic disorders. The key should be consulted to decode the symbols. Individuals are identified by their generation number and their order number in that generation. For example, **II-6** is the sixth person in the second generation. The arrow indicates the person through whom the pedigree was discovered (i.e. who reported the condition).

If the chart on the right were illustrating a human family tree, it would represent three generations: grandparents (I-1 and I-2) with three sons and one daughter. Two of the sons (II-3 and II-4) are identical twins, but did not marry or have any children. The other son (II-1) married and had a daughter and another child (sex unknown). The daughter (II-5) married and had two sons and two daughters (plus a child that died in infancy).

For the particular trait being studied, the grandfather was expressing the phenotype (showing the trait) and the grandmother was a carrier. One of their sons and one of their daughters also show the trait, together with one of their granddaughters.

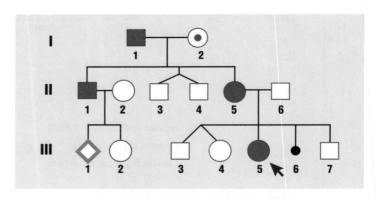

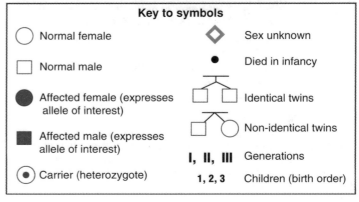

Key to symbols

◯ Normal female	◇ Sex unknown
☐ Normal male	● Died in infancy
⬤ Affected female (expresses allele of interest)	⬒ Identical twins
⬛ Affected male (expresses allele of interest)	⬔ Non-identical twins
	I, II, III Generations
◉ Carrier (heterozygote)	**1, 2, 3** Children (birth order)

    LS3.B P

©2018 **BIOZONE** International
ISBN: 978-1-927309-55-1
Photocopying Prohibited

EXPLORE: Pedigree charts are used to determine inheritance patterns

3. The pedigree chart right shows the inheritance of allele A in a flower that can be red or white. Red flower = affected:

   (a) Which color is produced by the dominant allele?

   _____ white _____

   (b) Write on the chart the genotype for each of the generation I individuals.

   (c) III4 is crossed with a white flower. What is the probability that any one offspring also has a white flower?

   _____

   _____

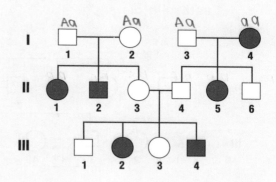

4. The pedigree chart right shows the inheritance of the allele B in a mammal that can have a coat color of black or white. Black coat = affected

   (a) Which color is produced by the dominant allele?

   _____ white _____

   (b) Explain how you know this: ___ 4 and 5 could have an offspring w/ a black coat, which shows that they were carriers.

   _____

   _____

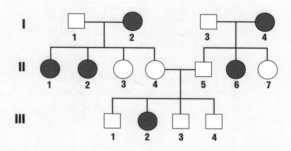

5. **Autosomal recessive traits**
   The sickle cell trait is inherited in an autosomal recessive pattern (although the alleles are codominant).

   (a) Write the genotype for each of the individuals I-1 & I-2 on the chart using the following letter codes:
   AA normal; **A-** normal (but unknown), **AS** carrier, **SS** sickle cell trait.

   (b) Why must the parents (II-5) and (II-6) be carriers:

   ___ Their child is affected with the recessive disease.

   _____

Sickle cell in humans

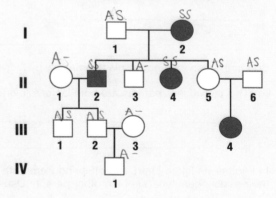

6. **Autosomal dominant traits**
   An unusual trait found in some humans is woolly hair (not to be confused with curly hair). Each affected individual will have at least one affected parent.

   (a) Write the genotype for each of the individuals on the chart using the following letter codes:
   WW woolly hair, **Ww** woolly hair (heterozygous), **W-** woolly hair, but unknown if homozygous, **ww** normal hair.

   (b) Describe a feature of this inheritance pattern that suggests the trait is the result of a **dominant** allele:

   _____

   _____

Woolly hair in humans

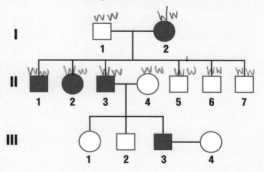

©2018 **BIOZONE** International
ISBN: 978-1-927309-55-1
Photocopying Prohibited

EXPLAIN: The inheritance of lactose intolerance

The pedigree chart below was one of the original studies to determine the inheritance pattern of lactose intolerance.

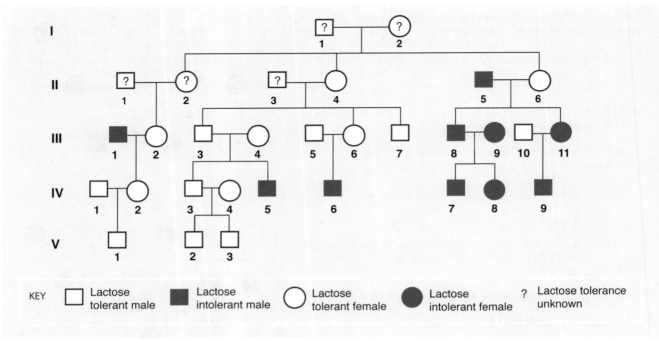

KEY | Lactose tolerant male | Lactose intolerant male | Lactose tolerant female | Lactose intolerant female | ? Lactose tolerance unknown

7. (a) Use the pedigree chart above to determine if lactose intolerance is a dominant trait or a recessive trait:

_____

(b) Provide evidence to support your answer: _____

_____

_____

_____

8. (a) On the pedigree above place an arrow by individual 6 in generation IV.

(b) This individual's parents are both lactose tolerant but he is lactose intolerant. Explain how this could have occurred?

_____

_____

9. In another pedigree chart, two offspring were both lactose tolerant. One of their parents was lactose intolerant. What are the possible genotype(s) of the other parent? Use the notation L = lactose tolerant, l = lactose intolerant.

_____

10. Using the examples in this activity, make up your own set of guidelines for interpreting pedigree charts. How do you identify an autosomal inheritance pattern? What are the features of autosomal recessive inheritance? Of autosomal dominant inheritance?

_____

_____

_____

_____

_____

_____

_____

_____

©2018 **BIOZONE** International
ISBN: 978-1-927309-55-1
Photocopying Prohibited

58 Environment Influences Phenotype

ENGAGE: Hydrangea flowers

Changes in the chemical environment influence flower color in hydrangeas (below). They have blue flowers when they are grown in acidic soil (pH <7.0) and pink flowers when grown in neutral to basic soils (≥ 7.0). The color change is a result of the mobility and availability of aluminum ions (Al^{3+}) at different pH. At low pH Al^{3+} is highly mobile. It binds with other ions and is taken up into the plant, reacting with the usually red/pink pigment in the flowers to form a blue color. In soil pH at or above 7.0, the aluminum ions combine with hydroxide ions to form insoluble and immobile aluminum hydroxide ($Al(OH)_3$). The plant doesn't take up the aluminum and remains red/pink. Other conditions (e.g. high phosphorus levels) can also affect aluminum mobility and availability.

Soil pH <7.0

Soil pH ≥ 7.0

1. Complete the table below to show what conditions cause color change in hydrangea flowers:

	Blue flower	Red/pink flower
pH		
Al^{3+} availability		

2. Litmus paper (right) is used to determine if a solution is acidic or basic. How is hydrangea flower color like litmus paper?

_____

_____

_____

3. Work in pairs or small groups to identify other examples to illustrate the role of environmental factors in determining phenotype and causing variation in a population. Hint: consider both abiotic and biotic factors.

_____

_____

_____

_____

_____

_____

_____

_____

©2018 **BIOZONE** International
ISBN: 978-1-927309-55-1
Photocopying Prohibited

CE LS3.B

EXPLORE: What factors influence phenotype?

▶ An organism's phenotype is influenced by the effects of the environment during and after development, even though the genotype remains unaffected.

▶ The phenotype is the product of the many complex interactions between the genotype, the environment, and the chemical tags and markers that regulate the expression of the genes (these are called epigenetic factors).

▶ Even identical twins have minor differences in their appearance due to epigenetic and environmental factors such as uterine environment before birth and diet after birth. Genes, together with epigenetic and environmental factors determine the unique phenotype that is produced.

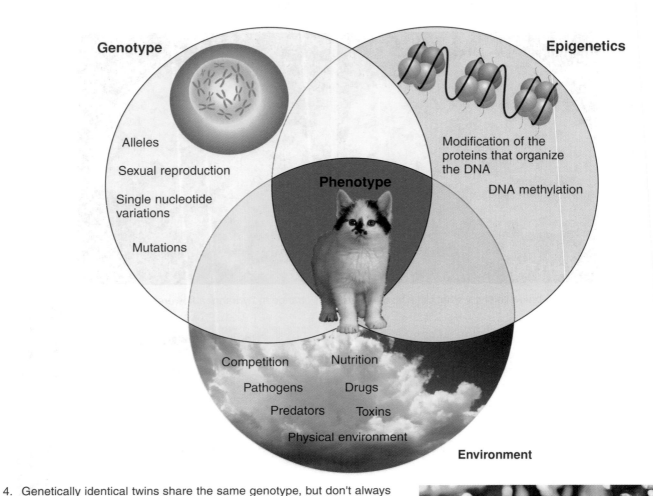

4. Genetically identical twins share the same genotype, but don't always look the same and they may have different outcomes in terms of their health and longevity. Explain how this can happen:

_____
_____
_____
_____
_____
_____
_____
_____
_____
_____
_____
_____

Marian and Vivian Brown were American actresses and identical twins. Although identical, Vivian the elder developed Alzheimer's disease and dies at age 85. Marian died later at age 87. Do you know any identical twins? In what ways are they different?

©2018 **BIOZONE** International
ISBN: 978-1-927309-55-1
Photocopying Prohibited

EXPLORE: How does environment alter phenotype?

Environmental factors can modify the phenotype encoded by genes without changing the genotype. This can occur both during development and later in life. Environmental factors that affect the phenotype of plants and animals include nutrients or diet, temperature, altitude or latitude, and the presence of other organisms.

The effect of temperature

Crocodiles hatching

Color pointing

The sex of some animals is determined by the incubation temperature during their embryonic development. Examples include turtles, crocodiles, and the American alligator. In some species, high incubation temperatures produce males and low temperatures produce females. In other species, the opposite is true. Temperature regulated sex determination may provide an advantage by preventing inbreeding (since all siblings will tend to be the same sex).

Siamese kittens are completely white when they are born. As they grow older they begin to change color. Dark coloration patterns develop in the cooler areas of the body while the rest of the body remains a pale or white color. This is called color pointing. It occurs because the enzyme responsible for producing the dark pigment (melanin) is temperature sensitive and can not work at normal body temperature.

The effect of other organisms

▶ The presence of other individuals of the same species may control sex determination for some animals. Some fish species, including Sandager's wrasse (right), show this characteristic. The fish live in groups consisting of a single male with their females and juveniles. In the presence of a male, all juvenile fish grow into females. When the male dies, the dominant female undergoes physiological changes to become a male. The male and female look very different.

Male
Female

▶ Some organisms respond to the presence of other, potentially harmful, organisms by changing their body shape. Invertebrates, such as some *Daphnia* species, grow a helmet and/or a longer tail spine when invertebrate predators are present. The helmet makes it more difficult for invertebrate predators (such as the phantom midge larva) to attack and handle *Daphnia*.

Non-helmeted *Daphnia*   Helmeted *Daphnia*

Chemical signal

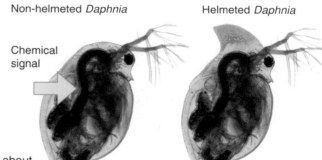

5. (a) Suggest why Siamese kittens are all born white. Hint: think about the environment they develop in.

_____

_____

(b) On the photo of the Siamese kitten (right), use arrows to indicate the cooler areas of the animal's body:

(c) What will the phenotype of the kitten be when it is an adult? Explain:

_____

_____

_____

©2018 **BIOZONE** International
ISBN: 978-1-927309-55-1
Photocopying Prohibited

The effect of altitude

Increasing altitude can stunt the phenotype of plants with the same genotype. In many tree species, such as Engelmann spruce (below) and mountain beech, plants at low altitude grow to their full genetic potential, but growth becomes progressively more stunted as elevation increases and the abiotic factors (e.g. temperature) change. Growth is gnarled and plants are shorter at the highest, most severe sites. Gradual change in phenotype over an environmental gradient is called a cline.

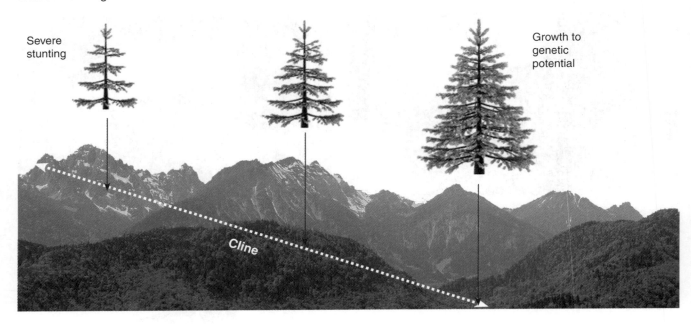

Severe stunting

Growth to genetic potential

Cline

6. (a) What is the environmental trigger for *Daphnia* to grow a helmet? _____

_____

(b) How does the helmet in *Daphnia* help them to survive in their environment?_____

_____

_____

_____

7. (a) Describe how the phenotype of the Englemann spruce changes with an increase in altitude: _____

_____

_____

(b) Physical (abiotic) factors change with altitude. Suggest what physical factors may influence the tree's phenotype:

_____

_____

8. Vegetable growers can produce enormous vegetables for competition (right). How could you improve the chance that a vegetable would reach its maximum genetic potential?

_____

_____

_____

_____

_____

_____

_____

©2018 **BIOZONE** International
ISBN: 978-1-927309-55-1
Photocopying Prohibited

EXPLAIN: Our ancestors' environment can have an effect of future generations

▶ Studies of heredity have found that the environment or lifestyle of an ancestor can have an effect on future generations. For example, certain environments or diets can affect the chemical tagging and packaging of the DNA (rather than the DNA itself) determining which genes are switched on or off at what time and so affecting the development of the individual. These effects can be passed on to offspring and even on to future generations. It is thought that these inherited effects may provide a rapid way to adapt to particular environmental situations (e.g. famine or prolonged stress).

▶ An example illustrating this was a study following the destruction of New York's Twin Towers (right) on September 11, 2001. These attacks traumatized thousands of people, including 1700 pregnant women. Some of the women suffered post-traumatic stress disorder (PTSD), whereas others did not. Studies showed that the mothers who developed PTSD had very low levels of cortisol in their saliva (a hormone that helps the body cope with stress). The children of these mothers also had much lower cortisol levels than those whose mothers had not suffered PTSD. The environment of the mother had affected the offspring.

Studying the effect of environmental triggers in generations of rats

The effect of the environment and diet of mothers on later generations exposed to a breast cancer trigger (a cancer-causing chemical) was investigated in rats fed a high fat diet or a diet high in estrogen. The length of time taken for breast cancer to develop in later generations after the trigger for breast cancer was given was recorded and compared. The data are presented below.

F_1 = daughters, F_2 = granddaughters, F_3 = great granddaughters.

	Cumulative percentage rats with breast cancer (rat mothers on high fat diet (HFD))					
	$F_1\%$		$F_2\%$		$F_3\%$	
Weeks since trigger	Mothers on high fat diet	Control	Mothers on high fat diet	Control	Mothers on high fat diet	Control
6	0	0	5	0	3	0
8	15	0	20	5	3	20
10	22	8	30	5	10	25
12	22	18	50	20	20	30
14	22	18	50	30	25	40
16	29	18	60	30	25	40
18	29	18	60	40	40	42
20	40	18	65	40	50	60
22	80	60	79	50	50	60

Data source: S. De Assis: Nature Communications 3 (Article 1053) (2012)

	Cumulative percentage rats with breast cancer (rat mothers on high estrogen diet (HED))					
	$F_1\%$		$F_2\%$		$F_3\%$	
Weeks since trigger	Mothers on high estrogen diet	Control	Mothers on high estrogen diet	Control	Mothers on high estrogen diet	Control
6	5	0	10	0	0	0
8	10	0	10	0	15	10
10	30	15	15	20	30	20
12	38	19	30	30	40	20
14	50	22	30	40	50	20
16	50	22	30	40	50	30
18	60	35	40	40	75	40
20	60	42	50	50	80	45
22	80	55	50	50	80	60

©2018 **BIOZONE** International
ISBN: 978-1-927309-55-1
Photocopying Prohibited

9. Use the data on the previous page to complete the graphs below. The first graph is done for you:

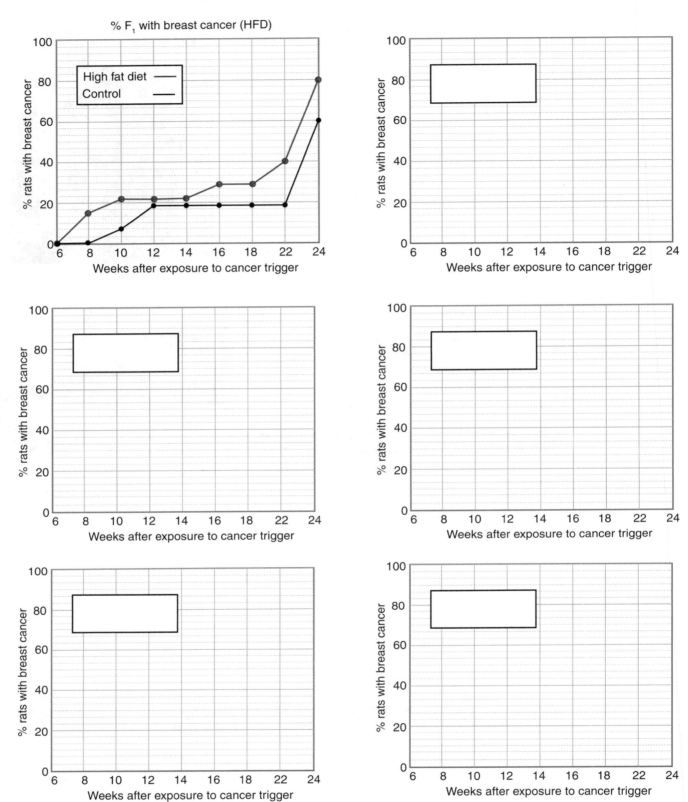

% F₁ with breast cancer (HFD)

10. (a) Which generations are affected by the original mother eating a high fat diet?_____

    (b) Which generations are affected by the mother eating a high estrogen diet? _____

    (c) Which diet had the longest lasting effect? _____

11. What do these experiments show with respect to diet and generational effects? _____

_____

©2018 **BIOZONE** International
ISBN: 978-1-927309-55-1
Photocopying Prohibited

59 Natural Selection Acts on Phenotype

ELABORATE: Natural selection in rock pocket mice

Recall from IS3 that natural selection acts on the phenotype of an individual. Variation in a population is important because individuals with phenotypes better suited to the environmental conditions at the time are more likely to survive and reproduce successfully than individuals with unfavorable phenotypes.

Rock pocket mice are found in the deserts of southwestern United States and northern Mexico. They are nocturnal, foraging at night for seeds, while avoiding owls (their main predator). During the day they shelter from the desert heat in their burrows. The coat color of the mice varies from light brown to very dark brown. Throughout the predominantly light colored desert environment where the mice live there are outcrops of dark volcanic rock. The presence of these outcrops and the mice that live on them present an excellent study in natural selection.

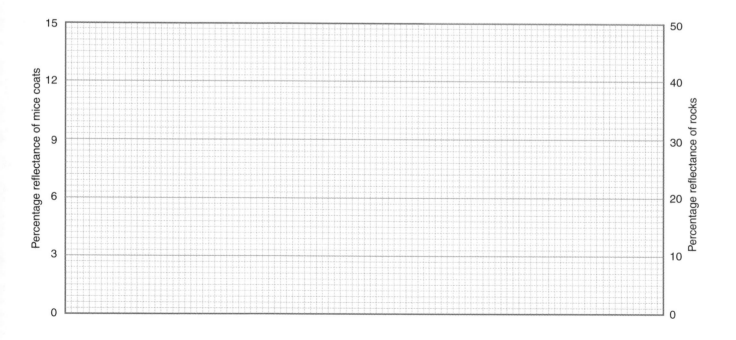

Site	Rock type (V volcanic)	Percent reflectance (%) Mice coat	Percent reflectance (%) Rock
KNZ	V	4	10.5
ARM	V	4	9
CAR	V	4	10
MEX	V	5	10.5
TUM	V	5	27
PIN	V	5.5	11
AFT		6	30
AVR		6.5	26
WHT		8	42
BLK	V	8.5	15
FRA		9	39
TIN		9	39
TUL		9.5	25
POR		12	34.5

▸ The coat color of the Arizona rock pocket mice is controlled by the Mc1r gene (in mammals this gene is commonly associated with the production of the pigment melanin). Coat color of mice in New Mexico is not related to the Mc1r gene.

▸ There are two alleles (versions) for the gene, D and d. Homozygous dominant (DD) and heterozygous mice (Dd) have dark coats, while homozygous recessive mice (dd) have light coats.

▸ 107 rock pocket mice from 14 sites were collected and their coat color and the rock color they were found on were recorded by measuring the percentage of light reflected from their coat (low percentage reflectance equals a dark coat). The data is presented right:

1. (a) What is the genotype(s) of the dark colored mice? _____

   (b) What is the genotype of the light colored mice? _____

2. Plot the data above as a dual column chart on the grid below. The y axes have been written in for you. Include a key:

©2018 BIOZONE International
ISBN: 978-1-927309-55-1
Photocopying Prohibited

CE P LS4.C LS4.B

3. (a) What do you notice about the reflectance of the rock pocket mice coat color and the reflectance of the rocks they were found on?

_____

_____

_____

(b) Suggest a cause for the pattern in 3(a). How do the phenotypes of the mice affect where the mice live?

_____

_____

_____

_____

(c) What are two exceptions to the pattern you have noticed in 3(a)? _____

_____

_____

_____

_____

(d) How might these exceptions have occurred? _____

_____

_____

_____

4. The rock pocket mice populations in Arizona use a different genetic mechanism to control coat color than the New Mexico populations. What does this tell you about the origin of the genetic mechanism for coat color?

_____

_____

_____

5. Imagine scientists moved 200 randomly selected rock pocket mice from an area with very few volcanic outcrops to an area where dark volcanic outcrops are very common.

(a) Predict the predominant coat color variant(s) after one generation: _____

_____

(b) Predict the predominant coat color variant(s) after 20 generations: _____

(c) Explain your predictions: _____

_____

_____

_____

_____

_____

_____

_____

_____

60 Pale and Interesting Revisited

The anchoring phenomenon for this chapter was albinism. Now you can apply the knowledge you have gained in this chapter to answer the following questions.

The most common form of albinism results from a mutation to the TYR gene. The TYR gene produces an enzyme called tyrosinase, which is needed to make a pigment called melanin. Melanin gives the skin, hair, and eyes their color. People with albinism have either a partial or complete lack of pigment. Medical conditions associated with albinism include skin cancer and vision problems.

1. In the US albinism occurs in around 1 in 17,000 people. In some regions of Africa it is far more common (1 in 3,000 people). One hypothesis suggests that albinism protects against a debilitating bacterial disease called leprosy (Hansen's disease), which has a high prevalence in some African countries. Some researchers have suggested that the bacterium is killed by UV light, which penetrates further into skin lacking melanin than skin with normal pigmentation. In 2011, the World Health Organization reported 12,673 new cases of leprosy in Africa, and 173 cases in the USA. New infection rates for leprosy are declining slowly worldwide following the WHO's resolution to reduce rates to < 1 per 10,000 by the year 2000 (achieved).

   (a) Do you think the data above support the hypothesis? Why or why not? _____

   _____

   _____

   (b) What other factors could account for the lower number of cases in the US? _____

   _____

   _____

   (c) If albinism protects against leprosy, predict what might happen to the rates of albinism in Africa as leprosy rates fall:

   _____

   (d) What assumption does this prediction make? _____

   _____

2. A pedigree for albinism in humans is shown on the right.

   (a) What type of inheritance pattern does albinism follow?

   _____

   (b) Explain how you came to your decision: _____

   _____

   _____

   _____

   (c) Write the genotype for each individual on the chart using the codes: **PP** normal skin color; **P-** normal (but unknown if they are homozygous), **Pp** carrier, **pp** albino.

   (d) Why must (II-3) and (II-4) be carriers of a recessive allele?_____

   _____

3. An unaffected homozygous male had children with a woman who is a carrier for albinism.

   (a) What is the genotype of each parent? _____

   (b) What is the probability they will have an albino child? _____

   (c) What is the probability they will have a child who is a carrier? _____

4. If an albino person has children with a person who is a carrier, what phenotypes are possible in their offspring?

   _____

©2018 **BIOZONE** International
ISBN: 978-1-927309-55-1
Photocopying Prohibited

61 Summative Assessment

1. The following dihybrid cross shows the inheritance of color and shape in pea seeds. Yellow (**Y**) is dominant over green (**y**) and a round shape (**R**) is dominant over the wrinkled (**r**) form.

   (a) Describe the phenotype of pea seeds with the genotype YyRr: _____

   (b) Complete the Punnett square below when two seeds with the YyRr genotype are crossed. Indicate the number of each phenotype in the boxes on the right.

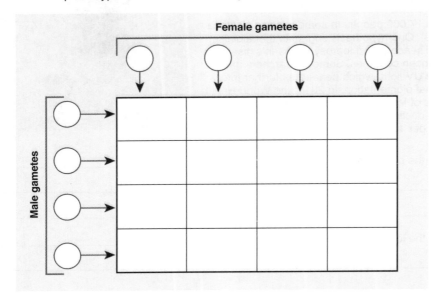

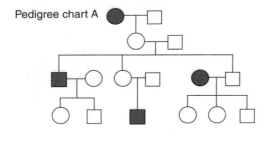

Yellow-round ☐

Green-round ☐

Yellow-wrinkled ☐

Green-wrinkled ☐

   (c) A plant breeder crossed two pea plants with known genotypes for three traits, seed color (yellow or green), seed shape (round or wrinkled), and stem length (tall or dwarf). For this cross YyRrTt x Yyrrtt, use probability rules to calculate the probability of getting offspring with the dominant phenotype in all three traits.

   _____

   _____

2. Study the two pedigree charts below.

Pedigree chart A

Pedigree chart B

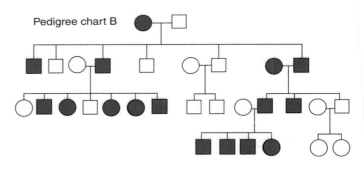

   (a) What type of inheritance pattern is shown in A? _____

   (b) Give a reason for your answer: _____

   _____

   _____

   _____

   (c) What type of inheritance pattern is shown in B? _____

   (d) Give a reason for your answer: _____

   _____

   _____

   _____

©2018 **BIOZONE** International
ISBN: 978-1-927309-55-1
Photocopying Prohibited

3. The peppered moth, *Biston betularia*, has two phenotypic variants. One is a gray mottled color and the other (melanic form) is a much darker color. The dark color is controlled by a dominant allele and follows a simple Mendelian inheritance pattern. The adult moths, which are night-active, rest on the trunks of trees during the day where they hide from predatory birds. In certain areas of England during the 1940s and 1950s, large amounts of coal were used to power industry. In areas where large amounts of coal was burned, the trees became covered in a dark soot. In these areas, the dark form quickly became more common than the lighter form.

Dark form
Genotype: MM or Mm

Gray form
Genotype: **mm**

(a) Suggest why the dark form became more common when large amounts of coal were being burned:

_____

_____

_____

(b) In terms of the inheritance pattern, suggest why the shift to dominance of the dark form happened very quickly:

_____

_____

_____

_____

4. Air quality improved between 1960-1980 because fewer coal burning factories were operating.

(a) Use the graph on the right to describe how this affected the frequency of the dark form of the peppered moth:

_____

_____

_____

_____

_____

Frequency of dark colored peppered moth related to air pollution

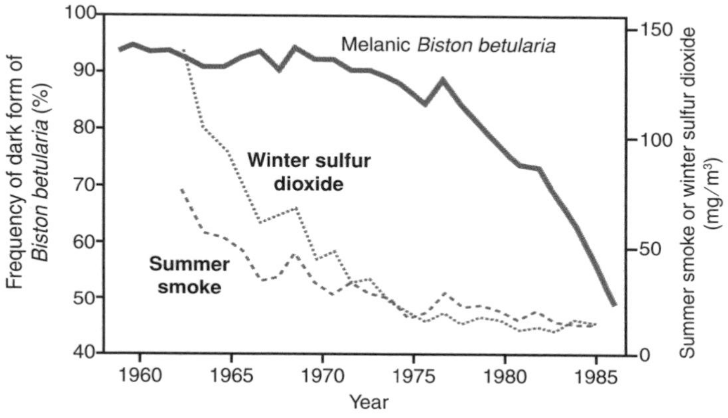

Melanic *Biston betularia*

Winter sulfur dioxide

Summer smoke

Frequency of dark form of *Biston betularia* (%)

Summer smoke or winter sulfur dioxide (mg/m^3)

Year

(b) Explain why the gray form would have higher fitness than the dark form as air quality improved:

_____

_____

_____

_____

(c) In the spaces below, draw the change in the frequencies of moth phenotypes over time given the scenarios described. Use filled circles to represent the dark form and open circles to represent the gray form:

High levels of coal burned, high levels of soot in the air.

Moderate levels of coal burned, soot levels decreasing.

Very low levels of coal burned, soot levels in the air very low.

Moderate levels of coal burned, soot pollution increasing.

©2018 **BIOZONE** International
ISBN: 978-1-927309-55-1
Photocopying Prohibited

5. Malaria is caused when a protozoan parasite called *Plasmodium* infects red blood cells. Humans become infected when they are bitten by an infected mosquito and the parasite passes into the blood infecting red blood cells. Over time, the red blood cells are destroyed. Untreated malaria can cause many serious health effects and can be fatal.

Malaria was eliminated from the US in 1951 following the successful completion of the National Malaria Eradication Program. However, malaria still occurs in many countries and, in 2016, caused 445,000 deaths worldwide. The graphs below show the occurrence of malaria and the frequency of the sickle cell allele.

Figure 1: Incidence of *P. falciparum* malaria

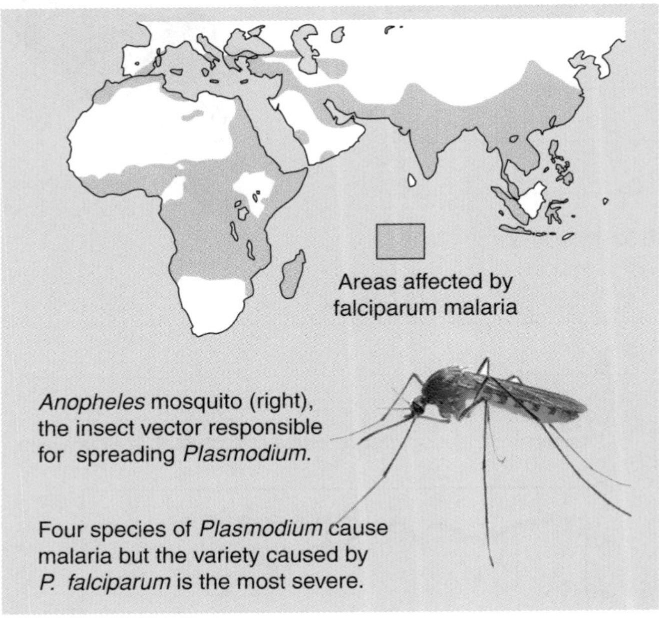

Figure 2: Frequency of the sickle cell allele (HbS)

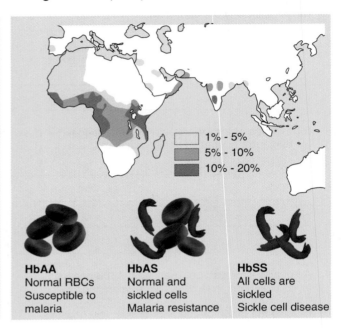

Areas affected by falciparum malaria

Anopheles mosquito (right), the insect vector responsible for spreading *Plasmodium*.

Four species of *Plasmodium* cause malaria but the variety caused by *P. falciparum* is the most severe.

1% - 5%
5% - 10%
10% - 20%

HbAA
Normal RBCs
Susceptible to malaria

HbAS
Normal and sickled cells
Malaria resistance

HbSS
All cells are sickled
Sickle cell disease

(a) Look at the occurrence of malaria around the world and the frequency of the sickle cell allele. What do you notice about the two?

_____

_____

_____

_____

(b) The sickle cell mutation is known to provide some protection against malaria because the parasite cannot enter the deformed cells. Explain when the mutation would be an advantage and when it would be a disadvantage:

_____

_____

_____

_____

(c) Which allele combination is likely to be most common in regions where malaria occurs? _____

(d) Explain why the other allele combinations are unlikely to provide an advantage: _____

_____

_____

_____

(e) How could malaria play a role in human evolution? _____

_____

_____

_____

©2018 **BIOZONE** International
ISBN: 978-1-927309-55-1
Photocopying Prohibited

The three-spined stickleback (*Gasterosteus aculeatus*) is a fish common to the Northern Hemisphere. It inhabits both coastal and fresh water and can live in either fresh, brackish, or salt water. Ancestral populations live in seawater and migrate to fresh or brackish water to breed but some populations are landlocked and remain in freshwater lakes. The different habitats are associated with different phenotypes.

Plates along the stickleback's sides protect against vertebrate predators such as birds.

One interesting feature of the species is that it has bony plates along its sides to protect it from predators. The number of these lateral plates present varies greatly among stickleback populations. A low plate form of stickleback typically has 4-7 plates, while the high plate form can have up to 36 plates.

Fish in seawater tend to have more lateral plates than fish in freshwater. In freshwater, a smaller number of plates is an advantage. Individuals with fewer plates grow more quickly (they quickly become bigger than their invertebrate predators) and they reach maturity faster. Too many plates are a disadvantage because they reduce speed and agility and provide surfaces for their main predator (dragonfly larvae) to attach to.

In 1982, the Loberg Lake population of sticklebacks was exterminated using poison so that trout could be introduced for recreational fishing. By 1990, sticklebacks had naturally recolonized the lake. Between 1990 and 2009, researchers at The Bell Lab (Stony Brook University) studied the Loberg Lake sticklebacks, counting how lateral plate numbers changed over time. Some of their data is presented below.

Dragonfly larvae have an extendable lower lip which is thrust out to capture unsuspecting prey, including sticklebacks.

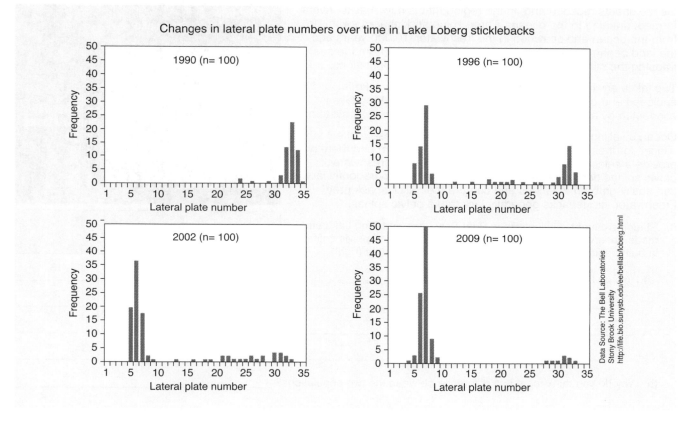

Changes in lateral plate numbers over time in Lake Loberg sticklebacks

Data Source: The Bell Laboratories
Stony Brook University
http://life.bio.sunysb.edu/ee/belllab/loberg.html

6. (a) Describe the lateral plate distribution in the 1990 data: _____

_____

_____

   (b) Where do you think the Lake Loberg colonizers probably came from? _____

_____

   (c) Explain your answer: _____

_____

_____

©2018 **BIOZONE** International
ISBN: 978-1-927309-55-1
Photocopying Prohibited

7. (a) Describe the change in lateral plate number over the sampling period: _____

_____

_____

_____

(b) What selection pressure is likely to have caused the change you described in (a)? _____

_____

_____

_____

(c) Why is a large number of lateral plates a disadvantage to freshwater sticklebacks? _____

_____

_____

_____

Pelvic spines in sticklebacks

North America was once covered in thick ice sheets. Eventually the ice sheets receded and, in the region that is now Alaska, rivers formed, draining to the oceans. Three spine sticklebacks swam up from the ocean and spawned in the freshwater. 10,000 years ago the land began to rise. Many lakes became isolated from the sea, trapping the sticklebacks and creating landlocked populations.

Two lakes are labeled on the photo (right). Bear Paw lake is enclosed and does not drain into another waterway. Frog lake is connected by a small stream to the surrounding waterway system.

Ocean dwelling sticklebacks have pelvic spines. These make it very difficult for their predators (fish and birds) to eat them and therefore provide a selective advantage against predation. In freshwater, pelvic spines decrease stickleback survival because dragonfly larvae can easily grab on to spines and capture their stickleback prey. Freshwater sticklebacks generally lack these pelvic spines.

Bear Paw Lake · Frog Lake · Knik Arm Waterway enters Gulf of Alaska · Glacial lakes around Houston, Alaska · Image: Google satellite

8. Students caught sticklebacks in Bear Paw Lake and Frog Lake and made observations about the presence or absence of pelvic spines in both populations. Their data is presented in the table (right).

(a) How do the two populations compare? _____

_____

_____

Pelvic spine description	Bear Paw Lake	Frog Lake
Fully present	0	18
Reduced	14	2
Absent	6	1

_____

(b) Why do you think there is a difference between the two populations? _____

_____

_____

9. Using the stickleback data as evidence, explain how the frequencies of particular traits can change from generation to generation in response to selection pressures (in this case the type of predator present in the environment):

_____

_____

_____

_____

_____

_____

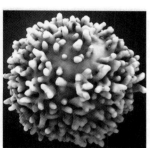
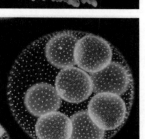
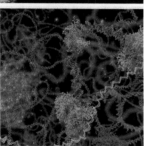
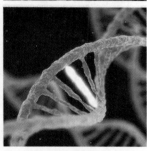
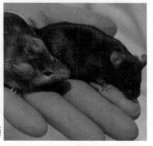

NIH

Instructional Segment 5

Structure, Function, and Growth

Activity number

Anchoring Phenomenon

Breast Cancer: Breast cancers are cell masses distinct from the tissue around them.   62 75

How do systems work in a multicellular organism (emergent properties) and what happens if there is a change in the system?

☐ 1 What do you understand by the term alive? What are your criteria for deciding if something is living or not? Is non-living the same as dead? Categorize objects as living or non-living according to your criteria.   63

☐ 2 Recognize cells as the fundamental unit of life. Categorize cells as either prokaryote or eukaryote based on the presence or absence of characteristic features. Recall when multicellular organisms arose on Earth (IS2) and what advantages multicellularity provided to those organisms. Using evidence from a model organism, explain how cells within a multicelluar organism are specialized to perform specific roles that contribute to the functioning of the organism as a whole.   63

☐ 3 Develop and use a model to show how a complex multicellular organism is produced and maintained through cell division and differentiation from stem cells. Use the model to explain how the expression of different genes during development leads to cells with different structure and function. Using skin as an example, construct an explanation for how differentiation of stem cells in the skin replenish the dead cells lost from the skin's surface. Use a model to explain how stem cells can be used to engineer new tissues such as skin.   64 76

☐ 4 Develop and use a model to show how the hierarchical organization of interacting systems (cells, tissues, organs, and organ systems) provides specific functions within multicellular organisms.   64 65 76

How does the structure of DNA affect how cells look and behave?

☐ 5 Recall that proteins are responsible for the traits we see in organisms and that these traits are encoded by genes carried on chromosomes. Describe the structure of proteins and explain how the variety of amino acid building blocks and their many possible arrangements enables a great protein diversity. Can you think why scientists in the early 20th century thought that proteins must carry the code?   66

☐ 6 Use a model to show how proteins are made by first transcribing (rewriting) the instructions in DNA and then translating them into a protein molecule using the cell's molecular machinery. Can you think of an analogy (comparison) for this process? Construct an explanation based on evidence for how the structure of DNA determines the structure of a protein.   66 76

☐ 7 In groups, come up with a list of the roles that proteins carry out in our bodies. Use examples and models (e.g. molecular models or drawings) to explain how the shape of a protein determines its function.   67

☐ 8 Describe the structure of the plasma (cell surface membrane), including the role that proteins play as part of its functional structure. Use a model to explain how different substances move through membranes by diffusion, osmosis, or active transport. Make the connection between these processes and the functioning of the cell and the organism as a whole.   68

☐ 9 **ENGINEERING CONNECTION: WATER PURIFICATION**   68
Explain how the usual movement of water molecules in osmosis can be reversed in order to treat or desalinate water. How could reverse osmosis be used to solve a water shortage problem? What is the cost?

☐ 10 Use what you have learned about the way proteins are made to predict the effect of DNA mutations of protein structure and function. Use a simple model to represent a body system to test your prediction. Explain how scientists determined the function and location of specific genes by looking at loss of function in mutant strains of organisms. Plan your own investigation into the effect of a mutation on the survival of an organism in different environments.   69

☐ 11 Using specific examples of human disorders, e.g. cystic fibrosis and PKU, explain how mutations lead to a loss of function and disruption of the system, e.g. disruption of membrane transport in cystic fibrosis or disruption of a metabolic pathway in PKU.   69

What happens if a cell in our body dies?

☐ 12 Has your skin ever peeled off after you have been sunburned? Or have you had a blister that has burst and left a flap of dead skin? Recall how organisms grow and cells are replaced by cell division and how cells differentiate into the different cell types that make up tissues and organs. Why must a cell duplicate its DNA before it divides into two identical cells? Compare different models of DNA replication and then model Meselson and Stahl's experiments that showed DNA replication is semi-conservative. Use these models to explain how the DNA is copied so that each cell contains the full set of genetic instructions for the organism. How are proteins (enzymes) involved in the process?

70 71

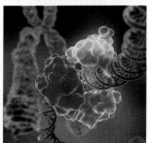

☐ 13 Recognize DNA replication as one part in the life cycle of a cell called the cell cycle. Describe the events and sequence of events in a model of the cell cycle. Use the model to explain how an organism grows and how dead cells are replaced in order to repair and maintain the organism. Show on your model how some cells may leave the cell cycle and stop dividing once they have differentiated or when they have become old and die. Use your model to predict the effect of mistakes in DNA replication or in the cell cycle process. Analyze data to show that environmental stimuli can affect events in the cell cycle.

71 72

How do organisms survive even when there are changes in their environment?

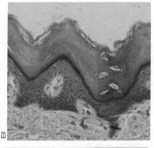

☐ 14 We have seen in this chapter how cells divide and differentiate to produce the tissues, organs, and organ systems that maintain the organism. Remembering back to your understanding about proteins, explain why organisms need to maintain a relatively stable state, called homeostasis, even though the environment around them may be fluctuating. Working as a group, describe examples to show how the body works as a set of interacting systems to maintain this stable state. Conduct an investigation of the body's response (e.g. change in heart rate) to a disturbance (e.g. exercise). Can you use your findings to explain the effect of exercise on the body and predict the response in different individuals.

73

☐ 15 Use diagrams or other conceptual models to explain the role of feedback mechanisms in homeostasis. Distinguish between negative (counterbalancing) feedback and positive (reinforcing) feedback and explain why negative feedback loops are more common in biological systems. Using humans as an example, describe how feedback mechanisms between the body's systems maintain a relatively constant temperature. Humans and other mammals use metabolism to generate the heat they need to maintain a constant body temperature (so they are called homeothermic endotherms). What is the cost of this process (called thermoregulation)? How is this different from the mechanisms used by ectotherms such as reptiles? Predict the effect of body size on the ability for large and small animals to maintain body temperature when the environmental temperature fluctuates.

73

☐ 16 Use an analogy to model the behavior of a system with interacting components. How is your analogy like the interacting systems of the human body? Now create your own model of interacting systems in the human body, showing how outputs are affected by how the interacting parts respond to inputs. Extend and refine your model to show how a disease such as cystic fibrosis affects the normal functioning of the body systems. Find out about how modern medicine has allowed us to diagnose and treat diseases or disorders so that homeostasis is restored. Make an evaluation of the feasibility of treatments and their success (considering costs and benefits) and communicate your findings as an oral report, poster, or infographic.

74 76

62 A Cancerous Creep

ANCHORING PHENOMENON: Breast cancers are cell masses distinct from the tissue around them

Breast cancer is caused by abnormal (and malignant) cell growth in breast tissue. The image on the right shows breast tissue that has been removed by mastectomy (removal of the breast and related tissues such as the lymph nodes). Note the large tumor occupying the upper left quadrant of the breast. Like all cancers, breast cancer can spread from the breast tissue to other organs and ultimately cause death.

Breast cancer accounts for 25% of all cancers in women worldwide, and around half a million deaths.

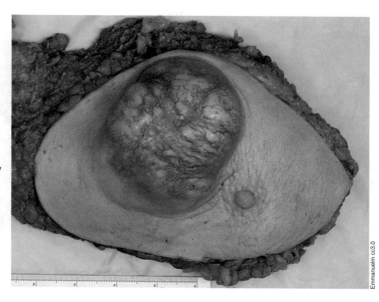

1. Answer the following questions about what you know about cancer as true or false:

    (a) Cancer can be spread from person to person: _____

    (b) Lifestyle choices over your lifetime can affect your chances of developing cancer: _____

    (c) Cancer cannot be treated: _____

    (d) Cancer is actually a group of over 100 diseases: _____

    (e) You can be vaccinated against cancer: _____

    (f) Abnormal gene function can cause cancer: _____

2. As a group discuss your answers above. Do you need to change any of your answers?

3. What else do you know about cancer? As a class or group discuss what you and others know. Summarize your ideas below:

    _____

    _____

    _____

    _____

    _____

    _____

    _____

4. Two genes that are often mentioned in conjunction with breast cancer are *BRCA1* and *BRCA2*. Research these two genes. What do they do and what percentage of breast cancer is related to mutations in them?

    _____

    _____

    _____

    _____

©2018 **BIOZONE** International
ISBN: 978-1-927309-55-1
Photocopying Prohibited

63 Cells and Life

ENGAGE: Classifying life

Lichen

Canary

Parvovirus

Bacterium (*E. coli*)

Insects in amber

Ladybug

Tunicate

Corn seeds

Stromatolite

1. Classify the nine images above as living or non-living:

    (a) Living: _____

    Reason: _____

    _____

    (b) Non-living: _____

    Reason: _____

    _____

EXPLORE: The cell theory

Cells are the fundamental unit of life: Study the images below:

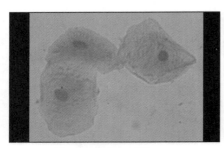

Human cheek cells

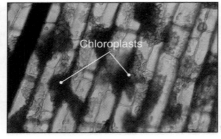

Chloroplasts

Plant leaf cells

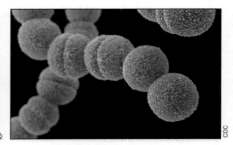

Bacterial cells dividing

Didinium capturing and eating *Paramecium*

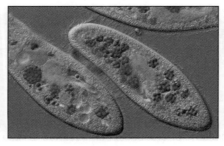

Organelles inside *Paramecium*

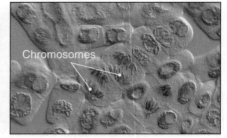

Chromosomes

Dividing plant cells

2. From the images above come up with three statements to describe the main features of the cell theory:

    (a) _____

    (b) _____

    (c) _____

   LS1.A SF

©2018 **BIOZONE** International
ISBN: 978-1-927309-55-1
Photocopying Prohibited

EXPLORE: Cell types

Life can be divided into two categories: prokaryotic and eukaryotic. Eukaryotic cells contain organelles and linear DNA molecules associated with proteins (chromosomes) within a nucleus. Prokaryotes lack membrane-bound organelles and the DNA is naked (not associated with proteins) and found as a single circular molecule free in the cytoplasm.

Prokaryotic cells

▶ Prokaryotic cells are bacterial cells.

▶ Prokaryotic cells lack a membrane-bound nucleus or any membrane-bound organelles.

▶ They are small (generally 0.5-10 μm) and usually single cells (unicellular or sometimes colonial).

▶ They are relatively basic cells with very little cellular organization (the DNA, ribosomes, and enzymes float free within the cell cytoplasm).

▶ Single, circular chromosome of naked DNA.

▶ Prokaryotes have a cell wall, but it is different to the cell walls that some eukaryotes have.

Eukaryotic cells

▶ Eukaryotic cells have a membrane-bound nucleus, and other membrane-bound organelles.

▶ Plant cells, animals cells, fungal cells, and protists are all eukaryotic cells.

▶ Eukaryotic cells are large (30-150 μm). They may exist as single cells or as part of a colonial or a multicellular organism.

▶ Multiple linear chromosomes consisting of DNA and associated proteins.

▶ They are more complex than prokaryotic cells, with more structure and internal organization.

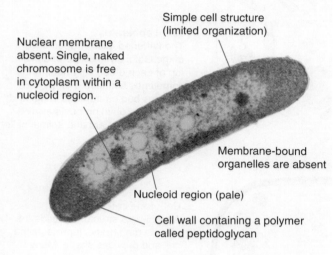

Nuclear membrane absent. Single, naked chromosome is free in cytoplasm within a nucleoid region.

Simple cell structure (limited organization)

Membrane-bound organelles are absent

Nucleoid region (pale)

Cell wall containing a polymer called peptidoglycan

A prokaryotic cell: *E.coli*

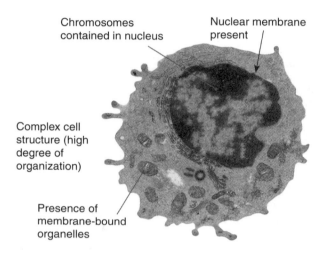

Chromosomes contained in nucleus

Nuclear membrane present

Complex cell structure (high degree of organization)

Presence of membrane-bound organelles

A eukaryotic cell: a human white blood cell

3. Study the photos below. Identify the cells shown as either prokaryotic or eukaryotic:

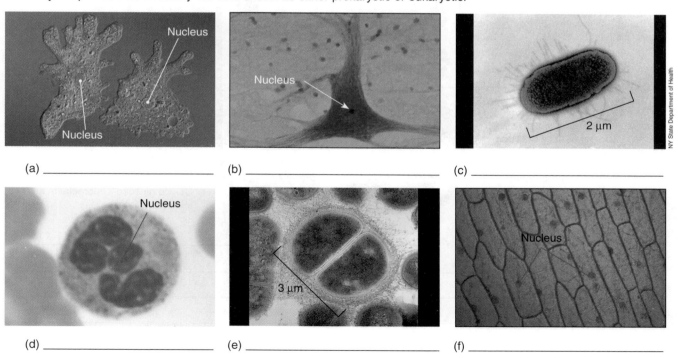

Nucleus

Nucleus

Nucleus

(a) _____

Nucleus

(b) _____

2 μm

NY State Department of Health

(c) _____

Nucleus

(d) _____

3 μm

(e) _____

Nucleus

(f) _____

©2018 **BIOZONE** International
ISBN: 978-1-927309-55-1
Photocopying Prohibited

EXPLORE: Cells

The diagrams below illustrate the features of a generalized plant cell and a generalized animal cell.

A generalized animal cell

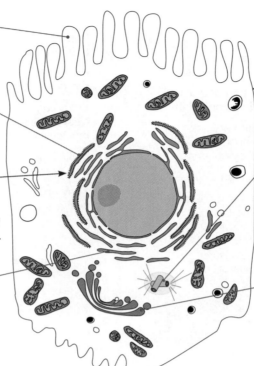

Microvilli
Small finger-like extensions which increase the cell's surface area (not all animal cells have these).

Rough endoplasmic reticulum (rough ER)
These have ribosomes attached to the surface. Proteins are made here in both plant and animal cells.

Ribosomes
Small structures that make proteins. They may be floating free in the cytoplasm or attached to the surface of rough endoplasmic reticulum (as here). They are also found in plant cells.

Smooth endoplasmic reticulum (smooth ER)
Its main role is to make lipids and phospholipids.

Plasma membrane
The boundary of all cells. The membrane is a semi-fluid phospholipid bilayer with embedded proteins. It separates the cell from its external environment and controls the movement of substances into and out of the cell.

Centrioles
Paired cylindrical structures contained within the centrosome (an organelle that organizes the cell's microtubules). The centrioles form the spindle fibers involved in nuclear division. Plant cells lack centrioles and the spindle is organized by the centrosome.

Golgi apparatus
The flattened, disc-shaped sacs of the Golgi are stacked one on top of each other, very near, and sometimes connected to, the ER. Vesicles bud off from the Golgi and transport protein products away. They occur in plant and animal cells.

A generalized plant cell

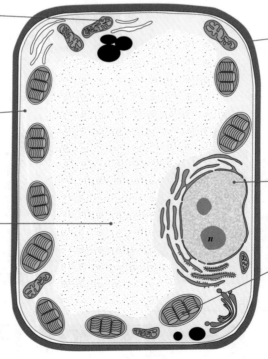

Mitochondrion
Organelles involved in the production of ATP (usable energy) in all eukaryotic cells.

Cytoplasm
A watery solution containing dissolved materials, enzymes, and the cell organelles (all eukaryotic cells).

Large central vacuole
Plant vacuoles contain cell sap. Sap is a watery solution containing dissolved nutrients, ions, waste products, and pigments. Functions include support (by pressing on the cell wall) storage, waste disposal, and growth. Vacuoles are found in animal cells but they are very small.

Cellulose cell wall
A semi-rigid structure outside the plasma membrane. It protects the cell and provides shape. Many materials pass freely through the cell wall. Animal cells lack cell walls.

Nucleus
A large organelle found containing most of the cell's DNA. Within the nucleus, is a denser structure called the **nucleolus** (*n*). The nucleus is a defining feature of eukaryotic cells.

Chloroplast
A specialized organelle containing the green pigment chlorophyll. They are the site for photosynthesis, in which light energy is used to convert CO_2 to glucose. Chloroplasts are not found in animal cells.

4. (a) Are plant and animal cells eukaryotic or prokaryotic? _____

    (b) Explain how you know? _____

    _____

    _____

    _____

©2018 **BIOZONE** International
ISBN: 978-1-927309-55-1
Photocopying Prohibited

5. Study the electron micrographs below. Identify the plant cell and the animal cell and give reasons for your identification:

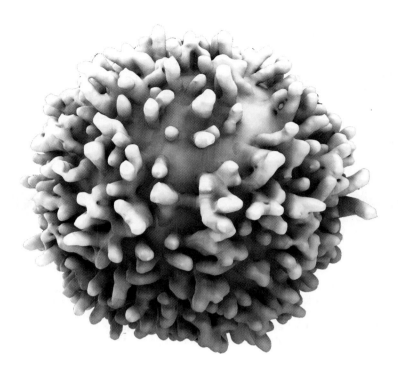

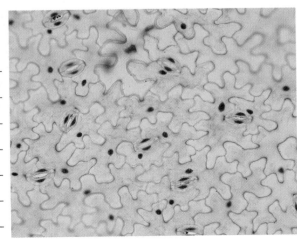

| **Scanning Electron Micrograph (false color)** | **Transmission Electron Micrograph** |

Cell type: _____

Reasons: _____

_____

_____

_____

_____

_____

Cell type: _____

Reasons: _____

_____

_____

_____

_____

_____

6. List three differences between plant and animal cells: _____

_____

_____

_____

7. The photograph on the right shows the surface cells of a plant leaf. Suggest why these cells have a jigsaw-like structure, instead of a block like structure as in many plant cells:

_____

_____

_____

_____

_____

_____

_____

©2018 **BIOZONE** International
ISBN: 978-1-927309-55-1
Photocopying Prohibited

EXPLAIN: Multicellularity

▶ Although there are many unicellular organisms in the world, being larger and multicellular clearly has advantages as the evolution of multicellular organisms shows. Why become multicellular? What advantages of multicellularity are there over being unicellular? Explore the information below and use it to explain the advantages of multicellularity.

Prokaryotic cells can work together

▶ Most prokaryotic cells 'behave' as individuals. However many form colonies, which may act in a rudimentary multicellular way. Some prokaryotes are able to form groups and some cells differentiate to perform specific tasks.

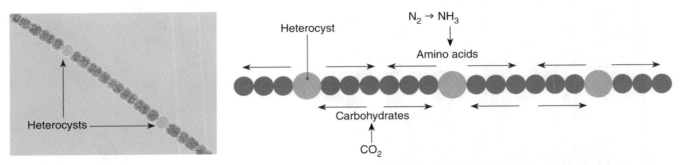

Cyanobacteria (e.g. *Anabaena* shown above) are a phylum of bacteria that are able to photosynthesize. The bacteria often form long filaments of individual cells joined together. Under low-nitrogen conditions, some of these cells will differentiate into **heterocysts**. These cells are able to fix nitrogen from molecular nitrogen (N_2). They show quite different gene expression from neighboring undifferentiated cells, most importantly in producing the enzyme **nitrogenase** and in being unable to photosynthesize. What's more, the heterocysts share the nitrogen they fix with neighboring cells and receive other nutrients from them, indicating basic cooperation.

Unicellular eukaryotes can behave as a single organism

▶ Some unicellular eukaryotes spend some of their life cycle as singular independent cells, but in certain conditions come together and behave as a single entity.

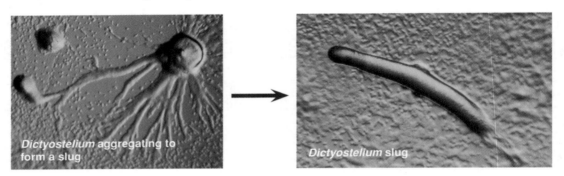

Cellular slime molds (e.g. *Dictyostelium*) spend much of their life cycle as individual ameboid-like cells, often living in the soil and feeding on bacteria. When food becomes scarce, the single cells group together to form a slug, which is capable of movement. The slug may move some distance before finding a suitable area to form a fruiting body. The fruiting body is held a few millimeters off the ground by a stalk. The cells in the stalk die whereas the cells in the fruiting body form spores, which will be dispersed to new areas.

Multicellularity means greater competitive advantage

▶ Some of the simplest multicellular organisms are species of the green alga, *Volvox*. Each *Volvox* is a hollow sphere of up to 50,000 cells arranged in a single layer. The internal space contains a clear fluid, called the extracellular matrix. There are two cell types. The majority are somatic cells with two flagella. The flagella beat in a coordinated way so that the colony can move in a specific direction. Embedded in this cell layer are a smaller number of germ cells, which can divide to form new somatic cells or a new colony.

▶ Living as a large colony gives *Volvox* an advantage over single celled algae such as *Chlamydomonas* (which are similar to individual *Volvox* cells). Phosphate is an important factor in algal growth. Importantly, it is the amount of phosphate that a cell can store that limits its growth (larger stores equal more growth). Volvox is able to store much more phosphate in its extracellular matrix than single cells can store internally. This gives the colonies an advantage when phosphate is scarce. In addition, the somatic cells serve as nutrient providers to the germ cells so that reproduction is more efficient.

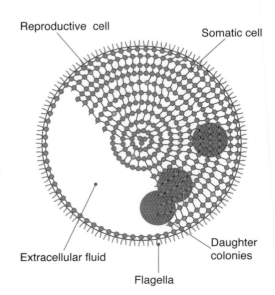

©2018 **BIOZONE** International
ISBN: 978-1-927309-55-1
Photocopying Prohibited

Multicellularity means specialization

▶ One advantage of multicellularity is being able to have cells that are specialized for a particular task. This results in those tasks being performed more efficiently and saves energy because unused organelles can be lost.

▶ Being multicellular also means an organism can become very large and acquire more resources.

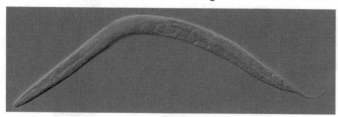

Caenorhabditis elegans is a simple multicellular organism. It is a free living nematode about 1 mm long. *Caenorhabditis elegans* has been very closely studied and the fate of all of its roughly 1000 cells is known. Even with such a small number of cells, it has cells that carry out specialized tasks. For example, 20 cells are specialized to form the intestine, while 95 cells are specialized as muscle cells for movement of the body.

Cell type	Number of cells
Hypodermis (part of body wall)	213
Nervous system	338
Body mesoderm (body cells)	122
Alimentary canal (intestines etc.)	143
Reproductive cells	143
Programmed cell deaths	131
Total	1090

8. Explain how *Anabaena* benefits by some individuals cells becoming heterocysts: _____

_____

_____

_____

9. What is the benefit of individual *Dictyostelium* cells forming a slug? _____

_____

_____

_____

10. Explain why the cells in a *Volvox* colony have an advantage over the individual free living *Chlamydomonas* cells:

_____

_____

_____

11. Explain how specialization in *Caenorhabditis elegans* allows it to live a highly active lifestyle:

_____

_____

_____

_____

_____

12. Use your answers to the questions above to summarize the advantages of multicellularity:

_____

_____

_____

_____

_____

_____

_____

©2018 **BIOZONE** International
ISBN: 978-1-927309-55-1
Photocopying Prohibited

64 Cells, Tissues, and Organs

ENGAGE: You start from just one cell

Early embryo: 100 cells

Zygote: 1 cell.

Adult: between 10 and 100 trillion cells (depending on how you estimate the number of cells). Latest estimates put the figure at 37.2 trillion.

Here's a question: Given that a newborn baby has about 2 trillion cells (it's going to be slightly different for everyone) how many times does the zygote (the original single cell) have to divide to reach this? The answer is only about 41 divisions (if we don't worry about programmed cell deaths or certain cells that divide more slowly or quickly that others). During those divisions all the 200 different types of cells from which you are made are formed.

NEED HELP?
See Activity 90

1. (a) The average mass of a cell is 1 nanogram (1 ng or 1×10^{-9} g). The average weight of an adult is about 70 kg. Use this information to calculate the number of cells in a 70 kg person:

    _____

    _____

    (b) The average volume of a cell is about 4 picoliters (4 pL or 4×10^{-12} L). The average human has a volume of about 66 liters. Use this information to calculate the number of cells in a 66 L person:

    _____

    _____

    (c) Why do you think both of these estimates are wrong? _____

    _____

    _____

    _____

EXPLORE: Differentiating cells

Every cell in your body has the same genetic information. How is it then that there are so many different types of cells in your body? The answer is that during different cell divisions only certain parts of the genetic information is used (much like reading only some books in the library), thus producing the many different cell types.

2. A hypothetical cell has 11 genes in its DNA. Each gene initiates certain processes in the cell as shown in the table below. Use these "genes" to fill in the boxes opposite showing the cell as it progresses through two divisions. Note if a gene is switched on, the cell follows the instruction for that gene.

Gene number	Instructions
1	Grow spikes (overrides gene 2)
2	Lose spikes
3	Grow by 20% in volume
4	Grow by 30% in volume
5	Keep single nucleus (overrides gene 6)
6	Divide nucleus in two to produce a cell with more than one nucleus
7	Grow 50% longer along y axis
8	Grow 50% longer along x axis
9	Remain as a singular cell (overrides genes 10 and 11)
10	Produce connections to join to parent cell
11	Grow microvilli along unattached border (only if gene 10 is on)

LS1.A LS1.B SF

©2018 **BIOZONE** International
ISBN: 978-1-927309-55-1
Photocopying Prohibited

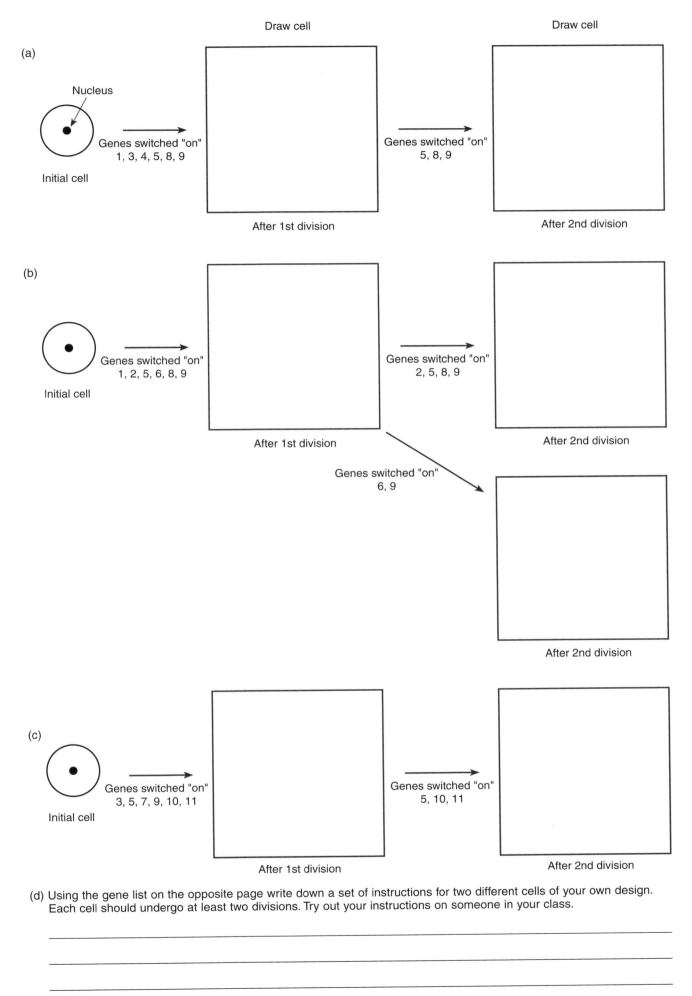

(a)

Nucleus

Initial cell

Genes switched "on"
1, 3, 4, 5, 8, 9

Draw cell

After 1st division

Genes switched "on"
5, 8, 9

Draw cell

After 2nd division

(b)

Initial cell

Genes switched "on"
1, 2, 5, 6, 8, 9

After 1st division

Genes switched "on"
2, 5, 8, 9

After 2nd division

Genes switched "on"
6, 9

After 2nd division

(c)

Initial cell

Genes switched "on"
3, 5, 7, 9, 10, 11

After 1st division

Genes switched "on"
5, 10, 11

After 2nd division

(d) Using the gene list on the opposite page write down a set of instructions for two different cells of your own design. Each cell should undergo at least two divisions. Try out your instructions on someone in your class.

©2018 **BIOZONE** International
ISBN: 978-1-927309-55-1
Photocopying Prohibited

EXPLAIN: Stem cells and differentiation

▸ **Cellular differentiation** is the transformation of unspecialized cells called **stem cells** into specialized cells that carry out a particular task in the body. Stem cells show the properties of self renewal and potency (below).

▸ Although each cell has the same genetic material (genes), different genes are turned on (activated) or off (inactivated) in different patterns during development in particular cell lines. The differences in gene activation control what type of cell forms (below). Once the developmental pathway of a cell is determined, it cannot alter its path and change into another cell type.

How stem cells give rise to different cell types

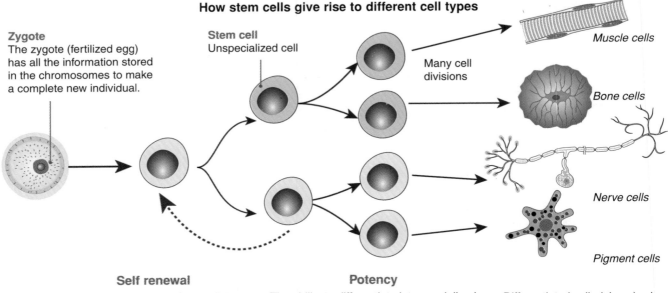

Zygote
The zygote (fertilized egg) has all the information stored in the chromosomes to make a complete new individual.

Stem cell
Unspecialized cell

Many cell divisions

Muscle cells

Bone cells

Nerve cells

Pigment cells

Self renewal
Stem cells have the ability to divide many times while maintaining an unspecialized state.

Potency
The ability to differentiate into specialized cells. Different types of stem cell have different levels of potency (below).

Differentiated cells (above) arise because genes are turned on or off in a particular sequence.

Categories of stem cells

Totipotent stem cells
These stem cells can differentiate into all the cells in an organism. Example: In humans, the zygote and its first few divisions. The meristematic tissue of plants (root and shoot tips) is also totipotent.

Pluripotent stem cells
These stem cells can give rise to any cells of the body, except extra-embryonic cells (e.g. placenta and chorion). Example: Embryonic stem cells (abbreviated ESC).

Multipotent stem cells
These adult stem cells (ASC) can give rise to a limited number of cell types, related to their tissue of origin. Examples: Bone marrow stem cells, epithelial stem cells, bone stem cells (osteoblasts).

3. Explain how so many different types of cells can be formed, even though all cells have the same DNA:

_____

_____

_____

4. (a) What are stem cells? _____

_____

(b) What are the two defining properties of stem cells?

i _____

ii _____

5. Name the cell from which all other cells in the body develop: _____

6. What is the difference between a multipotent cell and a totipotent cell? _____

_____

_____

©2018 **BIOZONE** International
ISBN: 978-1-927309-55-1
Photocopying Prohibited

ELABORATE: Stem cells and skin renewal

▶ The skin is the body's largest organ and acts as an essential physical barrier against infection and physical damage of the underlying tissues. It is made up of two parts: the epidermis, which forms a layered barrier, and the dermis, which supports and nourishes the epidermis.

▶ Different types of skin stem cells maintain the skin's epidermis and contribute to its healing after damage. As a result of constant renewal from the base and loss from the surface, the skin is completely renewed every 4 weeks.

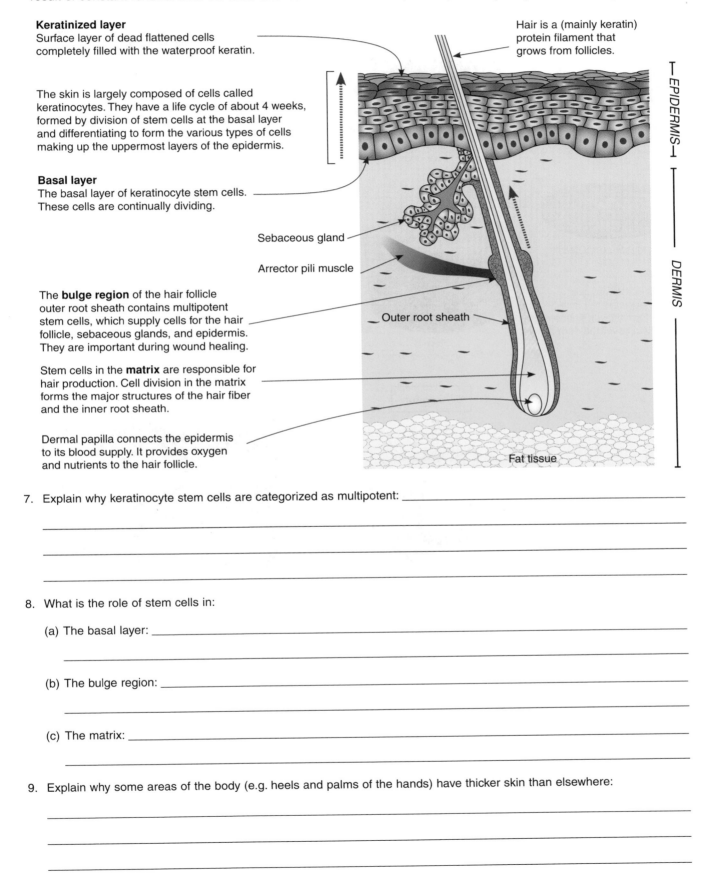

Keratinized layer
Surface layer of dead flattened cells completely filled with the waterproof keratin.

Hair is a (mainly keratin) protein filament that grows from follicles.

The skin is largely composed of cells called keratinocytes. They have a life cycle of about 4 weeks, formed by division of stem cells at the basal layer and differentiating to form the various types of cells making up the uppermost layers of the epidermis.

Basal layer
The basal layer of keratinocyte stem cells. These cells are continually dividing.

Sebaceous gland

Arrector pili muscle

The **bulge region** of the hair follicle outer root sheath contains multipotent stem cells, which supply cells for the hair follicle, sebaceous glands, and epidermis. They are important during wound healing.

Outer root sheath

Stem cells in the **matrix** are responsible for hair production. Cell division in the matrix forms the major structures of the hair fiber and the inner root sheath.

Dermal papilla connects the epidermis to its blood supply. It provides oxygen and nutrients to the hair follicle.

Fat tissue

EPIDERMIS

DERMIS

7. Explain why keratinocyte stem cells are categorized as multipotent: _____

_____

_____

_____

8. What is the role of stem cells in:

(a) The basal layer: _____

_____

(b) The bulge region: _____

_____

(c) The matrix: _____

_____

9. Explain why some areas of the body (e.g. heels and palms of the hands) have thicker skin than elsewhere:

_____

_____

_____

©2018 **BIOZONE** International
ISBN: 978-1-927309-55-1
Photocopying Prohibited

ELABORATE: Making synthetic skin

▶ Engineered skin is widely used to treat burns and other skin injuries. Stem cells from the skin are grown on a suitable scaffold or support material to produce a three-dimensional tissue.

Human dermal cells

Collagen

Day 0

Undifferentiated fibroblasts are combined with a gel containing collagen, the primary protein in skin. The fibroblasts move through the gel, rearranging the collagen to produce a fibrous, living matrix similar to the natural dermis.

Step 1
Form the lower dermal layer

Human epidermal cells

Day 6

Human epidermal cells (keratinocytes) are placed on top of the dermal layer. These cells multiply to cover the dermal layer.

Step 2
Form the upper epidermal layer

Air exposure

Day 10

Exposing the culture to air induces the epidermal cells to form the outer protective (keratinized) layer of skin. The final size of the product is about 75 mm. From this, many thousands of pieces of skin can be made.

Step 3
Form the outer layer

A real world lifesaver

Growing new skin from stem cells is useful for a burns victim or an otherwise healthy person needing a skin graft (e.g. after an infection or accident). But what if your stem cells carry a genetic disease? Or if your body's immune system rejects a skin graft produced from the cells of another person?

This scenario occurred in 2015 in Germany. A young child had a mutation in the *LAMB3* gene. His skin cells were not making the proteins needed to anchor the outer layers of his skin to the inner ones and as a result his skin was falling off. Doctors tried grafting skin from the boy's father but these grafts were rejected.

Finally the doctors sent a sample of the boy's skin to scientists in Italy. There the scientists used genetic engineering techniques to insert a working copy of the *LAMB3* gene into the stem cells in the sample of the boy's skin. The scientists then grew the genetically corrected stem cells on a protein and gauze gel to produce (in total) nearly one square meter of skin.

This was sent to the hospital in Germany in stages as it grew. There it was transplanted onto areas on the boy's body in sections, beginning with his arms and legs. The entire process took months but today the boy is able to live a normal life in completely new skin.

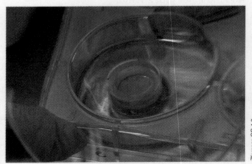

Growing human epidermal cells

Zhrgaas CC 4.0

10. Why are stem cells important in the generation of a new tissue? _____

_____

_____

_____

_____

11. Carry out some research about the scope of tissue engineering. What kinds of tissues can now be produced as three-dimensional ready to transplant tissues?

_____

_____

©2018 **BIOZONE** International
ISBN: 978-1-927309-55-1
Photocopying Prohibited

ELABORATE: Tissues work together

▶ A tissue is a collection of related cell types that work together to carry out a specific function. Different tissues come together to form organs. The cells, tissues, and organs of the body interact to meet the needs of the entire organism.

Muscle tissue	Epithelial tissue	Nervous tissue	Connective tissue
▶ Contractile tissue ▶ Produces movement of the body or its parts ▶ Includes smooth, skeletal, and cardiac muscle	▶ Lining tissue ▶ Covers the body and lines internal surfaces ▶ Can be modified to perform specific roles	▶ Receives and responds to stimuli ▶ Makes up the structures of the nervous system ▶ Regulates function of other tissues	▶ Supports, protects, and binds other tissues ▶ Contains cells in an extracellular matrix ▶ Can be hard or fluid

12. Research the type of tissue(s) that occur at the places in the body indicated below. State the types of tissues that occur in the spaces provided. Codes: E = epithelium, CT = connective tissue, SM = smooth muscle.

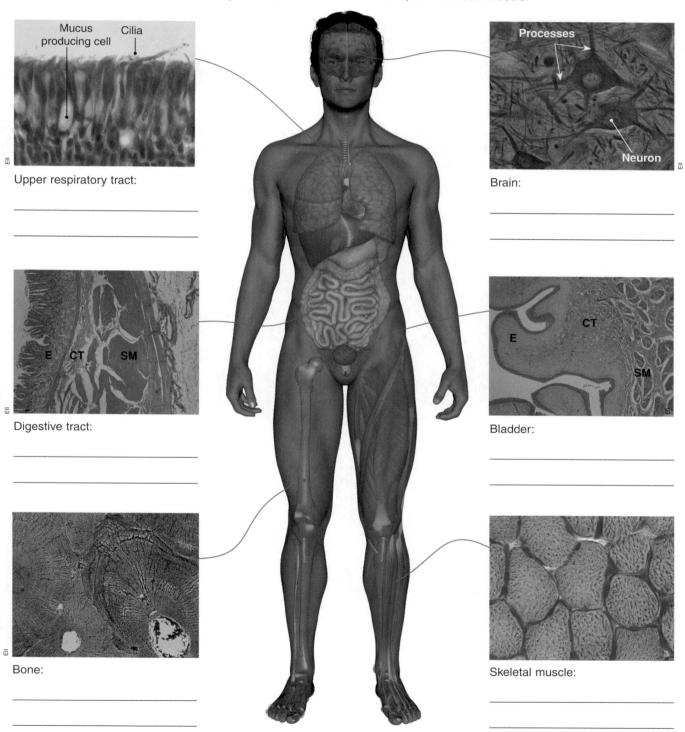

Upper respiratory tract:
_____
_____

Brain:
_____
_____

Digestive tract:
_____
_____

Bladder:
_____
_____

Bone:
_____
_____

Skeletal muscle:
_____
_____

©2018 **BIOZONE** International
ISBN: 978-1-927309-55-1
Photocopying Prohibited

Tissues make up organs

Different organs have different proportions of the four tissue types. Sometimes not all types are visible because they require special stains to be shown. The cells of the tissues, especially epithelial tissue, may be modified to carry out specific tasks:

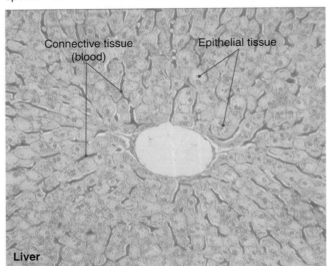

Liver

Nervous tissue is also present but not visible (not stained)

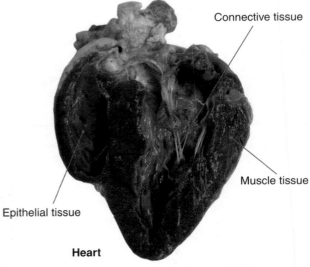

Heart

Nervous tissue is also present but not visible

EVALUATE: Developing a model for structural hierarchy in a multicellular organism

13. Draw a diagram below to show structural hierarchy in the body of a multicellular organism (i.e from the simplest components to the most complex). Label the diagram to show how the hierarchical arrangement leads to different interacting systems and how those system interactions result in the performance of new functions (e.g. movement). Properties that arise in a system as a result of increasing complexity are called emergent properties.

65 Interacting Systems

ENGAGE: Flex your muscles

Bones act with muscles to form levers that enable movement

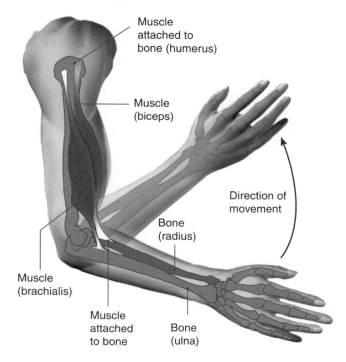

Muscle attached to bone (humerus)

Muscle (biceps)

Direction of movement

Bone (radius)

Muscle (brachialis)

Muscle attached to bone

Bone (ulna)

▸ An **organ system** is a group of organs that work together to perform a certain group of tasks.

▸ This example shows how the muscular and skeletal organ systems in humans work together to achieve movement in the arm. Of course the nervous system is involved too, providing the signal to move.

1. While you are sitting at your desk, lay your forearm on the desk, palm facing upwards. Now put your other hand on your upper arm and raise your resting arm towards you, bending at the elbow. You would have felt the muscles in your upper arm move.

(a) Describe how the shape of the muscle on your upper arm changes as you carry out the movement above:

_____

_____

(b) Which organ systems are involved in this movement? _____

_____

EXPLORE: Organ systems

▸ There are **11 organ** (body) systems in humans.

▸ Although each system has a specific job (e.g. digestion, reproduction, internal transport, or gas exchange) the organ systems must interact to maintain the functioning of the organism.

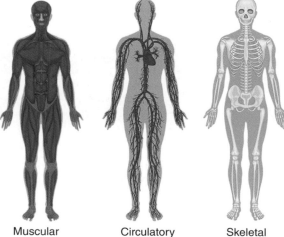

Muscular system   Circulatory system   Skeletal system

2. The image on the left shows three organ systems. Research the other eight organ systems in the human body. Write down what they are and some of the main organs involved:

(a) _____

(b) _____

(c) _____

(d) _____

(e) _____

(f) _____

(g) _____

(h) _____

©2018 **BIOZONE** International
ISBN: 978-1-927309-55-1
Photocopying Prohibited

EXPLAIN: Circulation and gas exchange

▶ The circulatory and respiratory systems interact to provide the body's tissues with oxygen and remove carbon dioxide.

Circulatory system

Function

Delivers oxygen (O_2) and nutrients to all cells and tissues. Removes carbon dioxide (CO_2) and other waste products of metabolism. CO_2 is transported to the lungs.

Components

▶ Heart
▶ Blood vessels:
  • Arteries
  • Veins
  • Capillaries
▶ Blood

Interaction between systems

In vertebrates, the respiratory system and cardiovascular system interact to supply oxygen and remove carbon dioxide from the body.

Respiratory system

Function

Provides surface for gas exchange. Moves fresh air into and stale air out of the body.

Components

▶ Airways:
  • Pharynx
  • Larynx
  • Trachea
▶ Lungs:
  • Bronchi
  • Bronchioles
  • Alveoli
▶ Diaphragm

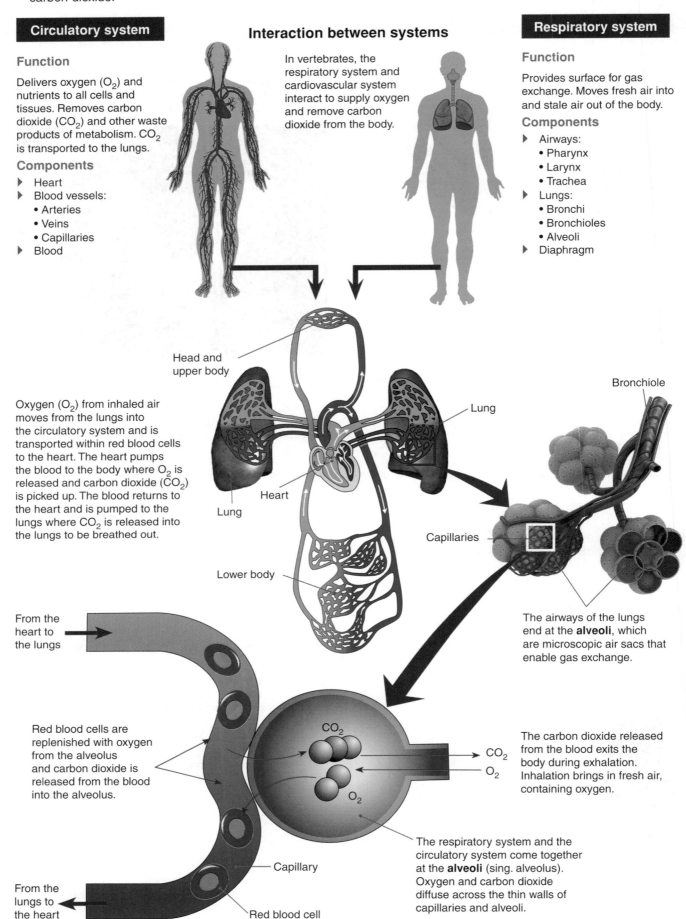

Head and upper body

Oxygen (O_2) from inhaled air moves from the lungs into the circulatory system and is transported within red blood cells to the heart. The heart pumps the blood to the body where O_2 is released and carbon dioxide (CO_2) is picked up. The blood returns to the heart and is pumped to the lungs where CO_2 is released into the lungs to be breathed out.

Lung

Heart

Lung

Lower body

Bronchiole

Lung

Capillaries

The airways of the lungs end at the **alveoli**, which are microscopic air sacs that enable gas exchange.

From the heart to the lungs

Red blood cells are replenished with oxygen from the alveolus and carbon dioxide is released from the blood into the alveolus.

CO_2

CO_2

O_2

O_2

The carbon dioxide released from the blood exits the body during exhalation. Inhalation brings in fresh air, containing oxygen.

Capillary

From the lungs to the heart

Red blood cell

The respiratory system and the circulatory system come together at the **alveoli** (sing. alveolus). Oxygen and carbon dioxide diffuse across the thin walls of capillaries and alveoli.

©2018 **BIOZONE** International
ISBN: 978-1-927309-55-1
Photocopying Prohibited

Responses to exercise

▶ During exercise, your body needs more oxygen to meet the extra demands placed on the muscles, heart, and lungs. At the same time, more carbon dioxide must be expelled. To meet these increased demands, blood flow must increase. This is achieved by increasing the rate of heart beat. As the heart beats faster, blood is circulated around the body more quickly, and exchanges between the blood and tissues increase.

▶ The arteries and veins must be able to resist the extra pressure of higher blood flow and must expand (dilate) to accommodate the higher blood volume. If they didn't, they could rupture (break). During exercise, the muscular, cardiovascular, and nervous systems interact to maintain the body's systems in spite of increased demands (right).

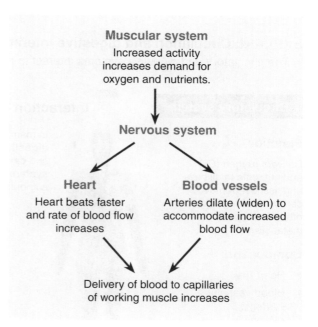

Muscular system
Increased activity increases demand for oxygen and nutrients.

Nervous system

Heart
Heart beats faster and rate of blood flow increases

Blood vessels
Arteries dilate (widen) to accommodate increased blood flow

Delivery of blood to capillaries of working muscle increases

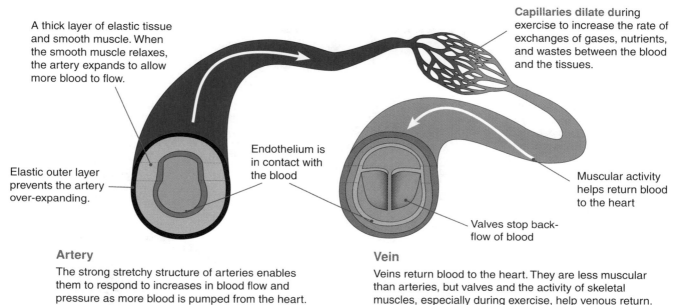

A thick layer of elastic tissue and smooth muscle. When the smooth muscle relaxes, the artery expands to allow more blood to flow.

Capillaries dilate during exercise to increase the rate of exchanges of gases, nutrients, and wastes between the blood and the tissues.

Elastic outer layer prevents the artery over-expanding.

Endothelium is in contact with the blood

Muscular activity helps return blood to the heart

Valves stop back-flow of blood

Artery
The strong stretchy structure of arteries enables them to respond to increases in blood flow and pressure as more blood is pumped from the heart.

Vein
Veins return blood to the heart. They are less muscular than arteries, but valves and the activity of skeletal muscles, especially during exercise, help venous return.

3. In your own words, describe how the circulatory system and respiratory system work together to provide the body with oxygen and remove carbon dioxide:

_____

_____

_____

_____

_____

_____

4. (a) What happens to blood flow during exercise? _____

(b) Explain how body systems interact to accommodate the extra blood flow needed when a person exercises?

_____

_____

_____

_____

©2018 **BIOZONE** International
ISBN: 978-1-927309-55-1
Photocopying Prohibited

EXPLAIN: Circulation and digestive interactions

▶ The circulatory and digestive systems interact to provide the body's tissues with nutrients.

Circulatory system

Function

Delivers oxygen (O_2) and nutrients to all cells and tissues. Removes carbon dioxide (CO_2) and other waste products of metabolism.

Components

▶ Heart
▶ Blood vessels:
  • Arteries
  • Veins
  • Capillaries
▶ Blood

Interaction between systems

In mammals, the **digestive system** and **cardiovascular system** interact to supply nutrients to the body.

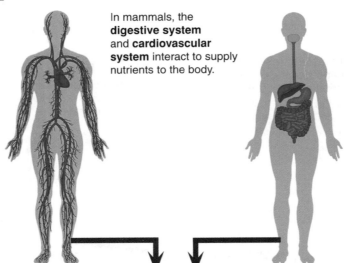

Digestive system

Function

Digest food and absorb useful molecules, reabsorb water and eliminate undigested material.

Components

▶ Mouth and pharynx
▶ Esophagus
▶ Stomach
▶ Liver and gall bladder (accessory organs)
▶ Pancreas (accessory organ)
▶ Small intestine
▶ Large intestine

Food is digested in the stomach and small intestine and the nutrients are then absorbed and passed to the circulatory system. The capillaries around the stomach and intestines collect nutrients and then drain to the hepatic portal vein, which carries the blood directly to the liver. The liver processes this nutrient-rich blood, e.g. glucose is stored as a larger storage molecule called glycogen. The hepatic vein then transports nutrients from the liver to supply the other tissues of the body.

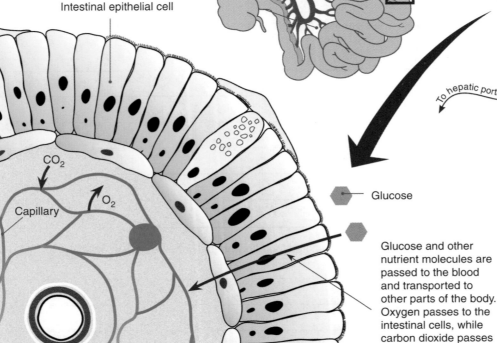

Intestinal epithelial cell

CO_2

O_2

Capillary

Glucose

Glucose and other nutrient molecules are passed to the blood and transported to other parts of the body. Oxygen passes to the intestinal cells, while carbon dioxide passes into the blood.

Hepatic vein

Liver

Stomach

Hepatic portal vein

Small intestine

Lumen (space inside the gut)

Villus

Capillaries

To hepatic portal vein

Villi project into the lumen (yellow) of the small intestine and absorb nutrients. Villi contain capillary networks which receive the nutrients and transport them to the hepatic portal system.

©2018 **BIOZONE** International
ISBN: 978-1-927309-55-1
Photocopying Prohibited

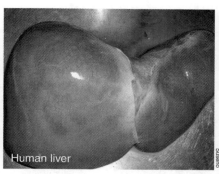
Human liver

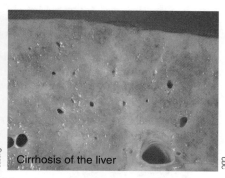
Cirrhosis of the liver

Blood flow to the digestive tract increases steadily after a meal and remains elevated for about 2.5 hours, reaching a maximum after about 30 minutes. During exercise, blood flow in the digestive tract is reduced as it is redirected to the muscles.

Nutrients, e.g. minerals, sugars, and amino acids, are transported in the blood plasma to the liver. The liver receives nutrient-rich deoxygenated blood from the digestive system via the hepatic portal vein and oxygen rich blood from the hepatic artery.

Scarring of the liver tissue (cirrhosis) can result in portal hypertension (high blood pressure). The scarred tissue obstructs the liver's blood flow causing pressure to build up in upstream blood vessels. This results in swelling and possible bleeding.

5. How are nutrients transported in the blood? _____

_____

6. Explain how a liver cirrhosis affects the circulatory system: _____

_____

_____

_____

7. (a) At which two points in the body do the digestive and circulatory systems directly interact? _____

_____

_____

(b) Explain what is happening at these points: _____

_____

_____

_____

8. (a) What happens to blood flow to the digestive tract after a meal? _____

_____

(b) Explain why it is often recommended that a person should exercise within 2.5 hours of eating, or eat within half an hour after exercising to gain the most benefit from the exercise (in terms of muscle development):

_____

_____

_____

_____

9. In your own words, describe how the circulatory and digestive systems work together to provide the body with nutrients:

_____

_____

_____

_____

_____

66 How Cells Make Proteins

ENGAGE: Proteins

▶ The body produces several different classes of proteins, e.g. those that provide shape and support (structural) or those that catalyze reactions (catalytic). Proteins work best under a certain set of conditions. Outside these conditions, the chemical bonds between the amino acid units making up the protein can be disrupted, causing the protein to lose its structure and thus its ability to carry out its task. This loss of protein function is called denaturation.

1. Proteins fall into seven broad categories: *Structural, contractile, storage, transport, immunological, signaling (regulatory), catalytic*. As a group, categorize each of the protein examples below (there is one example for each category):

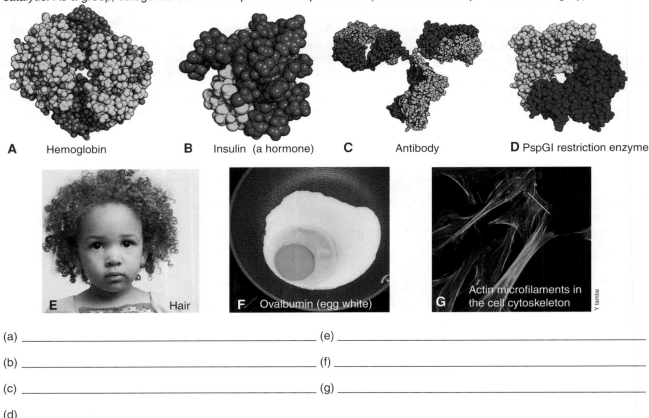

A Hemoglobin   **B** Insulin (a hormone)   **C** Antibody   **D** PspGI restriction enzyme

E Hair   **F** Ovalbumin (egg white)   **G** Actin microfilaments in the cell cytoskeleton

(a) _____

(b) _____

(c) _____

(d) _____

(e) _____

(f) _____

(g) _____

EXPLORE: Go boil your egg!

▶ Protein denaturation is a permanent event. Eggs are 13% protein and provide a good way to illustrate denaturation.

▶ Crack open a egg and place it in a 250 mL beaker that is about half full of water. Heat the water until it is nearly boiling. Record what the egg white looks like before and after boiling.

▶ Place about 100 mL of isopropyl alcohol (propan-2-ol or rubbing alcohol) into a 250 mL beaker. Crack open a egg and place it in the beaker. Stir the egg slowly and record the result.

2. (a) Describe what the egg white looked like before heating: _____

_____

(b) Describe what the egg white looked like after heating: _____

_____

(c) What has the heat done? _____

3. Describe what the egg white looks like after adding the isopropyl alcohol: _____

_____

_____

4. Which of the two methods above causes the most denaturation in egg white? _____

_____

_____

©2018 **BIOZONE** International
ISBN: 978-1-927309-55-1
Photocopying Prohibited

EXPLORE: Gene expression

▶ Cells make proteins in a process called gene expression. It involves **transcription** (rewriting) of the DNA making up a gene into mRNA and **translation** of the mRNA into protein. Translation is the job of ribosomes.

▶ Eukaryotic genes include non protein-coding regions called introns. These regions of intronic DNA must be removed before the mRNA is translated. Transcription of the genes and editing that primary transcript to form the mature mRNA occurs in the nucleus. Translation of the protein by the ribosomes occurs in the cytoplasm.

A summary of eukaryotic gene expression

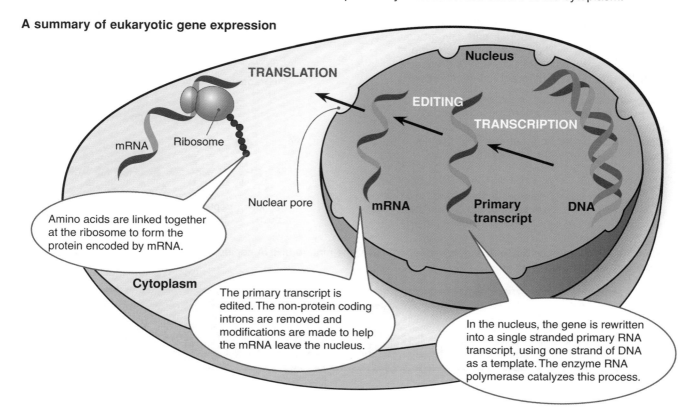

TRANSLATION

Nucleus

EDITING

TRANSCRIPTION

mRNA   Ribosome

Nuclear pore

mRNA

Primary transcript

DNA

Amino acids are linked together at the ribosome to form the protein encoded by mRNA.

Cytoplasm

The primary transcript is edited. The non-protein coding introns are removed and modifications are made to help the mRNA leave the nucleus.

In the nucleus, the gene is rewritten into a single stranded primary RNA transcript, using one strand of DNA as a template. The enzyme RNA polymerase catalyzes this process.

The diagram above shows a simple model of gene expression. Use it to answer the following questions:

5. (a) What are the three stages in gene expression in eukaryotes and where do they occur:

    (i) _____

    (ii) _____

    (iii) _____

    (b) What happens in each stage?

    (i) _____

    (ii) _____

    (iii) _____

6. This photograph shows an electron micrograph of a giant polytene chromosome. These chromosomes are common in the larval stages of flies, which must grow rapidly before changing to the adult form. Polytene chromosomes form as a result of repeated cycles of DNA replication without cell division. This creates many copies of genes. Within these chromosomes, visible 'puffs' show where there is active transcription of the genes.

    (a) What is the consequence of active transcription in a polytene chromosome?

    _____

    (b) Can you think why this might be useful in a larval insect? _____

    _____

    _____

©2018 **BIOZONE** International
ISBN: 978-1-927309-55-1
Photocopying Prohibited

EXPLORE: Amino acids

▶ Proteins are large molecules made up of many smaller units called amino acids joined together. The amino acids are joined together by peptide bonds (between the amine and carboxyl groups). The sequence of amino acids in a protein is determined by the order of nucleotides in DNA.

▶ All amino acids have a common structure (right) consisting of an amine group, a carboxyl group, a hydrogen atom, and an 'R' group. Each type of amino acid has a different 'R' group (side chain). Each "R" group has a different chemical property.

▶ The chemical properties of the amino acids are important because the chemical interactions between amino acids cause a protein fold into a specific three dimensional shape. The protein's shape helps it carry out its specialized role.

The order of amino acids in a protein is directed by the order of nucleotides in DNA (and therefore mRNA).

The general structure of an amino acid

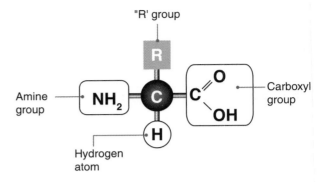

A polypeptide chain

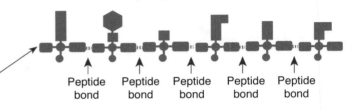

7. Examine the information below. It shows a length of DNA, the mature mRNA copied from the DNA (after editing), and the amino acid chain produced:

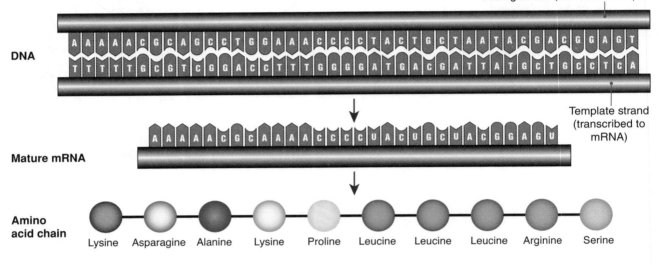

Use the information above to answer the following questions:

(a) What can you say about the matching of the letters that make up the DNA template strand with the letters that make up the mRNA strand?

_____

_____

_____

(b) Which mRNA letters were removed when the primary mRNA transcript was edited to form the mature mRNA?

_____

(c) Study the mRNA strand and the amino acid chain it coded for carefully. What are the rules for coding the amino acids from the mRNA?

_____

_____

_____

_____

©2018 BIOZONE International
ISBN: 978-1-927309-55-1
Photocopying Prohibited

EXPLORE: The genetic code

▶ The genetic code is the set of rules by which the genetic information in DNA (or mRNA) is translated into proteins.

▶ The genetic information for the assembly of amino acids is stored as three-base sequence. These three letter codes on mRNA are called codons.

▶ Each codon represents one of 20 amino acids used to make proteins. The code is effectively universal, being the same in all living things (with a few minor exceptions).

▶ The genetic code is summarized in a mRNA-amino acid table, which identifies the amino acid encoded by each mRNA codon.

▶ The code is degenerate, meaning there may be more than one codon for each amino acid. Most of this degeneracy is in the third nucleotide of a codon.

A triplet (three nucleotide bases) codes for a single amino acid. The triplet code on mRNA is called a codon.

mRNA-amino acid table

The table below decodes the genetic code from a given mRNA sequence to give a sequence of amino acids in a polypeptide chain. To work out which amino acid is coded for by a codon (triplet of bases) look for the first letter of the codon in the row label on the left hand side of the table. Then look for the column that intersects the same row from above matching the second base. Finally, locate the third base in the codon by looking along the row from the right hand side that matches your codon. The RNA base U (uracil) replaces the DNA base T (thymine) in RNA sequences. Example: Determine **CAG**: C on the left row, A on the top column, G on the right row. **CAG** is Gln (**glutamine**)

Read second letter here · **Second letter** · Read third letter here

Read first letter here

		U	C	A	G	
First letter	**U**	UUU Phe / UUC Phe / UUA Leu / UUG Leu	UCU Ser / UCC Ser / UCA Ser / UCG Ser	UAU Tyr / UAC Tyr / UAA STOP / UAG STOP	UGU Cys / UGC Cys / UGA STOP / UGG Trp	U C A G
	C	CUU Leu / CUC Leu / CUA Leu / CUG Leu	CCU Pro / CCC Pro / CCA Pro / CCG Pro	CAU His / CAC His / CAA Gln / CAG Gln	CGU Arg / CGC Arg / CGA Arg / CGG Arg	U C A G
	A	AUU Ile / AUC Ile / AUA Ile / AUG Met	ACU Thr / ACC Thr / ACA Thr / ACG Thr	AAU Asn / AAC Asn / AAA Lys / AAG Lys	AGU Ser / AGC Ser / AGA Arg / AGG Arg	U C A G
	G	GUU Val / GUC Val / GUA Val / GUG Val	GCU Ala / GCC Ala / GCA Ala / GCG Ala	GAU Asp / GAC Asp / GAA Glu / GAG Glu	GGU Gly / GGC Gly / GGA Gly / GGG Gly	U C A G

Third letter

Use the codon table above to answer the following questions:

8. What is the amino acid chain produced from the following mRNA sequences?

   (a) GAU CCG UAC GUA CGA ACA AUU ACC: _____

   (b) GGG UUU GCU UGG CAA AAC AGU GCA: _____

   (c) AAA CCC GGG GUA AUU CGC AAU GAU: _____

9. (a) If one RNA base (A, U, C, G) coded for one amino acid, how many amino acids could be coded for in total? _____

   (b) If two RNA bases coded for one amino acid, how many amino acids could be coded for in total? _____

10. (a) What is the effect (in general) of changing the last base in a codon? _____

   _____

   (b) How might this affect the chance of a mutation changing the final protein produced after gene expression?

   _____

   _____

©2018 **BIOZONE** International
ISBN: 978-1-927309-55-1
Photocopying Prohibited

EXPLAIN: How do we know the genetic code?

▶ In 1961, Marshall Nirenberg and Heinrich Matthaei developed an experiment to crack the genetic code.

▶ A cell free *E. coli* extract was produced for their experiment by rupturing the bacterial cells to release the cytoplasm. The extract had all the components needed to make proteins (except mRNA).

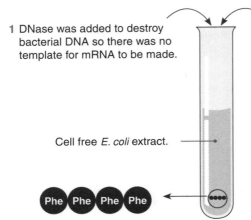

1 DNase was added to destroy bacterial DNA so there was no template for mRNA to be made.

Cell free *E. coli* extract.

2 Radio labeled amino acids and a synthetic mRNA strand containing only uracil (U) were added.

3 Once the mRNA was added, an amino acid was produced. The codon UUU produced the amino acid phenylalanine (Phe).

Nirenberg and Matthaei

11. Based on the method above briefly describe how you might eventually determine all the codons for the amino acids:

_____

_____

_____

_____

_____

EXPLAIN: Gene expression

12. Use the information in this activity to produce a diagram that shows the process of gene expression. In your diagram you should show how this process produces the following amino acid chain:

Histidine, isoleucine, threonine, proline, cysteine, alanine, phenylalanine, tyrosine

67 The Functions of Proteins

ENGAGE: Cosmetics and deadly toxins

▶ The botulinum toxin produced by the bacteria *Clostridium botulinum* is the most lethal toxin known. Just 2 ng (2 billionths of a gram) of the variant known as type H (discovered in 2013) will kill a human adult. There is no known antidote.

▶ The toxin is a protein that splits other proteins needed in order for nerve cells to function. This effectively shuts down the nervous system, leading to paralysis and ultimately death.

▶ A variant of the toxin known as type A is less lethal in its action (70 µg will kill an adult) and is used for medical and cosmetic procedures.

▶ Injecting the toxin under the skin relaxes the muscles and thus reduces the appearance of wrinkles. It will also paralyze those muscles for a time, which is why people sometimes can't raise their eyebrows after the procedure (right).

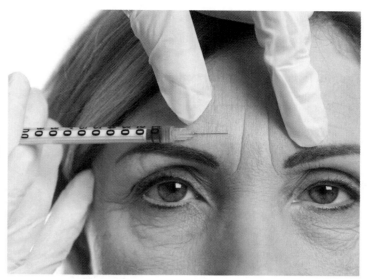

1. There are many organisms (e.g. venomous snakes and spiders) that produce proteins that are harmful. Some of the proteins they produce are now used in medical therapies, such as in treating muscle disorders. Research three proteins that are naturally deadly, but are now used in medicine.

(a) _____

_____

_____

(b) _____

_____

_____

(c) _____

_____

_____

EXPLORE: The shape of a protein determines its function

▶ The shape of a protein determines its role. As well as grouping proteins according to their function, they can be grouped according to their shape, i.e. as globular or fibrous.

▶ The amino acid sequence determines a protein's shape because it determines the interactions between the amino acids and therefore the way the protein folds up.

▶ A protein's shape is so critical to its function that if it is denatured it can no longer carry out its function.

Globular proteins

▶ Globular proteins are round and water soluble. Their functions include:

• Catalytic (e.g. enzymes)

• Regulatory (e.g. hormones)

• Transport (e.g. hemoglobin)

• Protective (e.g. antibodies)

DNase: a DNA cutting enzyme

Fibrous proteins

▶ Fibrous proteins are long and strong. Their functions include:

• Support and structure (e.g. connective tissue)

• Contractile (e.g. myosin, actin)

Collagen: a fibrous protein

2. Why are proteins important in organisms? _____

_____

_____

©2018 **BIOZONE** International
ISBN: 978-1-927309-55-1
Photocopying Prohibited

SF   LS1.A

EXPLORE: Protein functions

▸ In eukaryotic cells, most of a cell's genetic information (DNA) is found in a large membrane-bound organelle called the **nucleus**. The nucleus directs all cellular activities by controlling the synthesis of proteins.

▸ DNA provides instructions that code for the formation of proteins. Proteins carry out most of a cell's work. A cell produces many different types of proteins and each carries out a specific task in the cell.

▸ Proteins are involved in the structure, function, and regulation of the body's cells, tissues, and organs. Without functioning proteins, a cell can not carry out its specialized role and the organism may die.

The nucleus is the control center of a cell

The DNA within the nucleus provides instructions to a cell on how to carry out its functions to sustain essential life processes. This includes the production of proteins.

Different sections of DNA, called genes, code for specific proteins. A cell can control the type of protein it produces by only transcribing (rewriting) specific genes as their proteins are required.

The denser structure within the nucleus is the nucleolus (n), where ribosomes are synthesized.

The nuclear envelope is formed by a double-layered membrane. It keeps the DNA within the nucleus but has pores in it, which allow materials to move between the nucleus and the cytoplasm.

In eukaryotes, production of the protein is completed outside of the nucleus. Synthesis continues on ribosomes, which may be free in the cytoplasm or associated with the rough endoplasmic reticulum (rER).

An animal cell

3. Draw a line to match the protein function with its description. Provide examples of each:

Function	Description	Example(s)
Internal defense	Some proteins can function as enzymes, thereby controlling metabolism.	_____
Contractile	Proteins can function as chemical messenger molecules.	_____
Catalytic	Proteins can make up structural components of tissues and organs.	_____
Regulation	Some proteins can act as carrier molecules to transport molecules from one place to another.	_____
Structural	Some proteins form antibodies that combat disease-causing organisms.	_____
Transport	Some proteins form contractile elements in cells and bring about movement.	_____

4. For each of the examples on the next page, research and write down the protein's role or function (a)-(f) opposite:

5. Suggest what might happen to a protein's functionality if it was incorrectly encoded by the DNA. Explain your answer:

_____

_____

_____

©2018 **BIOZONE** International
ISBN: 978-1-927309-55-1
Photocopying Prohibited

Internal defense

Antibodies (also called immunoglobulins) are "Y" shaped proteins that protect the body by identifying and killing disease-causing organisms such as bacteria and viruses.

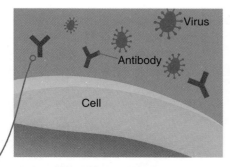

Cell, Virus, Antibody

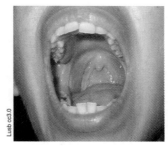
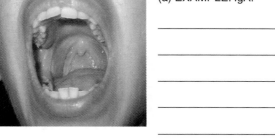

(a) EXAMPLE: IgA:

_____

_____

_____

_____

_____

Movement

Contractile proteins are involved in movement of muscles and form the cytoskeleton of cells.

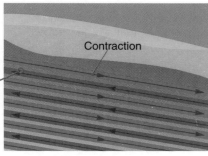

Contraction

(b) EXAMPLE: Actin / myosin:

_____

_____

_____

_____

_____

Catalytic

Thousands of different chemical reactions take place in an organism. Each chemical reaction is catalyzed by enzymes. The ending "ase" identifies a molecule as an enzyme.

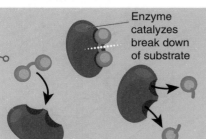

Enzyme catalyzes break down of substrate

(c) EXAMPLE: Amylase:

_____

_____

_____

_____

_____

Regulation

Regulatory proteins such as hormones act as signal molecules to control the timing and occurrence of biological processes and coordinate responses in cells, tissues, and organs.

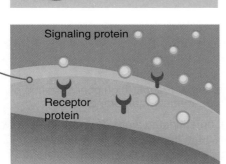

Signaling protein, Receptor protein

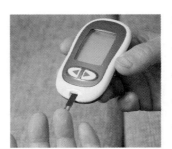

(d) EXAMPLE: Insulin:

_____

_____

_____

_____

_____

Structural

Structural proteins provide physical support or protection. They are strong, fibrous (thread like) and stringy.

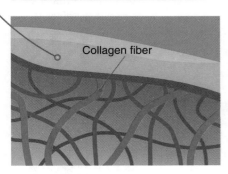

Collagen fiber

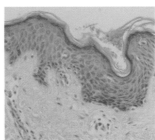

(e) EXAMPLE: Collagen:

_____

_____

_____

_____

_____

Transport

Proteins can carry substances around the body or across membranes. For example, hemoglobin (Hb) transports oxygen (left), and proteins in cell membranes help molecules move into and out of cells.

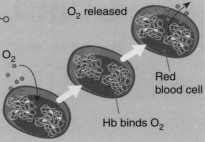

O_2 released, O_2, Hb binds O_2, Red blood cell

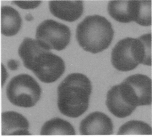
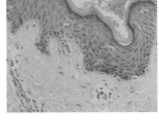

(f) EXAMPLE: Hemoglobin:

_____

_____

_____

_____

_____

©2018 **BIOZONE** International
ISBN: 978-1-927309-55-1
Photocopying Prohibited

68 Proteins are Part of Membranes

ENGAGE: Osmosis

▶ You may have carried out an investigation into osmosis in middle school. The image right shows the set up for osmosis using dialysis tubing. If you have not used this set up before, your teacher may set it up for you.

1. What happens to the water in the tube over time?

_____

_____

_____

2. Why does this happen? (Use the diagram on the right to help you if you're not sure):

_____

_____

_____

_____

3. Recall the plasma (cell surface) membrane surrounds a cell. How is it similar to the dialysis tubing in the osmosis investigation?

_____

_____

_____

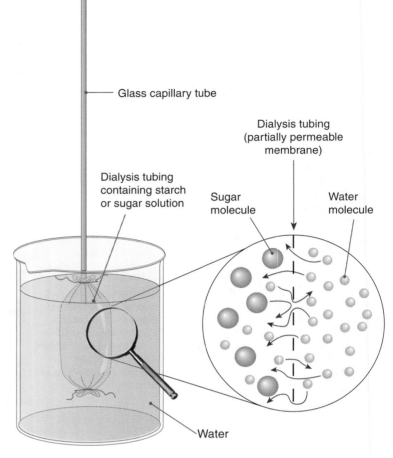

Glass capillary tube

Dialysis tubing (partially permeable membrane)

Dialysis tubing containing starch or sugar solution

Sugar molecule

Water molecule

Water

EXPLORE: Reversing osmosis to filter water

▶ Osmosis is a passive process in which water moves from an area of low solute concentration (high concentration of free water molecules) to an area of high solute concentration across a partially permeable membrane.

▶ Engineers have put partially permeable membranes to use in purifying water. Contaminated water (e.g. salt water or waste water) is forced under pressure through a partially permeable membrane in the opposite direction to which it would flow normally. This process is called reverse osmosis.

Reverse osmosis is commonly used in household water purification systems. Industrially it is used to treat and desalinate drinking water.

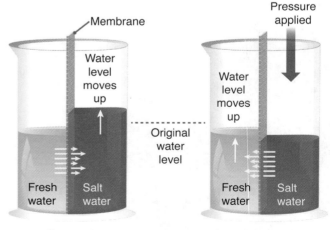

Membrane

Pressure applied

Water level moves up

Original water level

Water level moves up

Fresh water | Salt water

Fresh water | Salt water

Osmosis

Reverse osmosis

4. Why is energy needed to produce the reverse osmosis effect? _____

_____

_____

   LS1.A ETS1.B EM SF

©2018 **BIOZONE** International
ISBN: 978-1-927309-55-1
Photocopying Prohibited

EXPLORE: Cellular membranes

▶ The fluid-mosaic model of membrane structure (below) describes a phospholipid bilayer with proteins of different types moving freely within it. The double layer of phospholipids is quite fluid. It is a dynamic (constantly changing) structure and is actively involved in cellular activities.

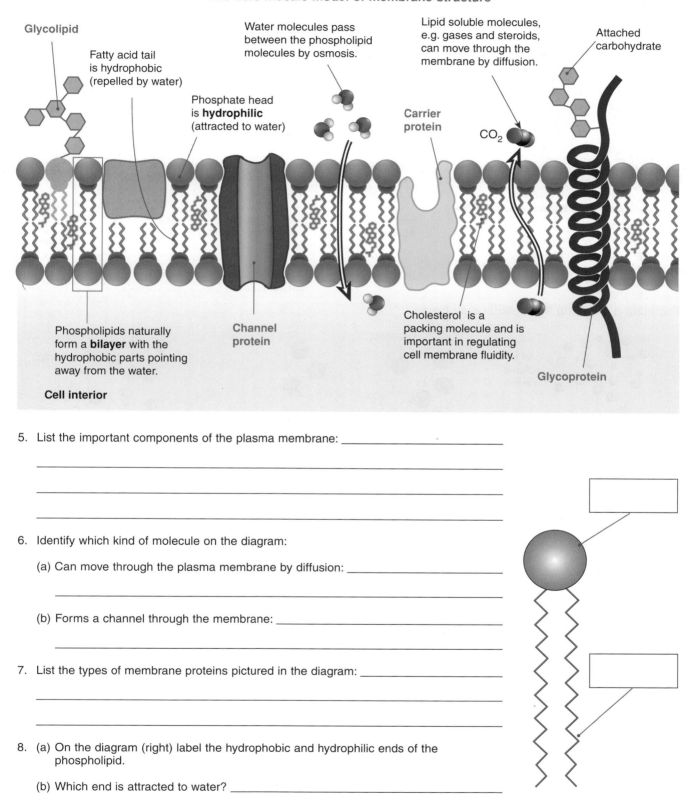

The fluid mosaic model of membrane structure

Glycolipid

Fatty acid tail is hydrophobic (repelled by water)

Phosphate head is **hydrophilic** (attracted to water)

Water molecules pass between the phospholipid molecules by osmosis.

Lipid soluble molecules, e.g. gases and steroids, can move through the membrane by diffusion.

Attached carbohydrate

Carrier protein

CO_2

Phospholipids naturally form a **bilayer** with the hydrophobic parts pointing away from the water.

Channel protein

Cholesterol is a packing molecule and is important in regulating cell membrane fluidity.

Glycoprotein

Cell interior

5. List the important components of the plasma membrane: _____

_____

_____

_____

6. Identify which kind of molecule on the diagram:

(a) Can move through the plasma membrane by diffusion: _____

_____

(b) Forms a channel through the membrane: _____

_____

7. List the types of membrane proteins pictured in the diagram: _____

_____

_____

8. (a) On the diagram (right) label the hydrophobic and hydrophilic ends of the phospholipid.

(b) Which end is attracted to water? _____

9. What would happen to the cell if the plasma membrane were to fail or split?

_____

_____

EXPLORE: Proteins facilitate diffusion in cells

▶ **Diffusion** is the movement of particles from regions of high concentration to regions of low concentration. Diffusion is a passive process, meaning it needs no input of energy to occur. During diffusion, molecules move randomly about, becoming evenly dispersed.

▶ Most diffusion in biological systems occurs across membranes. Simple diffusion occurs directly across a membrane, whereas facilitated diffusion involves helper proteins. Neither requires the cell to expend energy.

Factors that affect the rate of diffusion

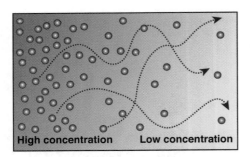

High concentration   **Low concentration**

Concentration gradient

If molecules can move freely, they move from high to low concentration (down a concentration gradient) until evenly dispersed. Net movement then stops.

Concentration gradient	Diffusion rate is higher when there is a greater concentration difference between two regions.
The distance moved	Diffusion occurs at a greater rate over shorter distances than over a larger distances.
The surface area involved	The larger the area across which diffusion occurs, the greater the rate of diffusion.
Barriers to diffusion	Rate of diffusion is slower across thick barriers than across thin barriers.
Temperature	Rate of diffusion increases with temperature.

$KMnO_4$ diffusing in a test tube

Facilitating diffusion in a cellular membrane

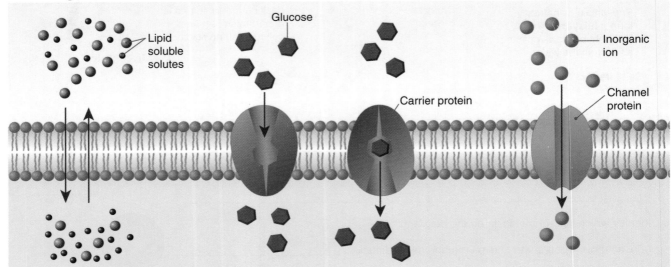

Glucose

Lipid soluble solutes

Carrier protein

Inorganic ion

Channel protein

Simple diffusion

Molecules move directly through the membrane without assistance and without any energy expenditure. Example: O_2 diffuses into the blood and CO_2 diffuses out.

Facilitated diffusion by carriers

Carrier proteins allow large lipid-insoluble molecules that cannot cross the membrane by simple diffusion to be transported into the cell. Example: the transport of glucose into red blood cells.

Facilitated diffusion by channels

Channel proteins (hydrophilic pores) in the membrane allow inorganic ions to pass through the membrane. Example: K^+ ions leaving nerve cells to restore membrane resting potential.

10. What is diffusion? _____

_____

_____

11. How are proteins involved in facilitated diffusion? _____

_____

_____

_____

©2018 **BIOZONE** International
ISBN: 978-1-927309-55-1
Photocopying Prohibited

EXPLAIN: Active transport

▶ Active transport is the movement of molecules (or ions) from regions of low concentration to regions of high concentration across a plasma membrane.

▶ Active transport needs energy to proceed because molecules are being moved against their concentration gradient.

▶ In a cell, the energy for active transport comes from ATP (adenosine triphosphate). Energy is released when ATP is hydrolyzed (water is added) forming ADP (adenosine diphosphate) and inorganic phosphate (P_i).

▶ Transport (carrier) proteins in the plasma membrane use energy to transport molecules across a membrane.

▶ Active transport can be used to move molecules into and out of a cell.

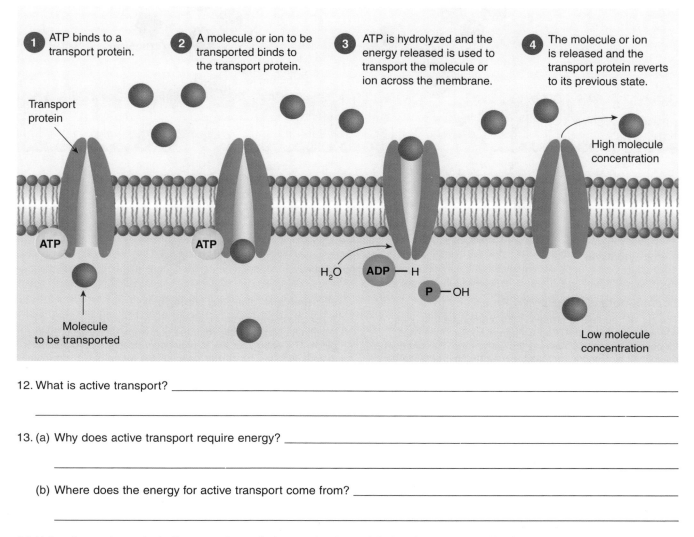

1. ATP binds to a transport protein.

2. A molecule or ion to be transported binds to the transport protein.

3. ATP is hydrolyzed and the energy released is used to transport the molecule or ion across the membrane.

4. The molecule or ion is released and the transport protein reverts to its previous state.

Transport protein

High molecule concentration

Molecule to be transported

Low molecule concentration

12. What is active transport? _____

_____

13. (a) Why does active transport require energy? _____

_____

(b) Where does the energy for active transport come from? _____

_____

14. Using the analogy of a ball on top of a wall, draw a simple model of active transport showing:
(a) The difference between active and passive transport
(b) Moving a particle over a barrier
(c) Using the passive transport of one particle to power the transport of a second particle against its concentration gradient (complete the next page first if you need help):

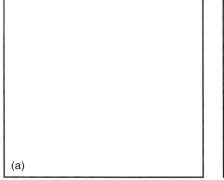
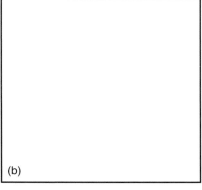
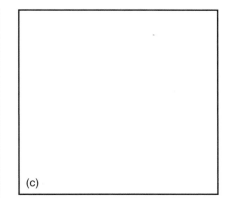

(a)   (b)   (c)

©2018 **BIOZONE** International
ISBN: 978-1-927309-55-1
Photocopying Prohibited

EXPLAIN: Coupling the work of transport proteins

▶ Cells can use the active transport of a molecule across the cell membrane in one direction to power the transport of a different molecule across the membrane in the opposite direction.

▶ The sodium-potassium pump (below, left) is found in almost all animal cells and is also common in plant cells. The concentration gradient created by ion pumps is often coupled to the transport of other molecules, such as glucose or sucrose, across the membrane (below right).

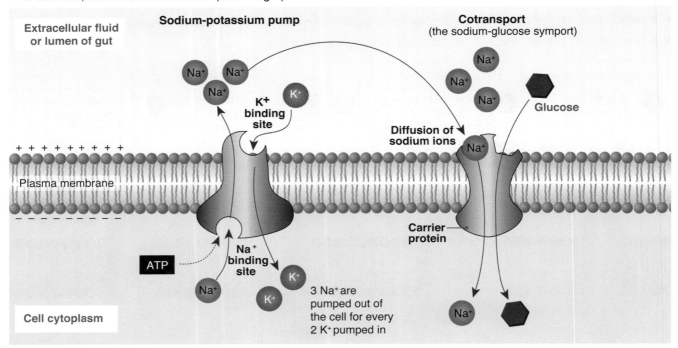

Sodium-potassium (Na^+/K^+) Pump

The Na^+/K^+ pump is a protein in the membrane that uses energy in the form of ATP to exchange sodium ions (Na^+) for potassium ions (K^+) across the membrane. The unequal balance of Na^+ and K^+ across the membrane creates large concentration gradients that can be used to drive transport of other substances (e.g. cotransport of glucose). The Na^+/K^+ pump also helps to maintain the right balance of ions and so helps regulate the cell's water balance.

Cotransport (coupled transport)

A specific carrier protein controls the entry of glucose into the intestinal epithelial cells from the gut where digestion is taking place. The energy for this is provided indirectly by a gradient in sodium ions. The carrier 'couples' the return of Na^+ down its concentration gradient to the transport of glucose into the cell. The process is therefore called cotransport. A low intracellular concentration of Na^+ (and therefore the concentration gradient for transport) is maintained by a sodium-potassium pump.

15. (a) Explain what is meant by cotransport: _____

_____

(b) How is cotransport used to move glucose into the intestinal epithelial cells? _____

_____

_____

16. During photosynthesis, the energy in excited electrons is used by proteins embedded in the thylakoid membranes in chloroplasts to pump H^+ ions (protons) across membrane. The protons flow back across the membrane via the protein ATP synthase, which uses their energy to power the production of ATP.

(a) Where does the energy to excite the electrons come from? _____

(b) How is the energy in the excited electrons used? _____

_____

(c) How is the result of (b) used? _____

_____

(d) How does the answer in (c) power ATP synthesis: _____

_____

_____

©2018 **BIOZONE** International
ISBN: 978-1-927309-55-1
Photocopying Prohibited

69 How Do We Know What Proteins Do?

ENGAGE: Missing a protein?

▸ Most of the proteins in your body do something useful, such as transporting molecules or signaling a cell to do something.

▸ Proteins are encoded by genes. If the gene that encodes a particular protein is not functioning, then that protein will not be produced.

▸ If a protein is not made then there is usually some physical result that we can record. For example, the mouse on the left of the photo is unable to make the protein leptin, a hormone that signals to the brain to stop eating. The mouse on the right has normal leptin production.

EXPLORE: Metabolic pathways

▸ A metabolic pathway is a series of biochemical reactions in which each subsequent step relies on the completion of the previous step. The reactions are usually mediated by enzymes (catalytic proteins) produced by specific genes. If a gene is not functioning, then the enzyme will not be produced and the metabolic pathway will stop.

▸ An interruption (failure) in a metabolic pathway often results in a specific phenotype because an intermediate in the pathway will accumulate. A study of the phenotype can show where in the pathway the interruption is located.

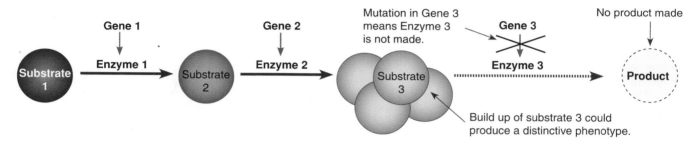

1. The diagram below shows a simple machine that carries out a several tasks. It represents a metabolic pathway. The gears labeled A, B, C, D, and E, are important for the machine to carry out its tasks.

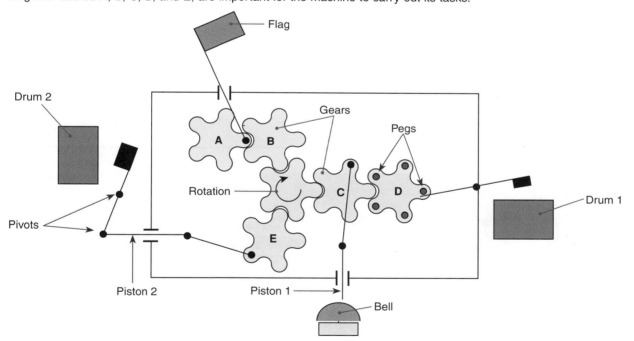

Work out which component(s) (A, B, C, D, or E) is non functional for each of the scenarios below.

(a) The bell does not sound: _____

(b) Drum 1 does not sound: _____

(c) Drum 2 does not sound: _____

2. Predict what would happen with a loss of function at gear B: _____

_____

©2018 BIOZONE International
ISBN: 978-1-927309-55-1
Photocopying Prohibited

SF CE LS1.A

EXPLORE: Animal models for defective genes

▶ We can find out the function of a gene and its associated protein by making the gene non-functional in a model animal and studying the associated effect.

▶ The technique pictured is called **gene knockout**. It produces a mouse with no functioning copies of a target gene and involves both genetic manipulation and selective breeding. Mice are used because they are the laboratory animal species most closely related to humans to which the gene knockout technique can easily be applied.

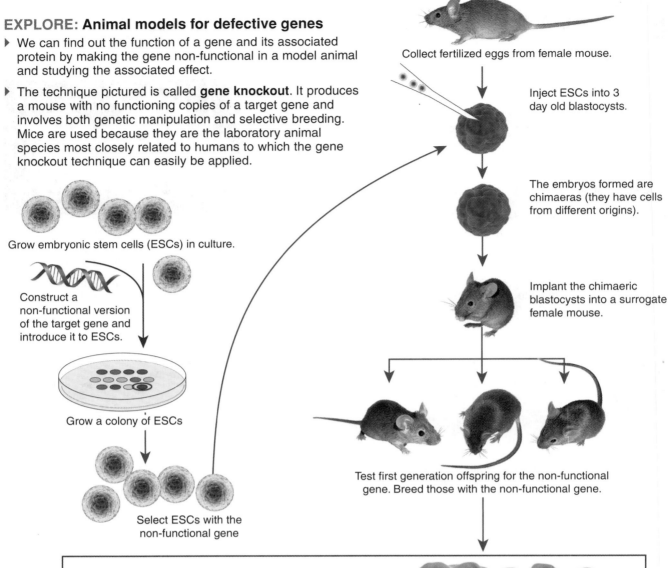

Collect fertilized eggs from female mouse.

Inject ESCs into 3 day old blastocysts.

The embryos formed are chimaeras (they have cells from different origins).

Implant the chimaeric blastocysts into a surrogate female mouse.

Grow embryonic stem cells (ESCs) in culture.

Construct a non-functional version of the target gene and introduce it to ESCs.

Grow a colony of ESCs

Select ESCs with the non-functional gene

Test first generation offspring for the non-functional gene. Breed those with the non-functional gene.

Second generation offspring are heterozygous for the non-functional gene. Selective breeding produces third generation offspring that are homozygous for the non-functional gene.

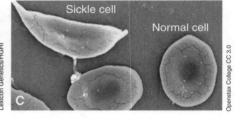

Many strains of gene knockout mice have been bred for research. The brown mouse has had a gene affecting hair growth knocked out. The black mouse has the normal gene (A). The knockout mouse above left was created as a model for obesity. There are over 50 strains of obesity-related gene knockout mice (B).

The BCL11A gene would usually cause sickle cell disease in mice. Knocking the gene out is helping to develop a drug treatment for sickle cell disease (C).

3. Why is gene knockout a useful technique for producing model animals? _____

_____

_____

4. Why are many strains of mice for the one phenotypic outcome (e.g. obesity) needed?_____

_____

_____

©2018 **BIOZONE** International
ISBN: 978-1-927309-55-1
Photocopying Prohibited

EXPLAIN: Finding the mutant gene

▶ Whenever a new mutation occurs there is always some part of the DNA responsible. Sometimes it is a single gene that is responsible, other times it is a large chunk of DNA involving multiple genes or even entire chromosomes.

▶ Being able to find the mutation in the DNA is useful because it can help us understand the processes affected in the body and the proteins involved when that process is working normally.

▶ Identifying the location of a DNA mutation can provide information on how to treat the disease it causes, especially now that technology provides ways to target (and edit) specific sections of DNA directly.

▶ The two common ways to find the location of a mutation are (i) by using crossover values from breeding experiments or (ii) by using marker probes in DNA analysis.

1. Crossover maps

▶ Genes found on the same chromosome are called linked genes, as they as tend to be inherited together. Crossing over occurs during meiosis (see activity 54) and results in linked genes being on separate chromosomes.

▶ Genes that are further away from each other cross over more often than genes that are close to each other. The amount of crossing over between genes is a direct measure of the distance between the genes on a chromosome. Once they are known, the distances between each gene can be used to produce a map of their relative positions.

▶ Crossover values have traditionally been obtained as a result of test crosses for the genes concerned.

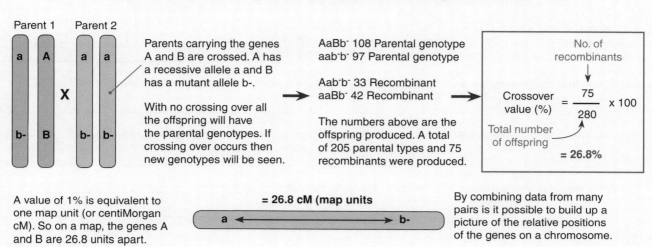

Parent 1   Parent 2

Parents carrying the genes A and B are crossed. A has a recessive allele a and B has a mutant allele b-.

With no crossing over all the offspring will have the parental genotypes. If crossing over occurs then new genotypes will be seen.

AaBb⁻ 108 Parental genotype
aab⁻b⁻ 97 Parental genotype

Aab⁻b⁻ 33 Recombinant
aaBb⁻ 42 Recombinant

The numbers above are the offspring produced. A total of 205 parental types and 75 recombinants were produced.

No. of recombinants

$$\text{Crossover value (\%)} = \frac{75}{280} \times 100$$

Total number of offspring

= 26.8%

A value of 1% is equivalent to one map unit (or centiMorgan cM). So on a map, the genes A and B are 26.8 units apart.

= 26.8 cM (map units

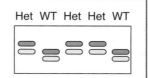

a ⟵⟶ b-

By combining data from many pairs is it possible to build up a picture of the relative positions of the genes on a chromosome.

2. Probing the DNA

▶ A second way to locate the position of a mutant gene is by looking at the DNA profiles of both the normal (wild type WT) and the mutant form. By comparing the banding pattern it is possible to identify the band that is likely to contain the mutant DNA. This part of the DNA can then be extracted and the DNA sequenced.

▶ Once the DNA is sequenced, a marker probe that binds to that DNA sequence is made. The DNA from each chromosome is mixed with the probe and if the probe binds to a particular site it is likely the DNA with the mutation is found there. This can be done with ever-increasing precision until the mutation's exact position on a chromosome can be determined.

Het WT Het Het WT

A mutation (Het) in the gene *Disc1* in mice leaves a distinctive banding pattern in the DNA profile (above).

DNA probes fluorescing to show the location of targeted genes (below).

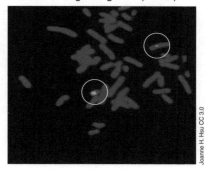

Joanne H. Hsu CC 3.0

5. Use the cross over values given to map the three genes onto the chromosome. Each division represents one map unit:

Gene map   Genes have the following crossover values:

X – G	18 %
B – G	3 %
B – X	15 %

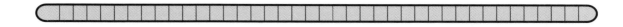

©2018 **BIOZONE** International
ISBN: 978-1-927309-55-1
Photocopying Prohibited

ELABORATE: Investigating the effect of a gene mutation

▶ *Arabidopsis thaliana* (thale cress) is a plant commonly used as a model organism in laboratory experiments due to its rapid life cycle and also its relatively small genome, which makes genetic analysis relatively simple.

Investigating loss of function in the *BON1* gene in *Arabidopsis thaliana*

▶ Both *BON1* and wild type plants were grown at 22°C under controlled conditions.

▶ After 4 weeks plants were measured for numerous features including stem length and diameter (table 1).

▶ A second set of plants were grown at 28°C. At this temperature there was no significant difference between the two strains of plants.

Measurements of wild type and *BON1* plants at 22°C

	Stem length	Stem diameter	Cell length	Cell width
Wild type	18 cm	1.0 cm	288 μm	18 μm
***BON1* mutant**	2.2 cm	0.38 cm	40 μm	12 μm

▶ Both wild type and *BON1* plants were grown at 22°C for three weeks and then at 28°C for 10 days. In this scenario, the wild type plant grew normally, whereas the *BON1* plant produced a tiny rosette but a normal length stem.

▶ Both strains of plant were then grown at 28°C for three weeks and then 22°C for 10 days. In this case, the wild type produced a normal sized rosette with a slightly shorter stem, whereas the *BON1* plant produced a normal sized rosette but a very short stem.

Stem —

Rosette

NASA

Arabidopsis thaliana

6. (a) What effect did growing the *BON1* plant at 22°C have on the morphology of the plant after four weeks?

_____

(b) What effect did growing the *BON1* plant at 28°C have on the morphology of the plant after four weeks?

_____

(c) What does this suggest about the product of the gene *BON1*? _____

_____

7. Which part of the plant develops first, the stem or the rosette? Explain how you know: _____

_____

_____

Planning your own investigation

You may want to investigate the effect of a certain mutation on the survival of an organism in different environments. Your teacher will provide you with the details of the organism and mutation involved. Suitable organisms include the plants *Brassica rapa* and *Arabidopsis thaliana*, the bacterium *E. coli*, and the fruit fly *Drosophila melanogaster*.

8. (a) What organism are you investigating? _____

(b) What is the name of the mutant gene you are investigating? _____

(c) What effect does the mutation have on your organism? _____

_____

(d) What conditions are you going to grow or raise your organism in (e.g. temperature range etc): _____

_____

_____

(e) What will you measure to quantify the effect of the mutation in those conditions? _____

_____

_____

(f) Attach any results from your investigation to this page.

©2018 **BIOZONE** International
ISBN: 978-1-927309-55-1
Photocopying Prohibited

ELABORATE: Cystic fibrosis

▶ Cystic fibrosis (CF) is an inherited disorder caused by a mutation of the CFTR gene. It is one of the most common lethal autosomal recessive conditions affecting white skinned people of European descent.

▶ The CFTR gene's protein product is a membrane-based protein that regulates chloride transport in cells. The ΔF508 mutation produces an abnormal CFTR protein, which degrades rapidly and so cannot take its position in the plasma membrane or perform its transport function. The ΔF508 mutation is the most common mutation causing CF.

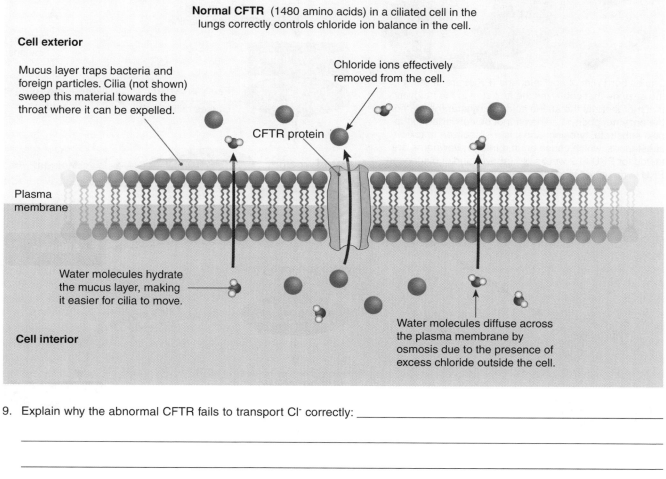

Normal CFTR (1480 amino acids) in a ciliated cell in the lungs correctly controls chloride ion balance in the cell.

Cell exterior

Mucus layer traps bacteria and foreign particles. Cilia (not shown) sweep this material towards the throat where it can be expelled.

Chloride ions effectively removed from the cell.

CFTR protein

Plasma membrane

Water molecules hydrate the mucus layer, making it easier for cilia to move.

Water molecules diffuse across the plasma membrane by osmosis due to the presence of excess chloride outside the cell.

Cell interior

9. Explain why the abnormal CFTR fails to transport Cl⁻ correctly: _____

_____

_____

10. Predict what will happen to chloride in the cell if a person has the ΔF508 mutation: _____

_____

_____

11. Predict what will happen to water movement into and out of the cell if a person has the ΔF508 mutation: _____

_____

_____

12. What will happen to the mucus layer if a person has the ΔF508 mutation? _____

_____

_____

13. Use your answer in 12 above to explain why CF patients have much higher incidences of bacterial lung infections:

_____

_____

14. The normal DNA sequence for the CFTR gene is CCG TGG TAA TTT CTT TTA TAG TAG AAA CCA CCA. The mutant sequence is CCG TGG TAA TTT CTT TTA TAG TAA CCA CCA. Identify the three bases deleted from the original DNA:

_____

©2018 **BIOZONE** International
ISBN: 978-1-927309-55-1
Photocopying Prohibited

EVALUATE: Metabolism of phenylalanine

▶ The metabolism of the essential amino acid phenylalanine is a well studied metabolic pathway. The effect of defective genes producing defective enzymes at each stage has been identified. A simplified pathway is shown below. The enzymes involved are in green ovals, with the result of defective enzymes shown in red.

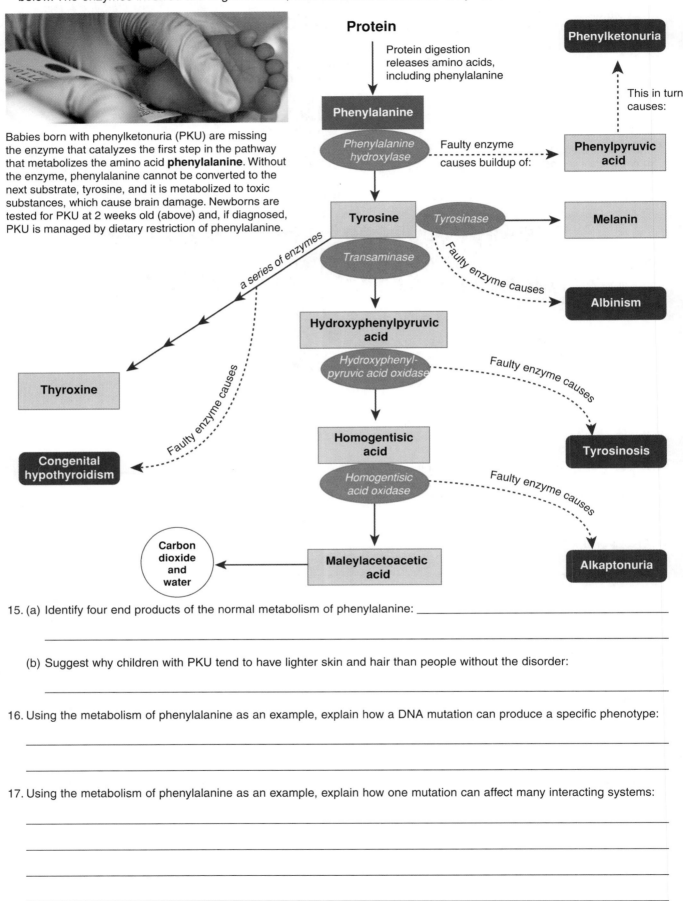

Babies born with phenylketonuria (PKU) are missing the enzyme that catalyzes the first step in the pathway that metabolizes the amino acid **phenylalanine**. Without the enzyme, phenylalanine cannot be converted to the next substrate, tyrosine, and it is metabolized to toxic substances, which cause brain damage. Newborns are tested for PKU at 2 weeks old (above) and, if diagnosed, PKU is managed by dietary restriction of phenylalanine.

15. (a) Identify four end products of the normal metabolism of phenylalanine: _____

_____

(b) Suggest why children with PKU tend to have lighter skin and hair than people without the disorder:

_____

16. Using the metabolism of phenylalanine as an example, explain how a DNA mutation can produce a specific phenotype:

_____

_____

17. Using the metabolism of phenylalanine as an example, explain how one mutation can affect many interacting systems:

_____

_____

_____

©2018 **BIOZONE** International
ISBN: 978-1-927309-55-1
Photocopying Prohibited

70 DNA Replication

ENGAGE: Models of DNA replication

▸ Like structural models of DNA, conceptual models of DNA replication can vary from simple to very complicated. These models have been built up after many years of study, with many different investigations.

1. (a) Using the internet, find models of DNA replication, print them out, and place them in the appropriate box

Simple replication showing a replication fork, the leading strand, and the lagging strand.

Replication showing replication fork, the leading strand, the lagging strand, Okazaki fragments, and DNA polymerase, primase, and helicase.

Replication showing replication fork, the leading strand, the lagging strand, Okazaki fragments, and DNA polymerase, primase, and helicase and a replication loop on the lagging strand.

(b) What do all three models of DNA replication show about the direction of replication? _____

_____

(c) How do the different models represent Okazaki fragments? _____

_____

_____

(d) Why is it useful for different models to have different levels of detail when showing DNA replication?

_____

_____

_____

©2018 **BIOZONE** International
ISBN: 978-1-927309-55-1
Photocopying Prohibited

EXPLORE: How DNA replicates

▶ In 1958 two scientists Matthew Meselson and Franklin Stahl carried out an experiment that showed how DNA replicates. At the time, there were several competing models of DNA replication. The two most viable models were conservative replication and semi-conservative replication.

▶ Meselson and Stahl grew bacteria in a solution containing a heavy nitrogen isotope (^{15}N) until their DNA contained only ^{15}N. The bacteria were then placed into a growth solution containing the nitrogen isotope (^{14}N), which is lighter than ^{15}N. After a set number of generation times, the DNA was extracted and centrifuged in a solution that provides a density gradient. Heavy DNA (containing only ^{15}N) sinks to the bottom, light DNA (containing only ^{14}N) rises to the top, and intermediate DNA (one light and one heavy strand) settles in the middle.

▶ In this activity, you will model the Meselson and Stahl experiment and determine the mode of DNA replication by comparing your results to the results of Meselson and Stahl.

2. In this part of the activity you will model the **semi-conservative** model of DNA replication. 2(a) is done for you:

(a) The DNA model below represents the DNA of the bacteria after growing in the solution containing the heavy nitrogen (generation 0). The relative mass of the DNA can be modeled by adding together the masses of the nitrogens.

(b) In the space below split the DNA along its center, then draw in the complementary base pairs to form two DNA strands. This represents the DNA after it has been grown in the solution containing the light nitrogen (^{14}N) for one generation.

(c) In the space below split the two DNA chains along their centers, then draw in the complementary base pairs. This represents the DNA after it has been growing in the solution containing the light nitrogen for two generations.

Generation 0

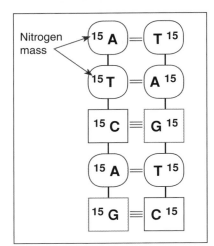

Mass of nitrogen in DNA strand:

_____150_____

> To help you record, label your DNA strands (i), (i, ii), (i, ii, iii, iv) as you replicate them.

Generation 1

Mass of nitrogen in DNA strands:

i: _____

ii: _____

Generation 2

Mass of nitrogen in DNA strands:

i: _____

ii: _____

iii: _____

iv: _____

(d) On the test tubes right mark a bar representing the mass of each of the DNA strands from (a), (b), and (c).

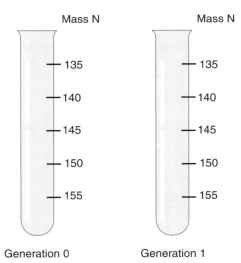

©2018 **BIOZONE** International
ISBN: 978-1-927309-55-1
Photocopying Prohibited

3. In this part of the activity you will model the **conservative** model of DNA replication:

(a) The DNA model of the below represents the DNA of the bacteria after it has grown in the solution containing the heavy nitrogen (generation 0). The relative mass of the DNA can be modeled by adding together the masses of the nitrogens.

(b) In the space below, redraw the original DNA strand and beside it draw a matching strand with bases using the light nitrogen. This represents the DNA after it has been grown in the solution containing the normal nitrogen (^{14}N) for one generation.

(c) In the space below, redraw the original DNA strand and the first generation strand. Beside these, draw matching strands with bases using the light nitrogen. This represents the DNA after it has been growing in the solution containing the normal nitrogen for two generations.

Generation 0

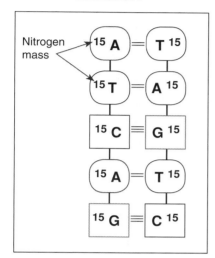

Nitrogen mass

Mass of nitrogen in DNA strand:

_____

To help you record, label your DNA strands (i), (i, ii), (i, ii, iii, iv) as you replicate them.

Generation 1

Mass of nitrogen in DNA strands:

i: _____

ii: _____

Generation 2

Mass of nitrogen in DNA strands:

i: _____

ii: _____

iii: _____

iv: _____

(d) On the appropriate test tubes right mark a bar representing the mass of each of the DNA strands from Generation 0, Generation 1, and Generation 2

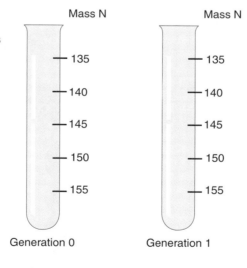

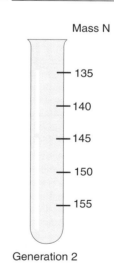

Mass N
— 135
— 140
— 145
— 150
— 155

Generation 0

Mass N
— 135
— 140
— 145
— 150
— 155

Generation 1

Mass N
— 135
— 140
— 145
— 150
— 155

Generation 2

4. In their experiment Meselson and Stahl obtained the following results: **Generation 0** = 100% "heavy DNA", **Generation 1** = 100%" intermediate DNA", **Generation 2** = 50%" intermediate DNA", 50% "light DNA".

From the results of your two modeling exercises, decide which matches the result of Meselson and Stahl. How does DNA replicate (conservatively or semi-conservatively)?

_____

_____

©2018 **BIOZONE** International
ISBN: 978-1-927309-55-1
Photocopying Prohibited

EXPLAIN: DNA replication

▸ Before a cell can divide, its DNA must be copied (replicated). **DNA replication** ensures that the two daughter cells receive identical genetic information.

▸ In eukaryotes, DNA is organized into structures called chromosomes in the nucleus.

▸ After the DNA has replicated, each chromosome consists of two chromatids, joined at the centromere. As you have seen already, the two chromatids will become separated during cell division to form two separate chromosomes.

▸ DNA replication takes place in the time between cell divisions. As you have seen in your model, the process is **semi-conservative**, and each chromatid contains half original (parent) DNA and half new (daughter) DNA.

DNA replication duplicates chromosomes

5. What is the purpose of DNA replication? _____

_____

_____

_____

_____

_____

DNA replication creates a chromosome with two identical chromatids

Parent chromosome

Replicated chromosome consists of two chromatids joined at the centromere.

Centromere links sister chromatids

Chromatid

The centromere keeps sister chromatids together in an organized way until they are separated before nuclear division.

6. What would happen if DNA was not replicated prior to cell division?

_____

_____

_____

_____

DNA replication is semi-conservative

3'   5'

Parent chromosome before replication. It is a double stranded DNA molecule.

Enzymes unzip the DNA and create a swivel point to unwind the DNA helix.

7. (a) What does a replicated chromosome look like?

_____

_____

(b) What is the purpose of the centromere?

_____

_____

_____

Original 'parent' DNA

New 'daughter' DNA

8. Explain what semi-conservative replication means:

_____

_____

_____

_____

_____

Enzymes catalyze the addition of bases during replication of the DNA strand. Enzymes proof-read and correct any copying errors made during DNA replication.

9. Why must the DNA be unzipped before it can be replicated?

_____

_____

_____

_____

3'   5'

DNA that will become one chromatid

3'   5'

DNA that will become the other chromatid

DNA replication is called **semi-conservative**. This is because each resulting DNA molecule is made up of one parent strand and one daughter strand of DNA.

©2018 **BIOZONE** International
ISBN: 978-1-927309-55-1
Photocopying Prohibited

Stages in DNA replication

▸ During DNA replication, new nucleotides (the units that make up the DNA molecule) are added at a region called the **replication fork**. The replication fork moves along the chromosome as replication progresses.

▸ Nucleotides are added in by complementary base-pairing: Nucleotide A is always paired with nucleotide T. Nucleotide C is always paired with nucleotide G.

▸ The DNA strands can only be replicated in one direction, so one strand has to be copied in short segments, which are joined together later.

▸ This whole process occurs simultaneously for each chromosome of a cell and the entire process is tightly controlled by enzymes.

10. How are the new strands of DNA lengthened?

_____

_____

_____

_____

11. What rule ensures that the two new DNA strands are identical to the original strand?

_____

_____

12. Why does one strand of DNA need to be copied in segments?

_____

_____

_____

_____

_____

_____

13. Describe three activities carried out by enzymes during DNA replication:

(a) _____

_____

(b) _____

_____

(c) _____

_____

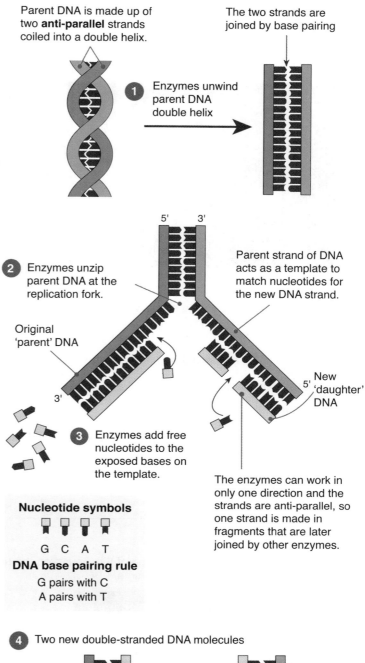

Parent DNA is made up of two **anti-parallel** strands coiled into a double helix.

The two strands are joined by base pairing

1 Enzymes unwind parent DNA double helix

5'　3'

2 Enzymes unzip parent DNA at the replication fork.

Parent strand of DNA acts as a template to match nucleotides for the new DNA strand.

Original 'parent' DNA

3'

3 Enzymes add free nucleotides to the exposed bases on the template.

5'　New 'daughter' DNA

The enzymes can work in only one direction and the strands are anti-parallel, so one strand is made in fragments that are later joined by other enzymes.

Nucleotide symbols

G　C　A　T

DNA base pairing rule
G pairs with C
A pairs with T

4 Two new double-stranded DNA molecules

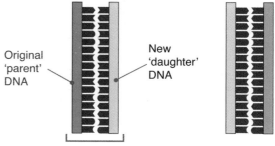

Original 'parent' DNA

New 'daughter' DNA

Enzymes are involved at every step of DNA replication. They unzip the parent DNA, add the free nucleotides to the 3' end of each single strand, join DNA fragments, and check and correct the new DNA strands.

14. **Group activity**: In a group using materials of your choice (e.g. paper clips and pipe cleaners, or Lego), develop a model to show how DNA replicates. Take photos or make drawings of the different stages of replication represented by your model. Use them to make a presentation of DNA replication for your classmates. You can choose your own method of presentation (e.g. video, poster, Powerpoint).

©2018 **BIOZONE** International
ISBN: 978-1-927309-55-1
Photocopying Prohibited

71 Growth and Repair of Cells

ENGAGE: Wounded!

▶ The photos below show a wound soon after it occurred, 12 days later, and 21 days later.

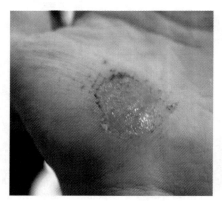

1 day after

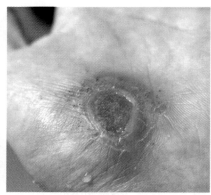

12 days after

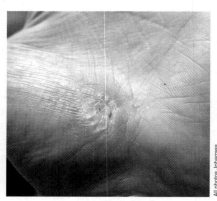

21 days after

All photos Jpbarrass

1. (a) Have you ever received a simple wound such as a graze or cut? _____

   (b) Did it heal quickly or slowly? _____

   (c) Did you use a bandage or wound dressing? _____

   (d) Did a scab form? _____

   (e) Was there any lasting scar? _____

2. (a) Looking at the images above, which part of the wound appears to heal first? _____

   _____

   (b) Suggest what is happening to the growth and division rate of the skin cells during the time the wound is healing:

   _____

EXPLORE: Healing

▶ The data below relates to the healing of the wound on the hand shown above. It shows the approximate area of the unhealed region of the wound over time:

Time (days)	Area unhealed (mm²)
0.02	350
0.66	350
1	340
2	252
12	171
13	135
17	72
18	64
21	16
30	0

3. Plot the data on the grid provided:

4. Describe the trend in the wound size as shown by the data:

   _____

   _____

   _____

     LS1.B SF SC

©2018 **BIOZONE** International
ISBN: 978-1-927309-55-1
Photocopying Prohibited

EXPLAIN: Cell growth and repair

▸ Mitotic cell division produces new genetically identical daughter cells from a parent cell. This type of cell division has three purposes:

▸ **Growth**: Multicellular organisms grow from a single fertilized cell into a mature organism. Depending on the organism, the mature form may consist of several thousand to several trillion cells. These cells, which form the building blocks of the body, are called somatic cells.

▸ **Repair**: Damaged and old cells are replaced with new cells.

▸ **Asexual reproduction**: Some unicellular eukaryotes (such as yeasts) and some multicellular organisms (e.g. *Hydra*) reproduce asexually by mitotic division.

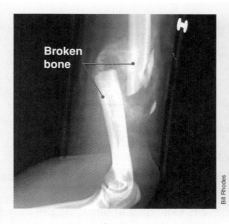

Zygote   *Embryo*   *Adult*

Matthias Zepper

Asexual reproduction

Some simple eukaryotes reproduce asexually by mitosis. Yeasts (e.g. baker's yeast) reproduce by budding. The parent cell buds to form a daughter cell (right). The daughter cell continues to grow, then separates from the parent cell. The production of CO_2 by the growing yeast is responsible for making the bread rise.

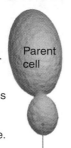

Parent cell

Daughter cell

Growth

Multicellular organisms develop from a single fertilized egg cell (zygote) and grow by increasing in cell numbers. Cells complete a cell cycle, in which the cell copies its DNA and then divides to produce two identical cells. When the organism is growing, the number of new cells produced exceeds the number of cells dying. Organisms, such as the 12 day old mouse embryo (above, middle), grow by increasing their total cell number and the cells become specialized to form tissues and organs as part of development. Cell growth is highly regulated. Once the mouse reaches its adult size, physical growth stops and the number of cell deaths equals the number of new cells produced.

Repair

Mitosis is vital in the repair and replacement of damaged cells. When you break a bone or graze your skin, new cells are generated to repair the damage. Some organisms, like the sea star (far right) are able to generate new limbs if they are broken off.

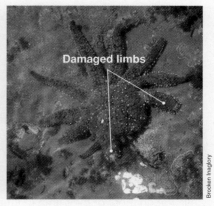

Broken bone

Damaged limbs

Bill Rhodes

Brocken Inaglory

5. Use examples to describe the role of mitotic cell division in:

   (a) Growth of an organism: _____

   _____

   (b) Replacement of damaged cells: _____

   _____

   (c) Asexual reproduction: _____

   _____

6. Explain the importance of mitotic cell division in repairing the wounded hand shown on the opposite page:

   _____

   _____

©2018 **BIOZONE** International
ISBN: 978-1-927309-55-1
Photocopying Prohibited

72 The Cell Cycle

ENGAGE: Cell division

▶ The image below shows cells dividing in the root tip of an onion

Dividing cell

Chromosomes

Cell

Nucleus

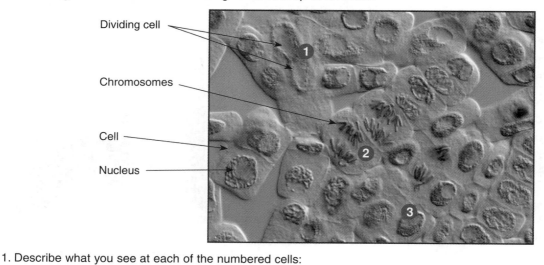

1. Describe what you see at each of the numbered cells:

(a) 1: _____

_____

_____

(b) 2: _____

_____

_____

(c) 3: _____

_____

_____

2. How many cells are in a similar stage to number 2 in the photograph? _____

EXPLORE: Stages of the cell cycle

▶ The life cycle of a eukaryotic cell is called the cell cycle. The cell cycle can be divided into two broad phases, called interphase and M-phase. Specific activities occur in each phase.

▶ Cells spend most of their time in interphase. During interphase, the cell increases in size, carries out its normal activities, and replicates its DNA in preparation for cell division. During M-phase the cell divides the replicated chromosomes equally (mitosis) and produces two new cells.

3. The onion cells in the photograph above have a cell cycle of about 24 hours. Use the photo to decide which phase, interphase or M-phase, appears to occur for the longer period of time in the cell cycle. Fill in the table and use it to justify your answer.

Stage	No. of cells	% of total cells	Estimated time in stage
Interphase			
M-phase			
Total		100	

_____

_____

_____

_____

_____

_____

_____

_____

_____

_____

    LS1.B SC

©2018 **BIOZONE** International
ISBN: 978-1-927309-55-1
Photocopying Prohibited

EXPLORE: Modeling the cell cycle

4. Use your answer and the information from the previous page to develop a simple model of the cell cycle:

Interphase

During interphase, the cell increases in size, carries out its normal activities, and replicates its DNA in preparation for cell division. Interphase is not a stage in mitosis. Interphase is divided into three stages:

▶ The first gap phase (G_1). Cell increases in size and makes the mRNA and proteins needed for DNA synthesis.

▶ The S-phase (S). Chromosome replication (DNA synthesis).

▶ The second gap phase (G_2). Rapid cell growth and protein synthesis. Cell prepares for mitosis.

Mitosis and cytokinesis (M-phase)

Mitosis (nuclear division) and cytokinesis occur during M-phase. Cytokinesis is distinct from nuclear division.

▶ During mitosis, the cell nucleus (containing the replicated DNA) divides in two, with equal distribution of chromosomes to each half.

▶ Cytokinesis occurs at the end of M-phase (although in animal cells it can be underway before nuclear division is complete). The cell cytoplasm divides and two new daughter cells are produced.

EXPLORE: Refining your model of the cell cycle

5. Produce a final model of the cell cycle by including the information above:

©2018 **BIOZONE** International
ISBN: 978-1-927309-55-1
Photocopying Prohibited

EXPLORE: The cell cycle is regulated

▶ Regulatory checkpoints are built into the cell cycle to ensure that the cell is ready to proceed from one phase to the next. The failure of these systems can lead to cancer.

Checkpoints during the cell cycle

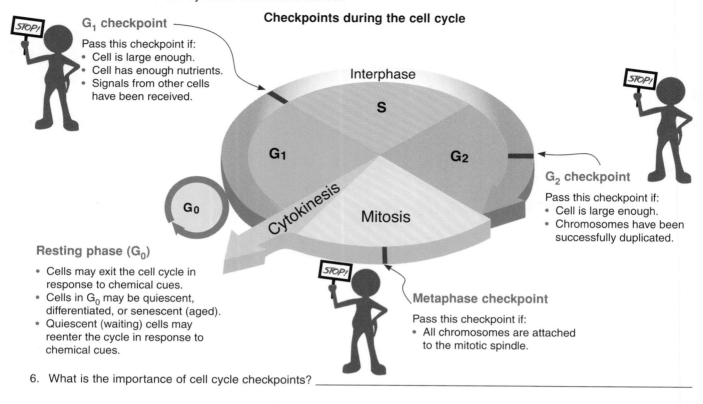

G_1 checkpoint

Pass this checkpoint if:
• Cell is large enough.
• Cell has enough nutrients.
• Signals from other cells have been received.

G_2 checkpoint

Pass this checkpoint if:
• Cell is large enough.
• Chromosomes have been successfully duplicated.

Resting phase (G_0)

• Cells may exit the cell cycle in response to chemical cues.
• Cells in G_0 may be quiescent, differentiated, or senescent (aged).
• Quiescent (waiting) cells may reenter the cycle in response to chemical cues.

Metaphase checkpoint

Pass this checkpoint if:
• All chromosomes are attached to the mitotic spindle.

6. What is the importance of cell cycle checkpoints? _____

_____

_____

EXPLORE: Mitosis is a stage of the cell cycle

▶ M-phase (mitosis and cytokinesis) is the part of the cell cycle in which the parent cell divides in two to produce two genetically identical daughter cells (right).

▶ Mitosis results in the separation of the nuclear material and division of the cell. It does not change the chromosome number.

▶ Mitosis is one of the shortest stages of the cell cycle. When a cell is not undergoing mitosis, it is said to be in interphase.

▶ In animals, mitosis takes place in the somatic cells. Somatic cells are any cell of the body except sperm and egg cells.

▶ In plants, mitosis takes place in the meristems. The meristems are regions of growth (where new cells are produced), such as the tips of roots and shoots.

7. Briefly outline the events in mitosis: _____

_____

_____

_____

_____

_____

8. A cell with 10 chromosomes undergoes mitosis.

(a) How many chromosomes does each daughter cell have? _____

(b) The genetic material of the daughter cells is the same as / different to the parent cell (delete one).

Mitosis produces identical daughter cells

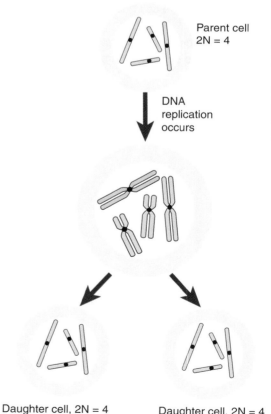

Parent cell
2N = 4

DNA replication occurs

Daughter cell, 2N = 4   Daughter cell, 2N = 4

©2018 BIOZONE International
ISBN: 978-1-927309-55-1
Photocopying Prohibited

EXPLORE: Stages in mitosis

Mitosis is a continuous process but it is helpful to recognize stages. Some of the key events are listed below:

▶ At the beginning of mitosis the nuclear membrane breaks down. The replicated chromosomes condense into the X structure that they are typically represented as (they only exist like this for a short time during mitosis).

▶ The condensed chromosomes line up along the center (equator) of the cell.

▶ In animal cells, the centrosome divides and a copy migrates to each end (pole) of the cell. Centrioles (within the centrosomes) form the spindle (protein fibers). Plant cells lack centrioles, and the spindle is organized by structures associated with the plasma membrane. The spindle fibers attach to the chromosomes.

▶ The spindle fibers contract to pull the individual chromatids of the condensed chromosomes apart.

▶ The chromatids move to opposite ends of the cells. Two new nuclei form.

▶ The cell divides producing two new identical cells (cytokinesis).

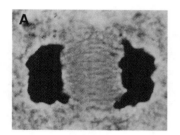

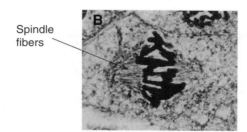

Spindle fibers

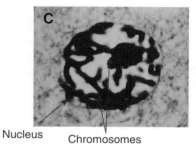

Nucleus   Chromosomes

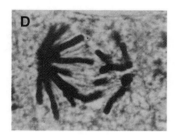

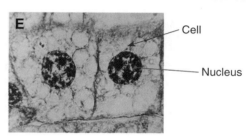

Cell

Nucleus

9. Use the description above to place the images of mitosis in the correct order: _____

Cytokinesis in animal cells

Animal cell cytokinesis (below left) begins shortly after the sister chromatids have separated during mitosis. A ring of microtubules assembles in the middle of the cell, next to the plasma membrane, constricting it to form a cleavage furrow. In an energy-using process, the cleavage furrow moves inwards, forming a region of separation where the two cells will divide once nuclear division has finished.

Cytokinesis in plant cells

In plant cells (below right), cytokinesis) involves construction of a cell plate (a precursor of the new cell wall) in the middle of the cell. The cell wall materials are delivered by vesicles derived from the Golgi. Once they have delivered their cell wall material, the vesicles join together to become the plasma membranes of the new cell surfaces.

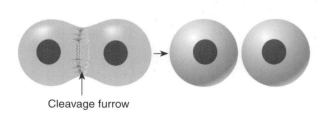

Cleavage furrow

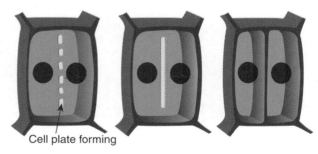

Cell plate forming

10. (a) What is the purpose of cytokinesis? _____

_____

(b) Describe the differences between cytokinesis in an animal cell and a plant cell: _____

_____

_____

_____

_____

©2018 **BIOZONE** International
ISBN: 978-1-927309-55-1
Photocopying Prohibited

EXPLAIN: Modeling mitosis

11. Use the information on the previous pages to model mitosis in an animal cell using pipe cleaners and yarn. Use four chromosomes for simplicity (2N = 4). Photo 1 (below) can be used as a starting point for your model. It represents a cell in interphase before mitosis begins. The circular structures are the replicated centrosomes. Photograph or film each stage. Photo 2 shows completion of mitosis. Fill in the space with photos as you model the process of mitosis:

Name the labeled structures in photo 1 before you begin. Remember to label your photos as your place them on the page.

A: _____

B: _____

C: _____

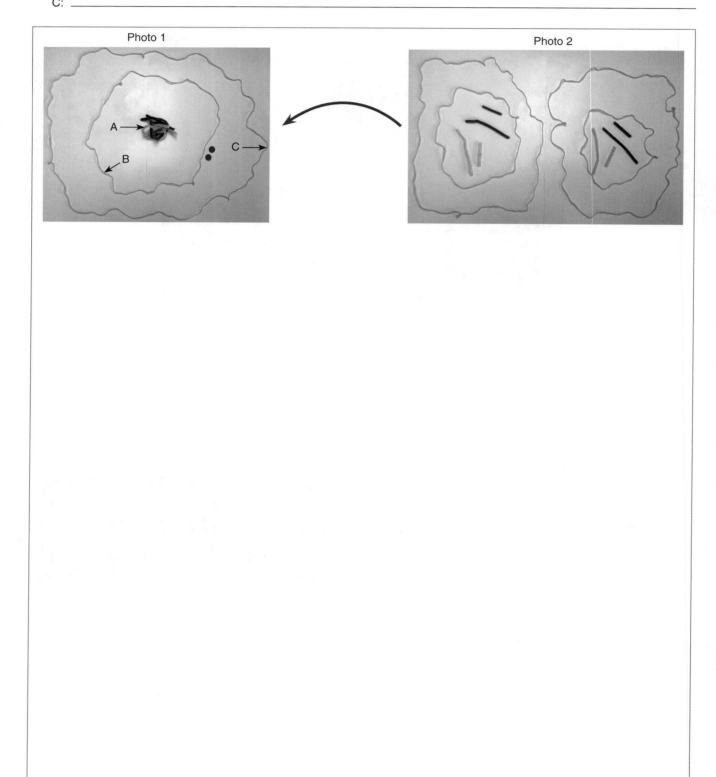

Photo 1

Photo 2

©2018 **BIOZONE** International
ISBN: 978-1-927309-55-1
Photocopying Prohibited

ELABORATE: **Factors that influence mitosis**

▸ The rate of mitosis can be influenced by environmental factors, such as the presence of carbohydrate-binding proteins called lectins. These have been shown to increase the rate of mitosis in blood cells.

▸ Some students wanted to determine if PHA-M (a lectin) would increase the level of mitosis in the root tip cells of onion plants. They exposed the roots of some onion plants to a 50 mg L^{-1} solution of PHA-M and others to a control solution (water).

▸ After an exposure period of 2 days, they prepared and stained the samples so they could count the number of cells in mitosis and interphase. The data they obtained is shown in table 1. The percentage of cells in interphase or mitosis for the control and treated samples are presented in table 2.

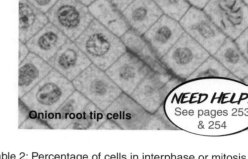

Onion root tip cells

NEED HELP?
See pages 253 & 254

Table 1: Class data

	Number of cells		
	Interphase	**Mitosis**	**Total**
Control	1292	302	1598
Treated	1406	306	1712

Table 2: Percentage of cells in interphase or mitosis

	Percentage of cells	
	Control	**Treated**
Interphase cells	81	82
Mitotic cells	19	18

12. Generate a null hypothesis and an alternative hypothesis for this experiment:

(a) Null hypothesis (H$_0$): _____

_____

(b) Hypothesis (H$_A$): _____

_____

▸ A chi-squared test (χ^2) was used to determine if the null hypothesis should be accepted or rejected at P=0.05 (see table 3). Expected values (E) were calculated by applying the control percentages in table 2 (blue column) to the treated total (i.e. 0.81 x 1712 = 1386). At the end of the calculation, a chi-squared value (Σ(O-E)^2 / E) is generated.

▸ The χ^2 value is compared to a critical value of χ^2 for the appropriate degrees of freedom (number of groups - 1). For this test it is 1 (2-1=1). For this data set, **the critical value is 3.84** (at P=0.05 and 1 degree of freedom). If χ^2 is < 3.84, the result in not significant and H$_0$ cannot be rejected. If χ^2 is ≥ 3.84, H$_0$ can be rejected in favor of H$_A$.

Table 3: Calculation of chi-squared value

	Observed (O)	**Expected (E)**	**(O - E)**	**(O - E)^2**	**(O -E)^2 / E**
Interphase cells	1406	1386			
Mitosis cells	306	325			
				Σ (O -E)^2 / E	

13. Calculate the chi-squared value by completing table 3. χ^2 = _____

14. (a) Should the null hypothesis be accepted or rejected?_____

(b) Is this the result you would have expected? Explain: _____

_____

_____

_____

15. Suggest further investigations to verify the results presented here. Use more paper and attach it here if you wish:

_____

_____

©2018 **BIOZONE** International
ISBN: 978-1-927309-55-1
Photocopying Prohibited

73 Keeping in Balance

ENGAGE: Changes during exercise

▸ The body must be able to respond to changes to maintain homeostasis. In the investigation below you will work in groups of three to see how exercise affects breathing and heart rates. Choose one person to carry out the exercise and one person each to record heart rate and breathing rate.

▸ Heart rate (beats per minute) is obtained by measuring the pulse (right) for 15 seconds and multiplying by four.

▸ Breathing rate (breaths per minute) is measured by counting the number of breaths taken in 15 seconds and multiplying it by four.

Measuring the carotid pulse

Gently press your index and middle fingers, not your thumb, against the carotid artery in the neck (just under the jaw) or the radial artery (on the wrist just under the thumb) until you feel a pulse.

Measuring the radial pulse

Procedure

Resting measurements

Have the person carrying out the exercise sit down on a chair for 5 minutes. They should try not to move. After 5 minutes of sitting, measure their heart and breathing rates. Record the resting data on the table (right).

Exercising measurements

Choose an exercise to perform. Some examples include step ups onto a chair, skipping rope, jumping jacks, and running in place.

Begin the exercise, and take measurements after 1, 2, 3, and 4 minutes of exercise. The person exercising should stop just long enough for the measurements to be taken. Record the results in the table.

Post exercise measurements

After the exercise period has finished, have the exerciser sit down in a chair. Take their measurements 1 and 5 minutes after finishing the exercise. Record the results on the table, right.

	Heart rate (beats/minute)	Breathing rate (breaths/minute)
Resting		
1 minute		
2 minutes		
3 minutes		
4 minutes		
1 minute after		
5 minutes after		

1. (a) Graph your results on separate piece of paper. You will need to use one vertical axis for heart rate and another for breathing rate. When you have finished answering the questions below, attach it to this page.

    (b) Analyze your graph and describe what happened to heart rate and breathing rate during exercise:

    _____

    _____

    _____

2. (a) Describe what happened to heart rate and breathing rate after exercise: _____

    _____

    _____

    (b) Why did this change occur? _____

    _____

    _____

    _____

3. Design an experiment to compare heart and breathing rates pre- and post exercise in students that do and do not regularly participate in a sport. Attach your design to this page. What do you predict will be the result and why?

        LS1.A SSM SC

©2018 **BIOZONE** International
ISBN: 978-1-927309-55-1
Photocopying Prohibited

EXPLORE: Significance of changes during exercise

▶ We have seen on the previous page that exercise affects heart rate. How do we know that the response we see is not just representative of the normal variation in a person's heart rate over time?

▶ We can evaluate the effect objectively by examining a larger group of people (a larger sample of the population) and comparing the pre- and post exercise heart rates objectively using a statistical measure called the 95% confidence interval of the mean (95% CI).

▶ The 95% CI is usually plotted either side of the mean (mean ± 95%CI). This tells you that, on average, 95 times out of 100, the true population mean will lie between these values.

x heart rate (beats per minute)	
Pre-exercise (A)	Post exercise (B)
72	102
116	175
79	96
97	100
90	132
67	158
115	152
82	141
95	113
82	136
77	130
$\bar{x}_A =$	$\bar{x}_B =$
$s_A =$	$s_B =$
$SE_A =$	$SE_B =$
$95\%CI_A=$	$95\%CI_B=$
$n_A = 11$	$n_B = 11$
$df = 10$	$df = 10$
$t_{(P=0.05)} = 2.228$	$t_{(P=0.05)} = 2.228$

▶ Students investigated how exercise affects heart rate by comparing their resting heart rate before exercise to their post exercise heart rate immediately after exercising.

▶ The students sat quietly for five minutes, then measured and recorded their heart rate to obtain their pre-exercise heart rate. They then performed star jumps as fast as possible for one minute. Immediately at the end of this, they measured their heart rate again. This was their post exercise heart rate. The class results are presented left:

Calculating 95% CI for pre- and post exercise data

Step 1: Calculate the sample mean ($\bar{x}$)

Calculate the sample mean for pre- and post exercise (A and B) and enter the values in the table left. If working longhand, attach your working to this page.

Step 2: Calculate standard deviation (s)

Calculate the sample standard deviation for pre- and post exercise and enter the values in the table left. If working longhand, attach your working to this page.

Step 3: Calculate standard error (SE)

Standard error is simply the standard deviation, divided by the square root of the sample size (n). Calculate the standard error of the mean for pre- and post exercise and enter the values in the table left. If working longhand, show your working below:

$$SE = \frac{s}{\sqrt{n}}$$

NEED HELP? See Activity 100

Step 4: Calculate 95% confidence interval and plot your data

The 95% CI of the mean is given by the standard error multiplied by the value of t at $P = 0.05$ (from a t table) for the appropriate degrees of freedom (df) for your sample (this is given by $n - 1$). Calculate the 95% CI of the mean for pre- and post exercise and enter the values in the table. The t value at $P = 0.05$ has been provided for you.

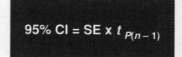

$$95\% \text{ CI} = SE \times t_{P(n-1)}$$

Plot the **mean values** for pre- and post exercise data as a column graph on the grid below. Add the 95% CI to each of the two columns as error bars, above and below the mean. Allow enough vertical scale to accommodate your error bars.

4. Complete steps 1-4 above (including a plot of the results).

5. (a) Do the error bars indicating the 95% confidence intervals of the two means overlap?

_____

(b) What does this tell you about the difference between the two data sets?

_____

_____

_____

(c) What can you conclude about the experiment from this? _____

_____

_____

_____

©2018 **BIOZONE** International
ISBN: 978-1-927309-55-1
Photocopying Prohibited

EXPLORE: Maintaining homeostasis

▶ An organism must constantly regulate its internal environment in order to carry out essential life processes, such as growing, respiring, feeding, removing wastes, and responding to the environment. Changes outside normal limits for too long can stop the body systems working properly and can result in illness or death.

▶ Homeostasis relies on monitoring all the information received from the internal and external environment and coordinating appropriate responses. This involves many different organ systems working together.

▶ Most of the time, an organism's body systems are responding to changes at the subconscious level, but sometimes homeostasis is achieved by changing a behavior (e.g. finding shade if the temperature is too high).

Negative feedback systems

Negative (counterbalancing) feedback is a control system that maintains the body's internal environment at a relatively steady state. It has a stabilizing effect by discouraging variations from a set point. When variations are detected, negative feedback returns internal conditions back to a steady state (right).

Changes are detected through receptors. The information is processed by the central nervous system and changes to the system are achieved through effectors (muscles and glands).

Most body systems achieve homeostasis through negative feedback. Body temperature, blood glucose levels, and blood pressure are all controlled by negative feedback mechanisms.

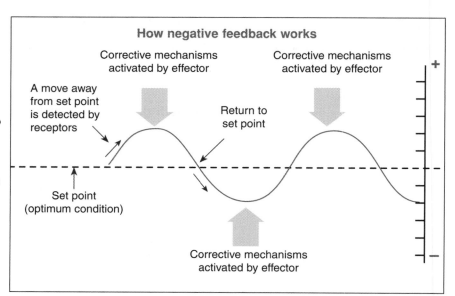

How negative feedback works

Corrective mechanisms activated by effector

Corrective mechanisms activated by effector

A move away from set point is detected by receptors

Return to set point

Set point (optimum condition)

Corrective mechanisms activated by effector

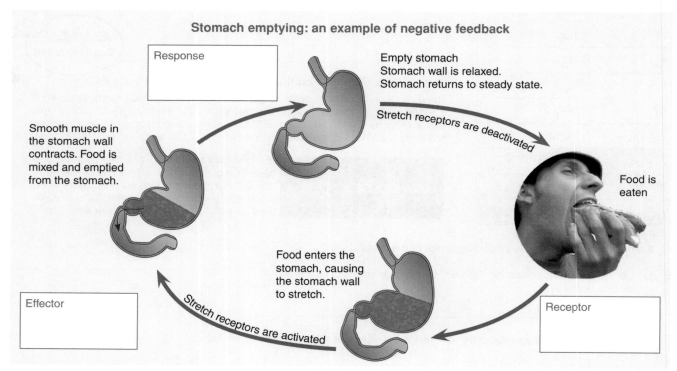

Stomach emptying: an example of negative feedback

Response

Empty stomach
Stomach wall is relaxed.
Stomach returns to steady state.

Stretch receptors are deactivated

Smooth muscle in the stomach wall contracts. Food is mixed and emptied from the stomach.

Food is eaten

Effector

Food enters the stomach, causing the stomach wall to stretch.

Stretch receptors are activated

Receptor

6. How does the behavior of a negative feedback system maintain homeostasis? _____

_____

_____

_____

7. (a) On the diagram of stomach emptying, name the receptor, effector, and response in the spaces provided.

(b) What is the steady state for this example? _____

©2018 **BIOZONE** International
ISBN: 978-1-927309-55-1
Photocopying Prohibited

Positive feedback systems

▶ Positive (reinforcing) feedback mechanisms amplify (increase) or speed up a physiological response, usually to achieve a particular outcome. Examples of positive feedback include fruit ripening, fever, blood clotting, childbirth (labor) and lactation (production of milk). A positive feedback mechanism stops when the end result is achieved (e.g. the baby is born, a pathogen is destroyed by a fever, or ripe fruit falls off a tree).

▶ Positive feedback is less common than negative feedback because it creates an escalation in response, which is unstable. This response can be dangerous (or even cause death) if it is prolonged.

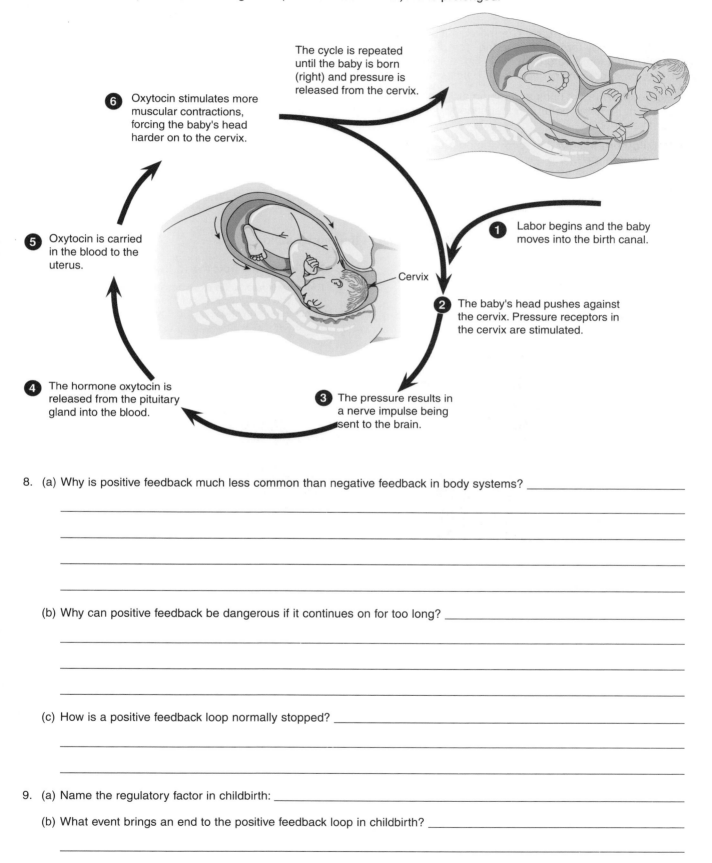

6 Oxytocin stimulates more muscular contractions, forcing the baby's head harder on to the cervix.

The cycle is repeated until the baby is born (right) and pressure is released from the cervix.

5 Oxytocin is carried in the blood to the uterus.

1 Labor begins and the baby moves into the birth canal.

Cervix

2 The baby's head pushes against the cervix. Pressure receptors in the cervix are stimulated.

4 The hormone oxytocin is released from the pituitary gland into the blood.

3 The pressure results in a nerve impulse being sent to the brain.

8. (a) Why is positive feedback much less common than negative feedback in body systems? _____

_____

_____

_____

_____

(b) Why can positive feedback be dangerous if it continues on for too long? _____

_____

_____

_____

(c) How is a positive feedback loop normally stopped? _____

_____

_____

9. (a) Name the regulatory factor in childbirth: _____

(b) What event brings an end to the positive feedback loop in childbirth? _____

_____

©2018 **BIOZONE** International
ISBN: 978-1-927309-55-1
Photocopying Prohibited

EXPLORE: Thermoregulation

Source of heat: Animals can generally be divided into two groups depending on *where* they get the energy to maintain their body temperature. These two groups are:

▶ **Ectotherms** depend on external sources of heat (from the environment) for their heat energy (e.g. heat from the sun).

▶ **Endotherms** generate most of their body heat from internal metabolic processes.

Constancy of temperature: Thermoregulation refers to the regulation of body temperature in the face of changes in environmental temperature. Animals can also be divided into two groups depending on *how they regulate* body temperature:

▶ **Homeotherms** (all birds and mammals) maintain a constant body temperature independently of environmental variation (right).

▶ **Poikilotherms** allow their body temperature to vary with the temperature of the environment (right). Most fish and all amphibians cannot regulate body temperature at all, but most reptiles use behavior both to warm up and to avoid overheating.

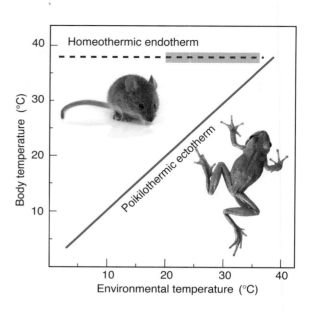

Mechanisms of thermoregulation

Homeothermic endotherm (mammal)

Always thermoregulate, largely by physiological mechanisms. This requires a large amount of energy, especially when outside the range of their normal body temperature.

Wool, hair, or fur traps air next to the skin providing an insulating layer to reduce heat loss and slow heat gain.

Panting and sweating cool through evaporation. Mammals usually sweat or pant but not both.

Heat can be generated by shivering.

In cold weather, many mammals cluster together to retain body heat.

Poikilothermic ectotherm (reptile)

Thermoregulate at the extremes of their temperature range. The energetic costs of this are much lower than for homeotherms because the environment provides heat energy for warming.

Increasing blood flow to the surface can help lose heat quickly.

Basking in the sun is common in lizards and snakes. The sun warms the body up and they seek shade to cool down.

Some lizards reduce points of contact with hot ground (e.g. standing on two legs instead of four) reducing heat uptake via conduction.

10. Why are the movements of many ectotherms slow in the early morning? _____

_____

_____

_____

11. (a) The graph (top) shows body temperature variations in a mammal and an amphibian with change in environmental temperature. Compare and explain their different responses:

_____

_____

(b) The blue shaded area on the graph marks the region where the energy cost of thermoregulation is lowest for the mouse. Why is the mouse using less energy to thermoregulate in this temperature range?

_____

_____

©2018 **BIOZONE** International
ISBN: 978-1-927309-55-1
Photocopying Prohibited

EXPLAIN: How is body temperature regulated?

▶ In humans, the temperature regulation center is a region of the brain called the **hypothalamus**. It has thermoreceptors that monitor core body temperature and has a 'set-point' temperature of 36.7°C.

▶ The hypothalamus acts like a thermostat. It registers changes in the core body temperature and also receives information about temperature change from thermoreceptors in the skin. It then coordinates the necessary nervous and hormonal responses to counteract the changes and restore normal body temperature, as shown in the diagram below.

▶ When normal temperature is restored, the corrective mechanisms are switched off. This is an example of a negative feedback regulation.

▶ Infection can reset the set-point of the hypothalamus to a higher temperature. Homeostatic mechanisms then act to raise the body temperature to the new set point, resulting in a fever (right).

Fever is an important defense against infection, but if the body temperature rises much above 42°C, a dangerous positive feedback loop can begin, making the body produce heat faster than it can get rid of it.

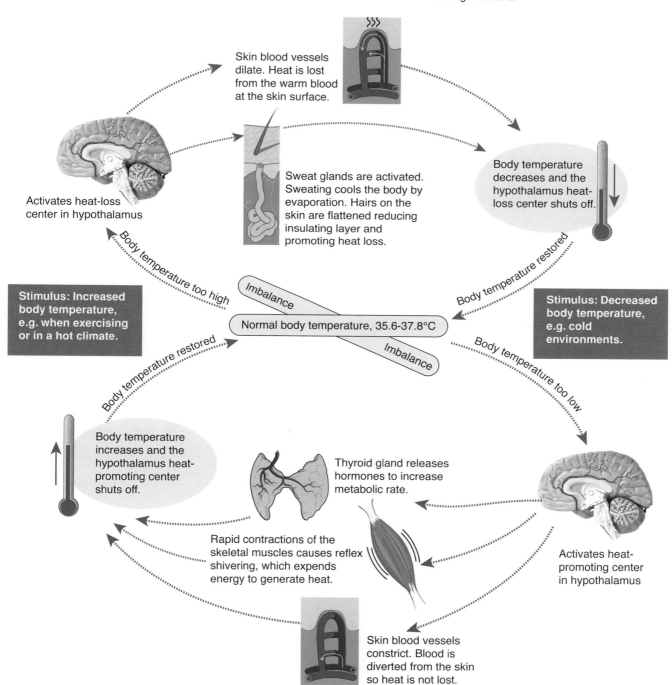

Skin blood vessels dilate. Heat is lost from the warm blood at the skin surface.

Sweat glands are activated. Sweating cools the body by evaporation. Hairs on the skin are flattened reducing insulating layer and promoting heat loss.

Body temperature decreases and the hypothalamus heat-loss center shuts off.

Activates heat-loss center in hypothalamus

Body temperature too high

Imbalance

Normal body temperature, 35.6-37.8°C

Imbalance

Body temperature restored

Stimulus: Increased body temperature, e.g. when exercising or in a hot climate.

Stimulus: Decreased body temperature, e.g. cold environments.

Body temperature restored

Body temperature too low

Body temperature increases and the hypothalamus heat-promoting center shuts off.

Thyroid gland releases hormones to increase metabolic rate.

Rapid contractions of the skeletal muscles causes reflex shivering, which expends energy to generate heat.

Activates heat-promoting center in hypothalamus

Skin blood vessels constrict. Blood is diverted from the skin so heat is not lost.

©2018 **BIOZONE** International
ISBN: 978-1-927309-55-1
Photocopying Prohibited

Thermoregulation in newborns

▸ Newborn babies cannot fully thermoregulate until six months of age. They can become too cold or too hot very quickly.

▸ Newborns minimize heat loss by reducing the blood supply to the periphery (skin, hands, and feet). This helps to maintain the core body temperature. Increased brown fat activity and general metabolic activity generates heat. Newborns are often dressed in a hat to reduce heat loss from the head, and tightly wrapped to trap heat next to their bodies.

▸ Newborns lower their temperature by increasing peripheral blood flow. This allows heat to be lost, cooling the core temperature. Newborns can also reduce their body temperature by sweating, although their sweat glands are not fully functional until four weeks after birth.

Newborns cannot shiver to produce heat.

A baby's body surface is three times greater than an adult's. There is greater surface area for heat to be lost from.

Heat losses from the head are high because the head is very large compared to the rest of the body.

Newborns have thin skin and blood vessels that run close to the skin. These features allow heat to be lost easily.

Newborns have very little white fat beneath their skin to insulate them against heat loss.

12. (a) Where is the temperature regulation center in humans located? _____

(b) How does it carry out this role? _____

_____

13. Describe the role of the following in maintaining a constant body temperature in humans:

(a) The skin: _____

_____

(b) The muscles: _____

_____

(c) The thyroid gland: _____

_____

14. How is negative feedback involved in keeping body temperature within narrow limits? _____

_____

_____

_____

15. (a) Why does infection result in an elevated core body temperature? _____

_____

(b) What is the purpose of this? _____

_____

(c) Explain why a prolonged fever can be fatal: _____

_____

16. (a) What features of a newborn cause it to lose heat quickly? _____

_____

(b) What mechanisms do newborns have to control body temperature? _____

_____

_____

©2018 **BIOZONE** International
ISBN: 978-1-927309-55-1
Photocopying Prohibited

ELABORATE: How does body shape help regulate temperature?

▶ It is often debated whether the dinosaurs were endo- or ectothermic and to what degree they were homeothermic. Some dinosaurs had large sail-like structures on their backs or heads. It is speculated that these might have been used in thermoregulation, suggesting the owners were ectothermic but regulated their temperature with behaviors similar to today's reptiles. However other dinosaurs did not have these. The sheer size of some dinosaurs (e.g. *Diplodocus* or *Brachiosaurus*) means that the environment would have had little effect on their body temperature so that maintaining a stable body temperature would have been relatively simple. Many other dinosaur genera had feathers or feather-like body coverings, suggesting their thermoregulation was somewhat independent of body shape.

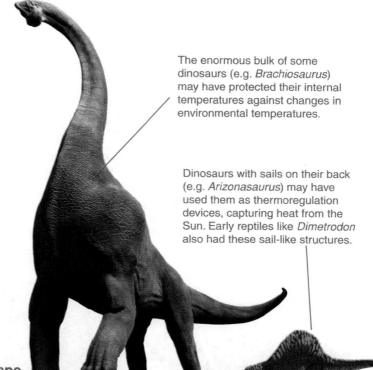

The enormous bulk of some dinosaurs (e.g. *Brachiosaurus*) may have protected their internal temperatures against changes in environmental temperatures.

Dinosaurs with sails on their back (e.g. *Arizonasaurus*) may have used them as thermoregulation devices, capturing heat from the Sun. Early reptiles like *Dimetrodon* also had these sail-like structures.

Investigating temperature gain and body shape
Method

For this investigation you will need a datalogger and temperature probe, a heat lamp (or a lamp with a high wattage incandescent bulb) and a roll of aluminum foil.

i.   Start by rolling an aluminum foil sphere about 3 cm in diameter. Push the temperature probe into the center of the ball. Place the ball 30 cm in front of the heat lamp. Record the temperature gain until the temperature stabilizes (or after a set amount of time) then switch the lamp off for two minutes, then switch it back on for two minutes.

Repeat this two more times then switch the lamp off and record the heat loss until the temperature levels off again. Record your results on a separate piece of paper and attach it here. Note that you can use a thermometer instead of a temperature probe and datalogger, but you will need to check the temperature regularly throughout.

ii. Now make a sphere of about 6 cm in diameter (or slightly smaller if you don't have enough aluminum foil). Repeat the method above with the large aluminum foil ball.

iii. Now make a fusiform (tapered) shape using the same amount of aluminum as you used to make the 5 cm sphere. Again place the probe into the middle of the fusiform body. Now place it (long) side on to the heat lamp at 30 cm distance and record the heat gain and loss as before.

iv. After this, turn the fusiform shape end on to the heat lamp and record the temperature gain and loss as before.

▶ Once you have collected data from these four shapes and orientations try some different shapes, include a sail shape on the fusiform body, try making other animal shapes (e.g. long snakes, coiled snakes etc).

▶ On the graph below make plots of internal temperature over time using appropriate axes.

17. Which shape heated up fastest?

_____

18. Which shape heated up slowest?

_____

19. Which shape had the most stable temperature during the phase of switching the lamp on and off?

_____

20. Does body shape or size have any effect on temperature regulation? Explain.

_____

_____

_____

_____

©2018 **BIOZONE** International
ISBN: 978-1-927309-55-1
Photocopying Prohibited

ELABORATE: Body shape influences heat loss

▶ Body shape influences how heat is retained or lost. Animals with a lower surface area to volume ratio will lose less body heat per unit of mass than animals with a high surface area to volume ratio.

▶ In humans, there is a negative relationship between surface area and latitude. Indigenous people living near the equator tend to have a higher surface area to volume ratio so they can lose heat quickly. In contrast, people living in higher latitudes near the poles have a lower surface area to volume ratio so that they can conserve heat.

▶ People from low latitudes (equatorial regions) have a taller, more slender body and proportionately longer limbs. Those at high latitudes are stockier, with shorter limbs.

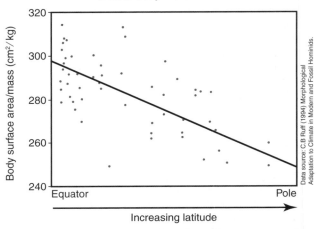

Relationship between ratio of surface area to body mass and latitude

Data source: C.B Ruff (1994) Morphological Adaptation to Climate in Modern and Fossil Hominids.

The Inuit people of Arctic regions are stocky, with relatively short extremities. This body shape is well suited to reducing the surface area over which heat can be lost from the body.

The indigenous peoples of equatorial Africa, such as these young Kenyan men, are tall and slender, with long limbs. This body shape increases the surface area over which heat can be lost.

The relationship holds true for fossil hominids. Neanderthals, which inhabited Eurasia during the last glacial, had robust, stocky bodies relative to modern humans as this reconstruction shows.

21. Why would having a reduced surface area to volume ratio be an advantage in a cold climate? _____

_____

_____

_____

22. Hypothermia is a condition that occurs when the body cannot generate enough heat and the core body temperature drops below 35°C. Prolonged hypothermia is fatal. Exposure to cold water results in hypothermia more quickly than exposure to the same temperature of air because water is much more effective than air at conducting heat away from the body.
In the graph (right), hypothermia resulting in death is highly likely in region 1 and highly unlikely in region 2.

(a) Which body shape has best survival at 15°C?

_____

(b) Explain your choice: _____

_____

_____

_____

Water exposure and survival times

Body shape: short and stocky

Average

Body shape: tall and thin

23. What body surface area per mass would you expect someone from 45° North to have? _____

©2018 **BIOZONE** International
ISBN: 978-1-927309-55-1
Photocopying Prohibited

74 Disease Affects Interactions

ENGAGE: A cascade

A cascade is when one action affects one or more actions, which affect other actions and so on. Consider the diagram below showing how a cascade involving the components of a home entertainment system can affect output.

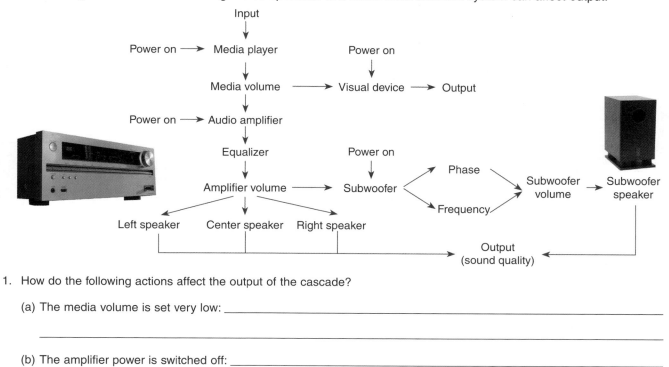

1. How do the following actions affect the output of the cascade?

   (a) The media volume is set very low: _____

   _____

   (b) The amplifier power is switched off: _____

   _____

   (c) The subwoofer frequency is changed: _____

   _____

▸ Cascades also occur in our bodies, both with cells (e.g. signal cascades) or in the interactions between various organs and parts of the body (e.g. high muscle activity affects heart beat and breathing rates).

2. (a) In the box below draw a simple cascade for the following body system: Exercise (muscle movement) (input) affects heart rate and breathing rate. Heart rate affects blood supply to the muscles. Breathing rate affects oxygen availability to the blood. Output is the rapidity or strength of the movement. Output is affected by the ability of the heart and lungs to respond to the input:

   (b) From your diagram, explain how a disease affecting the lungs (e.g. emphysema) would affect your diagram's output:

   _____

   _____

   _____

©2018 BIOZONE International
ISBN: 978-1-927309-55-1
Photocopying Prohibited

SC CE LS1.A

EXPLORE: Diseases may have many effects
Cystic fibrosis

▶ Recall the cause of cystic fibrosis is the deletion of three DNA base pairs resulting in a defective transport protein. This one defect causes a cascade of problems, some of which are directly caused by the mutation, others are secondary results that follow on from the initial problems.

3. In groups of three or four, research how the cystic fibrosis ΔF508 mutation affects the organs listed below. Each person should research the effect on a different organ. When each of you has done this, return to your group and describe your findings to the others in the group.

Lungs: _____

_____

_____

_____

_____

Pancreas and digestive tract: _____

_____

_____

_____

_____

Other organs or body systems: _____

_____

_____

_____

_____

4. Draw a cascade diagram to illustrate the "normal" body processes without cystic fibrosis. Show on the diagram the effect on the cascade of the ΔF508 mutation and how it results in the disorders your group has investigated:

EXPLAIN: Medical technology and disease

▶ Medical technology is constantly advancing, from simple X-rays to advanced MRIs. Technology is used to help diagnose and treat various diseases or disorders of the body.

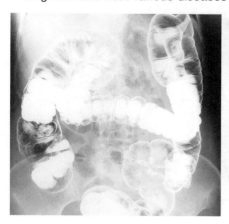

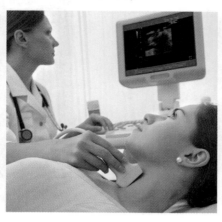

X-ray images are produced by X-rays passing through a patient onto an X-ray sensitive film. Bone and hardened structures block the X-rays, making it possible to see bones, internal scarring, or dense material that may correspond to a developing tumor or cell mass.

Ultrasound uses high frequency sound waves to build up an image of the tissue below the transducer. The technique is non invasive and extremely safe. Ultrasound is used in a similar way to X-rays in that it can be used to obtain images of dense tissues such as tumors.

MRI (magnetic resonance imaging) Uses strong magnetic and electric fields and radio waves to generate highly detailed internal images of the body. MRI is able to build detailed 3D images of the part of the body being studied, making diagnosis of specific diseases or disorders possible.

▶ Medical technology can treat many diseases and disorders. However many are still not treatable. Treatments range from drugs (chemotherapy) to surgery.

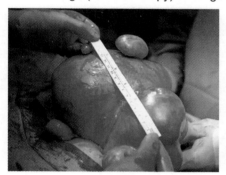

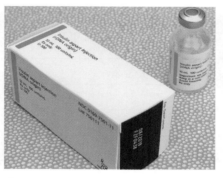

Surgery has multiple uses in treating disease. Diseased tissue may be removed (e.g. a tumor) or it may be use to repair damaged tissues (e.g. ligament damage). Surgery is the only option when an organ needs to be replaced (transplant).

There are a wide range of drugs available to treat diseases. Antibiotics are used to treat bacterial infections but drugs can also be used to treat cancers. Indeed, most people associate the term "chemotherapy" with cancer treatment.

Radiation therapy is an extreme example of medical technology. Targeted gamma rays are used to kill malignant tumor cells. Physiotherapy treats muscle and joint disorders by manipulation and training exercises.

5. As a group, research a common disorder, its diagnosis, and its treatment. You may want to research treatments for cystic fibrosis. Present your research as a poster or oral report:

_____

_____

_____

_____

_____

_____

_____

_____

_____

©2018 **BIOZONE** International
ISBN: 978-1-927309-55-1
Photocopying Prohibited

75 A Cancerous Creep Revisited

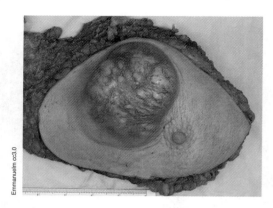

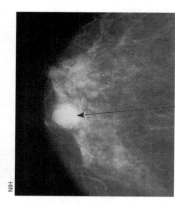

Tumor seen in mammogram

1. From your understanding of DNA, proteins, cells, and interacting systems, write an explanation of how gene mutations such as *BRCA1* and *BRCA2* can cause breast cancer.

   Your explanation should include an explanation of the link between DNA, proteins, and phenotype. There should be references to the cell cycle, including DNA replication and mitosis.

   You should make reference to system interactions and homeostasis in the body and how these can be affected by diseases such as cancer.

   A brief discussion of the treatment of breast cancer should be included.

©2018 **BIOZONE** International
ISBN: 978-1-927309-55-1
Photocopying Prohibited

76 Summative Assessment

1. (a) Plant cells are eukaryotic cells. Label the following diagram of a plant cell and explain how the organelles pictured contribute to the functioning of the cell as a whole. You may use extra paper if required. In your answer you should:

   • Label the structures and organelles
   • Briefly state the functions of the structures and organelles labeled
   • Explain how the organelles contribute to the functioning of the cell as a whole.

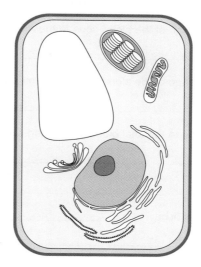

_____

_____

_____

_____

_____

_____

_____

(b) Which organelle shown in the plant cell above is not found in an animal cell?

_____

2. Describe the features of the plasma membrane and discuss how its structure contributes to the overall functioning of the cell (and the organism). You should include reference to membrane transport processes. Use extra paper if required.

_____

_____

_____

_____

_____

_____

_____

_____

_____

_____

_____

_____

©2018 **BIOZONE** International
ISBN: 978-1-927309-55-1
Photocopying Prohibited

    SF EM SSM LS1.B LS1.A

3. The graph (right) compares the change in cardiac output (a measure of total blood flow in L) during rest and during exercise. The color of the bars indicates the proportion of blood flow in skeletal muscle relative to other body parts.

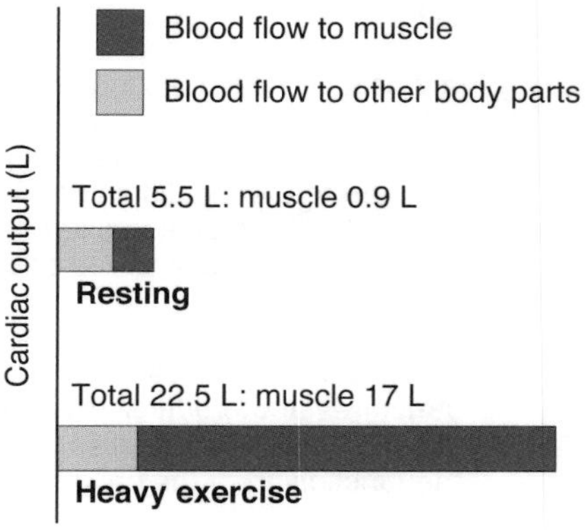

Blood flow to muscle

Blood flow to other body parts

Total 5.5 L: muscle 0.9 L

Resting

Total 22.5 L: muscle 17 L

Heavy exercise

(a) What percentage of the blood goes to the muscles at rest?

_____

(b) What percentage of the blood goes to the muscles during exercise?

_____

(c) What happens to the total blood flow during exercise compared to at rest?

_____

_____

(d) Why does this occur? _____

_____

(e) What would be happening to heart and breathing rates during this time? Explain: _____

_____

_____

4. Using investigative techniques called microarrays, which show the extent of gene activity, scientists mapped the genes involved in the differentiation of a stem cell into a β islet cell in the pancreas. Study the diagram below. Each step represents a cell division:

Gene control of cell differentiation

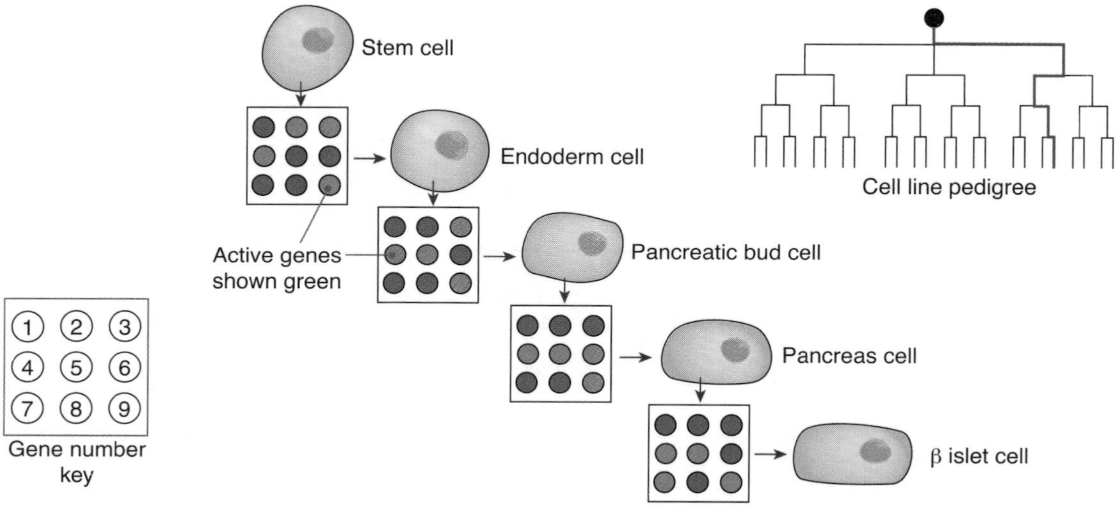

Stem cell

Active genes shown green

Endoderm cell

Cell line pedigree

Pancreatic bud cell

Pancreas cell

β islet cell

①	②	③
④	⑤	⑥
⑦	⑧	⑨

Gene number key

(a) Which gene(s) are active at every cell division (use the key to identify)? _____

(b) What might happen if the activity of the gene(s) was inhibited at any of the cell divisions? _____

_____

_____

(c) Suggest how this technique would be able to help scientists treat a disorder in a β islet cell:_____

_____

_____

_____

_____

_____

5. Pawnee Farm Arlinda Chief was a breeding bull of the Holstein cattle breed. He sired as many as 16,000 daughters and his chromosomes account for almost 14% of the genome of the current U.S.A. Holstein cattle population. However, it has been discovered that a mutation in Arlinda Chief's APAF1 gene (which is involved in the development of the central nervous system) leads to reduced fertility in his offspring via spontaneous abortion of the fetus. The mutation is recessive, so a fetus must have two copies of the mutation to be affected.

(a) Explain why the effect of the mutation is not seen in calves that survive to birth:

_____

_____

The DNA sequences below show the relevant part of the coding strand for the unaffected sequence and the relevant part of the coding strand for the mutant sequence for Arlinda Chief's mutation:

Unaffected DNA sequence: TCA GAG GTT TAT CGG CAA GCT AAG CTG CAG

Mutant DNA sequence:　　TCA GAG GTT TAT CGG TAA GCT AAG CTG CAG

(b) Convert the DNA sequences to template DNA strands and then to mRNA:

Unaffected DNA template strand: _____

Unaffected mRNA sequence: _____

Mutant DNA template strand: _____

Mutant mRNA sequence: _____

(c) Use the mRNA table on page 297 to work out the amino acid sequence for the unaffected mRNA sequence:

_____

(d) Use the mRNA table on page 297 to work out the amino acid sequence for the mutant mRNA sequence:

_____

(e) What was the change to the amino acid sequence caused by the mutation? _____

_____

(f) How would this affect the protein produced? _____

_____

_____

(g) Use this example to explain how DNA determines the structure of proteins, which carry out the functions of life:

_____

_____

_____

_____

_____

_____

_____

(h) Explain why the mutation does not affect live born cattle: _____

_____

_____

_____

©2018 **BIOZONE** International
ISBN: 978-1-927309-55-1
Photocopying Prohibited

6. Explain the links between DNA replication, mitosis and the cell cycle. Use a diagram to explain each process and how they relate to each other. Describe how these processes help to maintain a multicellular organism:

©2018 **BIOZONE** International
ISBN: 978-1-927309-55-1
Photocopying Prohibited

7. The two graphs below show two different feedback mechanisms. Identify the type of mechanism its mode of action and describe an example of its operation in a living organism:

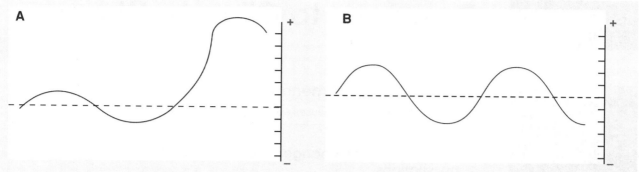

Type of feedback mechanism:

_____

Mode of action:

_____

_____

_____

_____

Biological example of this mechanism:

_____

_____

_____

Type of feedback mechanism:

_____

Mode of action:

_____

_____

_____

_____

Biological example of this mechanism:

_____

_____

_____

8. (a) Why does a human expend a large amount of energy when exposed to cold temperatures for a long period of time?

_____

_____

(b) How is this different to a lizard exposed to the same conditions? _____

_____

_____

9. Draw a diagram to show how the biological systems in a human react to a decrease in ambient environmental temperature in order to maintain thermal homeostasis:

©2018 **BIOZONE** International
ISBN: 978-1-927309-55-1
Photocopying Prohibited

USCG

Instructional Segment 6

Ecosystem Stability and the Response to Climate Change

Activity number

Anchoring Phenomenon

Weather whiplash: What causes extreme weather events?

77 84

What causes the changes in ecosystems that ultimately affect populations?

☐ 1 Describe examples of ecosystems that appear stable over time and those that seem to change dramatically. Using evidence from real ecosystems, construct an argument for the stability or change in ecosystems over time. Describe factors that might cause an ecosystem to change so dramatically that it cannot recover its original state. Use evidence from populations to evaluate the effect of ecosystem change on populations.

78 85

☐ 2 Develop and use a model to show how populations might change in response to a shift in climate. Explain how some individuals within a population might respond differently to changes in temperature and explain how this might influence the ability of these populations to adapt to climate change (recall IS3).

78

What are the changes that are happening to climate and what effects are those having on life?

☐ 3 Describe evidence that documents changes in the Earth's climate. Create and use simple models of the Earth's energy budget to illustrate the flows of energy into, within, and out of the Earth system. Explain the Greenhouse Effect and understand its significance to life on Earth. Analyze and interpret data relating to the increase in greenhouse gases in the atmosphere and construct an evidence-based argument for the relationship between increases in greenhouse gases and climate change and the role of humans in this.

79 85

☐ 4 Describe how the cause and effect relationship between greenhouse gases and global temperature is complicated by negative (counterbalancing) and positive (reinforcing) feedback loops within and between the Earth's various systems. Create and use a simple model to show how an increase in the Earth's surface temperature can create a reinforcing feedback loop and then investigate this using a model of ice sheet melting.

80

☐ 5 Explain how scientists develop, use, and test climate models to predict the effects of increased greenhouse gases on the Earth's systems. Analyze and interpret data from climate models to make a forecast about the likely impact of global climate change on the Earth's systems, e.g. sea level changes or changes to the distribution of crops or diseases. Evaluate the evidence for changes in populations predicted under different climate change scenarios.

81 85

How are human activities affecting Earth's systems and life on Earth?

☐ 6 Recall (IS1) how humans, by altering the availability of resources and changing the landscape (including through climate change), might cause changes to ecosystems. Analyze and interpret data about the effects of climate change on sea levels in coastal parts of the US. Explain how modeling can help us to understand and predict future sea level changes and mitigate their effects on the organisms affected.

83

☐ 7 Use a computational representation to illustrate the relationships among Earth's systems and how human activities are modifying them.

83

How can humans mitigate their negative effect on the environment?

☐ 8 Use examples to illustrate how humans, through their ability to model and predict changes in the Earth's climate, are finding out more about how the Earth's systems interact and are altered by human activity.

82 83

☐ 9 Design, evaluate, and refine a solution for reducing the effect of human-induced climate change of biodiversity.

82 83

77 Weather Whiplash

ANCHORING PHENOMENON:

What causes extreme weather events?

The sudden swing in weather conditions from one extreme to the another is called weather whiplash. California's climate naturally fluctuates between dry summers and wet winters. Historically, large swings between the extremes have been quite rare, only occurring every 100-200 years.

The Great Flood of 1862 is a well documented example, and affected several states including California. So much rain fell on the Central Valley and the Los Angeles Basin over 45 days that a large area was turned into an inland sea. An area 480 km long, averaging 32 km in width and covering 13,000 to 16,000 km² was under water. Towns and infrastructure were swept away and agricultural lands were ruined.

The United States Geological Survey (USGS) have studied and mapped the flood to predict the effect of another such megastorm on modern day California. The predicted flood event is officially called the ARkStorm or unofficially called "the other big one'.

Some researchers predict weather extremes like the one causing The Great Flood will become more common in California as the influences of global warming are felt.

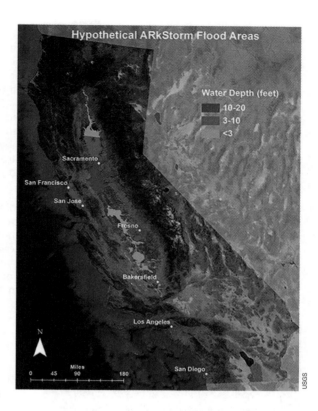

Hypothetical ARkStorm Flood Areas

Water Depth (feet)
10-20
3-10
<3

California has experienced many droughts, including one that lasted from 2012-2017. This period was the driest on record since monitoring was started. This drought was followed by the wettest winter ever recorded in Northern California. The rainfall was greater than the conditions that caused The Great Flood of 1862. Water reservoirs were replenished, but the flooding caused widespread damage. The Fresno River is normally dry at the construction site of the Fresno River Viaduct, but was filled with water during the 2017 flood. The images on the right illustrate this.

1. Would you consider the events described for California (2012-2017) as a weather whiplash? Explain why or why not:

_____

_____

_____

_____

_____

_____

Fresno River Viaduct 2016

Fresno River Viaduct 2017

Images: California High-Speed Rail Authority (public domain)

2. In small groups, research the occurrence of weather whiplash events in California. Has there been a change in their frequency? If so, explain the changes observed:

_____

_____

_____

_____

_____

_____

_____

©2018 **BIOZONE** International
ISBN: 978-1-927309-55-1
Photocopying Prohibited

78 Ecosystem Dynamics

ENGAGE: The vine that ate the South

Human activities can have unintended consequences on ecosystems. For example, humans may deliberately introduce a new species into an area and only later find that it has a negative effect on the ecosystem. An example is the introduction of kudzu into the US from Japan in the 1800s. Kudzu was intended as an ornamental plant for shade porches, as a high protein food source for cattle, and as ground cover to stop erosion. Kudzu was promoted by The Soil Erosion Service, which recommended it to control slope erosion. The government funded the distribution and planting of 85 million seedlings.

However, as kudzu became established it became obvious that it would damage ecosystems. Its growth is so rapid that it climbs and grows over other plants so rapidly that they don't receive any sunlight and are killed. A forest can be overrun in 2-3 years, greatly reducing biodiversity (photo, bottom, left). Kudzu grew virtually unchecked in the climate of the Southeastern US and was finally listed as a weed in 1970.

CA EP&Cs V: The process of making decisions about resources and natural systems, and how the assessment of social, economic, political, and environmental factors has changed over time (Vb).

Before the 20th century many species were actively introduced without government oversight. Today policies are in place to prevent the introduction of invasive species.

▶ Policies including The Lacey Act of 1900, and the The Alien Species Prevention and Enforcement Act of 1992 help to regulate the introduction of new species.

▶ These acts help the United States Fish and Wildlife Service protect sensitive habitats or endangered species.

▶ The United States Department of Agriculture helps prevent the introduction of invasive species through the Animal and Plant Health Inspection services.

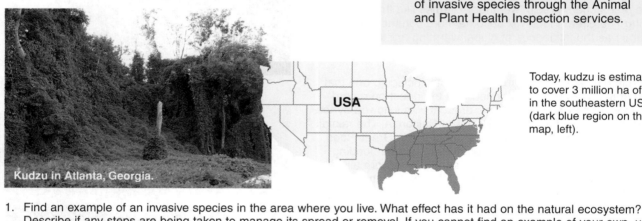

Scott Ehardt, Public Domain

Kudzu in Atlanta, Georgia.

USA

Today, kudzu is estimated to cover 3 million ha of land in the southeastern US (dark blue region on the map, left).

1. Find an example of an invasive species in the area where you live. What effect has it had on the natural ecosystem? Describe if any steps are being taken to manage its spread or removal. If you cannot find an example of your own, use **BIOZONE's Resource Hub** to select an invasive organism in California.

_____

_____

_____

_____

_____

_____

_____

_____

2. Very little government control regarding the introduction of new species was in place when kudzu was introduced. How has the introduction of legislation and the involvement of specific agencies helped to safeguard natural ecosystems against the introduction of invasive species:

_____

_____

_____

_____

_____

    LS2.C LS4.C SC

©2018 **BIOZONE** International
ISBN: **978-1-927309-55-1**
Photocopying Prohibited

EXPLORE: Ecosystems are dynamic

▶ Ecosystems are dynamic in that they are constantly changing. Many ecosystem components, including the seasons, predator-prey cycles, and disease cycles, are cyclical. Some cycles may be short term, such as the change of seasons, or long term, such as the expansion and retreat of deserts.

▶ Although ecosystems may change constantly over the short term, they may be relatively static over longer periods, with their essential characteristics remaining unchanged. Small scale disturbances, such as seasonal fires, may temporarily alter an ecosystem but recovery to its previous state is swift.

▶ Ecosystems are less likely to recover from large scale disturbances because a greater proportion of the ecosystem is affected than when the disturbance occurs on a small scale (see photos below).

Forest fire

Open cast mining

An ecosystem may remain stable for many hundreds or thousands of years provided that the interacting components within it remain stable.

Small scale changes usually have minimal long term effect on an ecosystem. Fire or flood may destroy some parts, but enough is left for the ecosystem to return to is original state relatively quickly.

Large scale disturbances such as volcanic eruptions, sea level rise, or large scale open cast mining remove all components of the ecosystem, changing it forever.

3. (a) What do you think is meant by the term dynamic ecosystem? _____

   _____

   _____

   (b) Some ecosystems appear not to change. Do you think this is likely or unlikely to be true? Explain your reasoning:

   _____

   _____

   _____

   _____

4. Suggest why ecosystems are more likely to recover from a small scale disturbance than a large scale one: _____

   _____

   _____

   _____

5. Decide if the following scenarios describe large scale or small scale disturbance events:

   (a) A major volcanic eruption destroys a mountain's vegetated slopes: _____

   (b) An area is experiencing a seasonal drought: _____

   (c) A landslide measuring a few meters across tumbles down a hillside: _____

   (d) Humans drain an extensive wetland area for cropping: _____

   (e) An extensive forest fire destroys an old growth forest: _____

   (f) Sea level rise causes land to become permanently inundated: _____

   (g) A windstorm brings down several large trees in a forest: _____

©2018 BIOZONE International
ISBN: 978-1-927309-55-1
Photocopying Prohibited

EXPLORE: What is ecosystem stability?

Stability refers to how an ecosystem can deviate very little from its average state despite changes in environmental conditions. Ecosystem stability has various components, including **inertia** (the ability to resist disturbance) and **resilience** (ability to recover from external disturbances).

Ecosystem stability is closely linked to biodiversity, with more diverse systems being more stable. It is hypothesized that this is because greater diversity results in a greater number of biotic interactions and few (if any) vacant niches. The system is buffered against change because it is resistant to invasions and there are enough species present to protect ecosystem functions (such as productivity or nutrient cycling) if a species is lost.

This hypothesis is supported by experimental evidence but there is uncertainty as to what level of biodiversity provides stability (because some species contribute more to ecosystem function than others). It is also not at all clear what factors will overstretch an ecosystem's resilience.

The stability of an ecosystem can be illustrated by a ball in a tilted bowl. Given a slight disturbance the ball will eventually return to its original state (**line A**). However given a large disturbance the ball will roll out of the bowl and the original state with never be restored (**line B**).

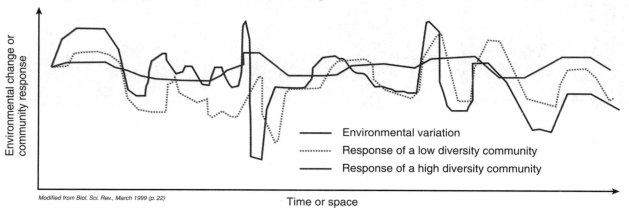

Community response to environmental change

Environmental change or community response

— Environmental variation
.... Response of a low diversity community
— Response of a high diversity community

Modified from Biol. Sci. Rev., March 1999 (p. 22)

Time or space

6. Compare the responses of the high and the low diversity communities (above) to environmental variations:

_____

_____

7. A hypothetical community consists of three different plant species. Each species is adapted to grow best within a different temperature range.

   (a) Complete the model below by coloring in the circles to represent changes in the community composition in response to climate change over a 20 year period.
   Use blue to represent a species that grows best at temperatures -4–10°C.
   Use black to represent a species that grows best at temperatures 0–14°C.
   Use red to represent a species that grows best at temperatures 6–20°C.

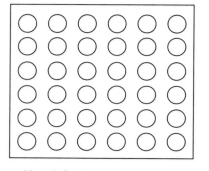

Year 0: Cool climate (-4–10°C)

Year 10: Moderate climate (4–16°C)

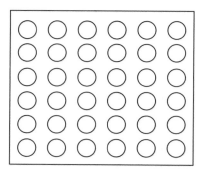

Year 20: Warm climate (14–24°C)

   (b) What has happened to community composition over time? _____

_____

_____

_____

Adapted from Biodiversity and Ecosystem Stability
Cleland, E. E. (2011) Biodiversity and Ecosystem Stability. Nature
Education Knowledge 3(10):14

EXPLORE: Ecosystems can be resilient

Resilience is the ability of the ecosystem to recover after disturbance and is affected by three important factors: diversity, ecosystem health, and frequency of disturbance. Some ecosystems are naturally more resilient than others. Some of the factors affecting ecosystem resilience are described below.

Ecosystem biodiversity

The greater the diversity of an ecosystem the greater the chance that all the roles (niches) in an ecosystem will be occupied, making it harder for invasive species to establish and easier for the ecosystem to recover after a disturbance.

Ecosystem health

Intact ecosystems are more likely to be resilient than ecosystems suffering from species loss or disease.

Disturbance frequency

Single disturbances to an ecosystem can be survived, but frequent disturbances make it more difficult for an ecosystem to recover. Some ecosystems depend on frequent natural disturbances for their maintenance, e.g. grasslands rely on natural fires to prevent shrubs and trees from establishing. The various grass species have evolved to survive frequent fires.

Fishing

Resilience and harvesting

It is important to consider the resilience of an ecosystem when harvesting ecosystem resources. For example, logging and fishing remove organisms from an ecosystem for human use. Most ecosystems will be resilient enough to withstand the removal of a certain number of individuals. However, excessive removal may go beyond the ecosystem's ability to recover. Examples include the overfishing of North Sea cod or deforestation in the Amazon basin.

8. How might over-harvesting affect an ecosystem's resilience? _____

_____

_____

_____

EXPLAIN: Case studies in ecosystem resilience

A case study in resilience: The Great Barrier Reef

A study of coral and algae cover at two locations in Australia's Great Barrier Reef (below) showed how ecosystems recover after a disturbance.
At Low Isles, frequent disturbances (e.g. from cyclones) made it difficult for corals to reestablish, whereas at Middle Reef, infrequent disturbances made it possible for coral to reestablish its dominant position in the ecosystem.
The image right shows a crown of thorns starfish (red) attacking a coral (white).

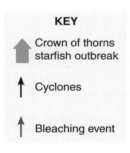

KEY

Crown of thorns starfish outbreak

Cyclones

Bleaching event

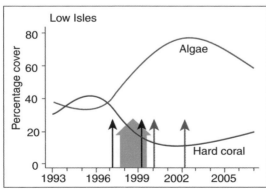
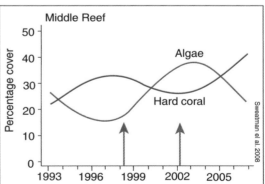

9. Study the graphs above. Suggest why the coral at Middle Reef remained abundant (high numbers) from 1993 to 2005 whereas the coral at Low isles did not:

_____

_____

_____

©2018 **BIOZONE** International
ISBN: 978-1-927309-55-1
Photocopying Prohibited

A case study in resilience: Spruce budworm and balsam fir

A case study of ecosystem resilience is provided by the spruce-fir forest community in northern North America. Organisms in the community include the spruce budworm and balsam fir, spruce, and birch trees. The community fluctuates between two extremes:

▶ Between spruce budworm outbreaks the environment favors the balsam fir.

▶ During budworm outbreaks the environment favors the spruce and birch species.

Balsam fir

Spruce budworm

1 Under certain environmental conditions, the spruce budworm population grows so rapidly it overwhelms the ability of predators and parasites to control it.

2 The budworm feeds on balsam fir (despite their name), killing many trees. The spruce and birch trees are left as the major species.

3 The population of budworm eventually collapses because of a lack of food.

4 Balsam fir saplings grow back in thick stands, eventually out-competing the spruce and birch. Evidence suggests these cycles have been occurring for possibly thousands of years.

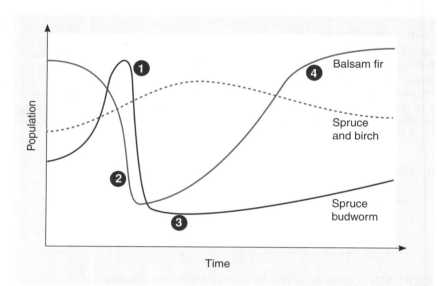

10. (a) Why could predators not control the budworm population? _____

_____

_____

_____

(b) What was the cause of the budworm population collapse after its initial rise? _____

_____

_____

11. Under what conditions does the balsam fir out-compete the spruce and birch? _____

_____

_____

12. In what way is the system resilient in the long term? _____

_____

_____

_____

_____

©2018 **BIOZONE** International
ISBN: 978-1-927309-55-1
Photocopying Prohibited

EXPLAIN: Why are some species more important to ecosystem stability than others?

Some species have a disproportionate effect on the stability of an ecosystem. These species are called **keystone species** (or key species). The term keystone species comes from the analogy of the keystone in a true arch. If the keystone is removed, the arch collapses.

The role of the keystone species varies from ecosystem to ecosystem, but if they are lost, the ecosystem can rapidly change or collapse completely. The pivotal role of keystone species is a result of their influence in some aspect of ecosystem functioning, e.g. as predators, prey, or processors of biological material.

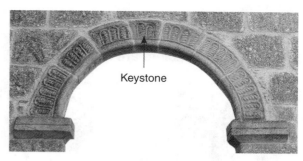

Keystone

An archway is supported by a series of stones, the central one being the keystone. Although this stone is under less pressure than any other stone in the arch, the arch collapses if it is removed.

Case study: Sea otters are a keystone species

▶ Sometimes the significant keystone effects of a species can be indicated when a species declines rapidly to the point of near extinction. This is illustrated by the sea otter example described below.

▶ Sea otters live along the Northern Pacific coast of North America and have been hunted for hundreds of years for their fur. Commercial hunting didn't fully begin until about the mid 1700s when large numbers were killed and their fur sold to overseas markets.

▶ The drop in sea otter numbers had a significant effect on the local marine environment. Sea otters feed on shellfish, particularly sea urchins. Sea urchins eat kelp, which provides habitat for many marine creatures. Without the sea otters to control the sea urchin population, sea urchin numbers increased and the kelp forests were severely reduced.

istockphotos.com

Sea otters are critical to ecosystem function. When their numbers were significantly reduced by the fur trade, sea urchin populations exploded and the kelp forests, on which many species depend, were destroyed.

The effect can be seen on Shemya and Amchitka Islands. Where sea otters are absent, large numbers of sea urchins are found, and kelp are almost absent.

13. (a) What effect do sea otters have on sea urchin numbers? _____

_____

(b) What effect do sea urchins have on kelp cover? _____

_____

(c) What evidence is there that the sea otter is a keystone species in these Northern Pacific coastal ecosystems?

_____

_____

_____

_____

©2018 **BIOZONE** International
ISBN: 978-1-927309-55-1
Photocopying Prohibited

79 Climate Change

ENGAGE: The glaciers are retreating!

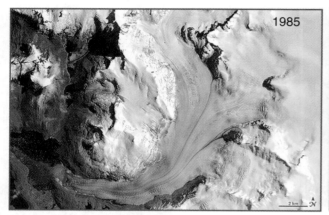
1985

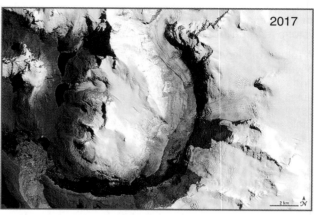
2017

NASA Earth Observatory images by Joshua Stevens

The Southern Patagonian Ice Field is located in Chile and Argentina. It extends approximately 350 kilometers and has an area of 12,363 km². The North Patagonian Ice field is experiencing ice thinning and rapid recession, with the fastest rates of recession recorded between 2001 to 2011. The ice field contains many glaciers, including the HPS-12 glacier shown above. The images were taken in 1985 (left) and again in 2017 (right), and document the rapid retreat of HPS-12 over this period. The red bars mark the glacier's end point in each case. In 1985, HPS-12 measured 26 km long and several glaciers contributed to it. In the 2017 image, the glacier was less than 13 km long. In addition, three of the contributing glaciers were cut off and isolated from it. Climate change is thought to have contributed to the loss.

1. Climate change is thought to have contributed to the loss of ice fields and glaciers such as HPS-12. Describe any factors that you think may be contributing to the loss of ice:

   _____

   _____

   _____

2. Can you think of any problems that may be caused by glaciers melting? _____

   _____

   _____

   _____

   _____

3. One third of California's surface water is stored as snow and ice in the Sierra Nevada, which accumulates as a result of winter precipitation. Between April and July the snow melts at a fairly constant rate and is used to help meet water demands throughout the state. For example, meltwater from the Lyell and Maclure glaciers feeds into the Tuolumne River and the Hetchy Reservoir, which supplies water to 2.4 million residents in the Bay area.

   Red Slate Mountain is part of the Sierra Nevada Range.

   Poppy CC3.0

   (a) Suggest how a decrease in snow and ice could affect the security of California's water supply:

   _____

   _____

   _____

   (b) What could the consequences of this be? _____

   _____

   _____

   _____

        ESS3.C ESS3.D SPQ SC

©2018 **BIOZONE** International
ISBN: 978-1-927309-55-1
Photocopying Prohibited

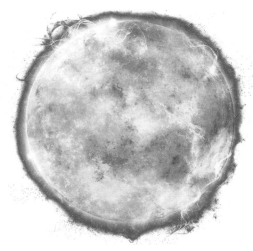

EXPLORE: Creating models of the Earth's energy budget

The Sun (right) ultimately provides the energy required to power all life on Earth. The Sun produces almost unimaginable amounts of energy (174 petawatts, or 174 quadrillion joules per second). To put this in context, the world's most powerful lasers can produce power of 1.25 PW and only keep this up for one picosecond (1×10^{-12} seconds).

Not all of the solar radiation reaching Earth is retained, and several factors determine how much solar radiation is retained and how much is radiated back into space.

In this section you will develop a series of simple drawings (models) with increasing complexity to account for the factors contributing to the Earth's energy budget. Use simple shapes and arrows of different sizes and colors to illustrate energy movements between the Sun and Earth.

4. (a) Imagine the Earth is a dark sphere and absorbs all incoming energy from the Sun. In the space below draw a diagram to represent this.

(b) The Earth does not absorb all the incoming solar energy but has regions where some is reflected back from light areas on the surface. This is called albedo. Draw a diagram to represent this.

(c) Add an atmosphere to your model. The atmosphere is continuous around the Earth. It allows most of the solar energy through, reflects some back, and also absorbs some energy. Draw a diagram to represent this in the space below.

(d) Now add clouds to your model. Clouds reflect some incoming solar energy but also absorb it in roughly equal amounts. Clouds are patchy and do not occur in a continuous layer around the Earth. In the space below draw a diagram to represent this.

5. (a) What do you think would happen to the Earth's surface temperature in 4(a)? _____

_____

(b) How do you think the addition of the other components of the model affect Earth's surface temperature? _____

_____

_____

_____

(c) Suggest why these components are important to maintaining life on Earth: _____

_____

_____

©2018 **BIOZONE** International
ISBN: 978-1-927309-55-1
Photocopying Prohibited

The diagram below represents a more complex model of the Earth's energy budget. A large amount of incoming solar radiation is absorbed by the atmosphere or reflected off clouds or the Earth's surface. About 51% of the incoming solar radiation reaches the Earth's surface. Some (0.023%) of this is used by photosynthesis in plants to build organic molecules. The rest drives atmospheric winds and ocean circulation and is eventually radiated back into space.

6. Use the information provided to complete the calculations:

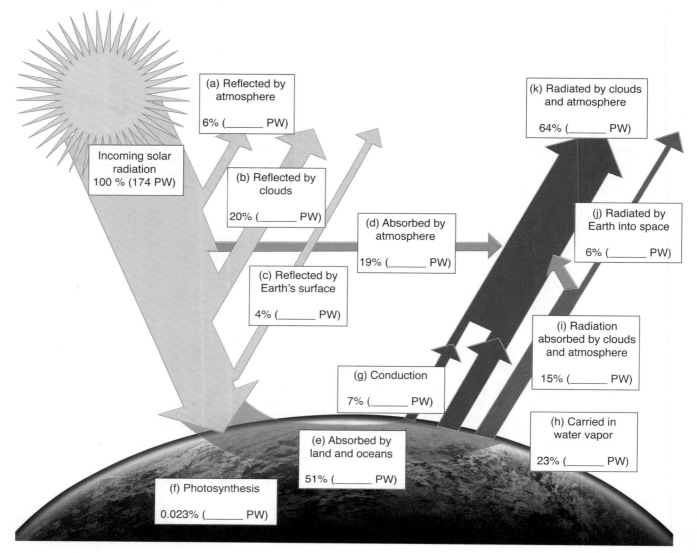

(a) Reflected by atmosphere

6% (_____ PW)

(b) Reflected by clouds

20% (_____ PW)

(c) Reflected by Earth's surface

4% (_____ PW)

(d) Absorbed by atmosphere

19% (_____ PW)

Incoming solar radiation
100 % (174 PW)

(k) Radiated by clouds and atmosphere

64% (_____ PW)

(j) Radiated by Earth into space

6% (_____ PW)

(i) Radiation absorbed by clouds and atmosphere

15% (_____ PW)

(h) Carried in water vapor

23% (_____ PW)

(g) Conduction

7% (_____ PW)

(e) Absorbed by land and oceans

51% (_____ PW)

(f) Photosynthesis

0.023% (_____ PW)

Energy is not evenly distributed on Earth

The energy from the Sun is not distributed evenly about the globe. Because the Earth is spherical, the poles receive less energy per square kilometer than the equator. The Earth's angle of rotation further influences the uneven distribution of the energy received.

Below the Arctic and Antarctic circles, the Earth receives only about 40% of the solar energy that is received at the equator.

The tropics receive the full amount of sunlight and energy available. This causes heating, which carries water into the air, creating a hot, wet climate.

Differential heating between the tropics and poles drives air currents from the tropics towards the poles. This is because air rises at the equator and falls at the poles.

©2018 **BIOZONE** International
ISBN: 978-1-927309-55-1
Photocopying Prohibited

EXPLORE: The greenhouse effect

▶ The Earth's atmosphere is made up of a mix of gases including nitrogen, oxygen, and water vapor. Small quantities of carbon dioxide (CO_2) and methane, and a number of other trace gases (e.g. helium) are also present.

▶ Water vapor, CO_2, and methane are called **greenhouse gases** because they produce a warming or 'greenhouse' effect, which moderates the surface temperature of the Earth.

▶ The **greenhouse effect** describes the natural process by which heat is retained within the atmosphere by greenhouse gases trapping the heat that would normally radiate back into space. About 75% of the natural greenhouse effect is due to water vapor. The next most significant agent is CO_2.

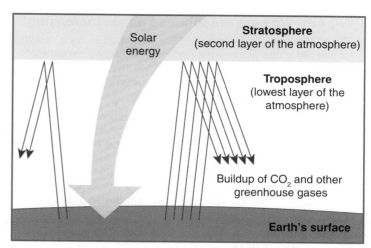

The greenhouse effect results in the Earth having a mean surface temperature of about 15°C, which is 33°C warmer than it would have without an atmosphere.

7. A common analogy for the greenhouse effect is a person wrapped up in a blanket.

   (a) What part of this model represents the Earth and what part represents the atmosphere?

   _____

   _____

   (b) What would be the effect of adding another blanket to this model?

   _____

   _____

   (c) Explain how an increase in greenhouse gases may be analogous to adding an extra blanket to your model:

   _____

   _____

   _____

Use the experiment outlined below to see how the addition of a simple "atmosphere" can influence temperature.

▶ Take two empty soda cans and add 50 mL of tap water to each. Place a thermometer in each can and after two minutes record the temperature in the initial temperature column on the table below.

▶ Place a zip lock bag around one of the cans and its thermometer. Leave the other can uncovered. Place both into a sunny spot. After one hour record the temperature (this is the final temperature) on the table below.

	Initial temperature (°C)	Final temperature (°C)	Change in temperature (°C)
Can with bag			
Can without bag			

8. (a) What does the plastic bag represent? _____

   (b) Was there a difference between the two cans? If yes, what do you think caused the difference? _____

   _____

   _____

©2018 **BIOZONE** International
ISBN: 978-1-927309-55-1
Photocopying Prohibited

EXPLAIN: Why are greenhouse gases increasing?

▶ Fluctuations in the Earth's surface temperature as a result of climate shifts are normal and have occurred throughout the history of the Earth. However since the mid 20th century, the Earth's surface temperature has been increasing. This phenomenon is called **global warming**.

▶ Most scientists attribute accelerated global warming (and associated climate change) to the increase in atmospheric levels of greenhouse gases emitted into the atmosphere as a result of human activity (e.g. burning fossil fuels). The 1800s saw the Industrial Revolution gathering momentum in Europe. During the Industrial Revolution, manual manufacturing techniques were replaced with machine driven techniques, many of which were powered by burning coal.

▶ Combustion releases both CO_2 and water. Although water vapor is the greatest greenhouse gas, CO_2 stays in the atmosphere for much longer. The effects of water vapor on warming are also very complex and the feedbacks involved with water in the Earth's climate are still poorly understood.

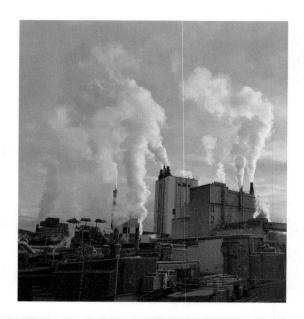

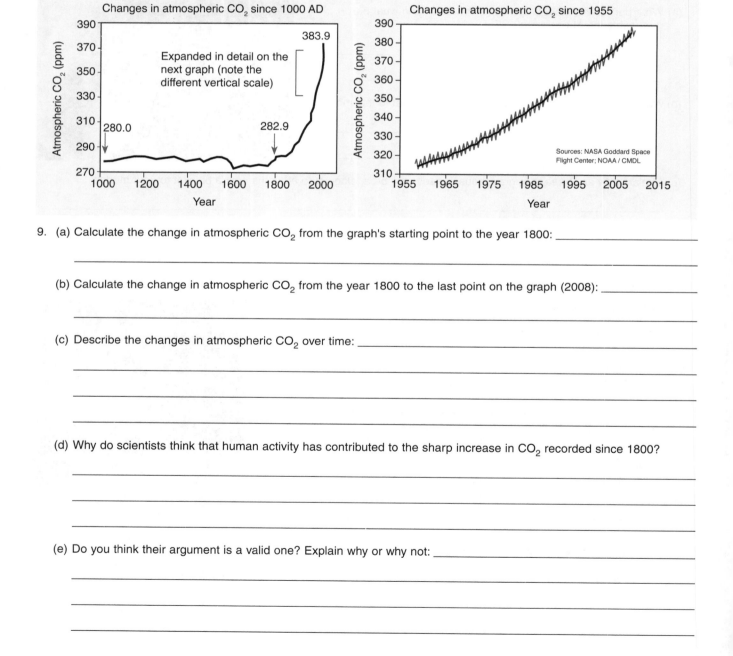

Changes in atmospheric CO_2

Changes in atmospheric CO_2 since 1000 AD

383.9

Expanded in detail on the next graph (note the different vertical scale)

280.0

282.9

Changes in atmospheric CO_2 since 1955

Sources: NASA Goddard Space Flight Center; NOAA / CMDL

9. (a) Calculate the change in atmospheric CO_2 from the graph's starting point to the year 1800: _____

_____

(b) Calculate the change in atmospheric CO_2 from the year 1800 to the last point on the graph (2008): _____

_____

(c) Describe the changes in atmospheric CO_2 over time: _____

_____

_____

_____

(d) Why do scientists think that human activity has contributed to the sharp increase in CO_2 recorded since 1800?

_____

_____

_____

(e) Do you think their argument is a valid one? Explain why or why not: _____

_____

_____

_____

©2018 **BIOZONE** International
ISBN: **978-1-927309-55-1**
Photocopying Prohibited

EXPLAIN: What is the relationship between greenhouse gases and global warming?

▸ There is considerable evidence that the Earth's atmosphere is experiencing a period of accelerated warming.

▸ Fluctuations in the Earth's surface temperature as a result of climate shifts are normal and the current period of warming climate is partly explained by warming after the end of the last glacial that finished 12,000 years ago.

▸ However, fifteen of the sixteen warmest years on record (since 1880) have been since the year 2000. Global surface temperatures in 2015 set a new record, being 0.9°C above the long term average.

▸ As we have already seen, many researchers attribute the accelerated increase to the increased levels of CO_2 and other greenhouse gases entering the atmosphere from human activity.

▸ The potential of a greenhouse gas to warm the climate is given as a value relative to CO_2 (right).

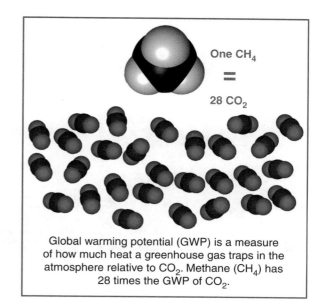

Global warming potential (GWP) is a measure of how much heat a greenhouse gas traps in the atmosphere relative to CO_2. Methane (CH_4) has 28 times the GWP of CO_2.

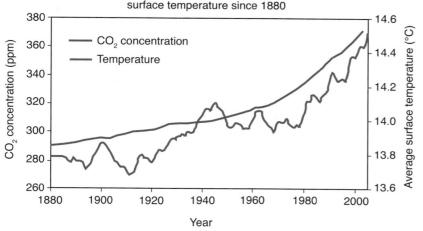

Atmospheric CO_2 concentration and surface temperature since 1880

World temperatures

Rank	Year	Anomaly (°C)
1	2015	+0.90
2	2014	+0.74
3	2010	+0.70
4	2013	+0.66
5	2005	+0.65

Some of the warmest temperatures on record have been since 2000. The table above shows departure from the long term average (anomaly).

10. Describe the relationship between atmospheric CO_2 levels and temperature shown in the graph above: _____

_____

_____

_____

_____

11. Based on the information in this activity so far, predict what could happen to the Earth's average surface temperature if atmospheric carbon dioxide levels continue to rise at an accelerated rate:

_____

_____

12. The greenhouse effect is essential to life on Earth. Why then are scientists concerned about the continuing rise in atmospheric carbon dioxide? What dangers might increasing carbon dioxide levels present to humans and ecosystems?

_____

_____

_____

_____

_____

_____

©2018 **BIOZONE** International
ISBN: 978-1-927309-55-1
Photocopying Prohibited

80 Feedback Systems

ENGAGE: Sea ice in retreat

The series of photos above shows the progressive melting of sea ice in the same area over time.

1. (a) What do you notice about the relative change in sea ice (white patches) to water and land (dark patches) over time:

_____

_____

_____

   (b) Recall from the model you developed earlier that lighter areas of the Earth reflect back some of the Sun's energy. This is called albedo. Predict how a decrease in sea ice could affect the amount of energy reflected back:

_____

_____

_____

EXPLORE: Feedback in Earth's systems

Feedback occurs when the output of a system is used as input in that same system forming a circuit or loop. Any change continues to influence the activity of the system. There are many feedback loops on Earth, both negative and positive, operating at the same time. As you have seen in the earlier chapter, negative feedback tends to stabilize a system around a mean (average or stable condition) whereas positive feedback tends to increase departures from the mean. Positive feedback is sometimes called reinforcing feedback because of the way it acts.

Negative feedback in nature

Feedback within and between the Earth's systems are complex because changes in one part of the system can have multiple effects (for example, more water vapor traps more heat and enhances the warming effect of CO_2 but it also increases cloud cover, which reduces heat absorption). The diagram on the right illustrates a simplified negative feedback loop involving the production of clouds. Clouds reflect incoming sunlight back into space so have the effect of lowering the Earth's surface temperature.

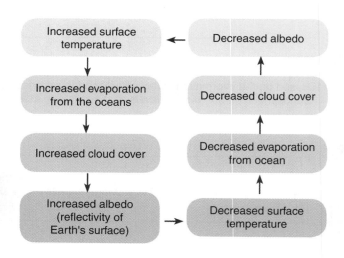

Negative feedback loops help to stabilize the Earth's climate. The evaporation of water from the oceans is affected by temperature, which may be influenced by an increase in solar output or carbon dioxide. The negative feedback of cloud production keeps the cloud cover of the Earth relatively constant.

NASA Earth Observatory

ESS3.D   CE   SSM   SC

©2018 **BIOZONE** International
ISBN: 978-1-927309-55-1
Photocopying Prohibited

Positive feedback in nature

Positive feedback loops on Earth tend to drive large scale changes to environments and the climate. The current increase in CO_2 in the atmosphere is driving numerous positive feedback loops. The diagram on the right illustrates the effect of methane (a potent greenhouse gas) release from permafrost. Permafrost is a subsurface layer of soil that remains frozen throughout the year. As the Earth warms, the permafrost melts, releasing methane which in turn causes the Earth to warm further.

Several positive feedback loops acting at the same time can potentially cause large changes to the climate. Although these are balanced to some extent by counterbalancing negative feedbacks, it is likely there will eventually be a "tipping point" at which a runaway climate change event will occur.

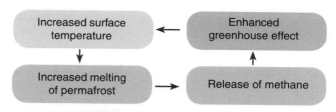

Melting permafrost, Hudson Bay, Canada

Steve Jurvetson cc2.0

2. Increased surface temperatures also increase the amount of ice melting and so decrease the Earth's albedo. In the space below, draw a simple feedback model (similar to the one above) to show how an increased surface temperature can cause a positive feedback loop:

Visualizing the albedo effect

A model where sea ice is retained

Arctic sea-ice summer minimum (white area) **1980**: 7.8 million km²

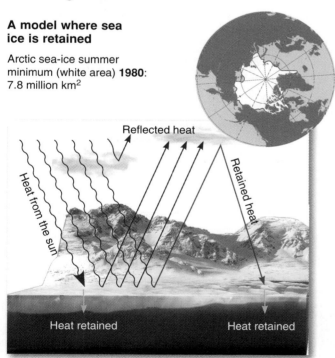

The high **albedo** (reflectivity) of sea-ice helps to maintain its presence. Thin sea-ice has a lower albedo than thick sea-ice. More heat is reflected when sea-ice is thick and covers a greater area. This helps to reduce the sea's temperature.

A model where sea ice is decreasing

Arctic sea-ice summer minimum (white area) **2012**: 3.41 million km²

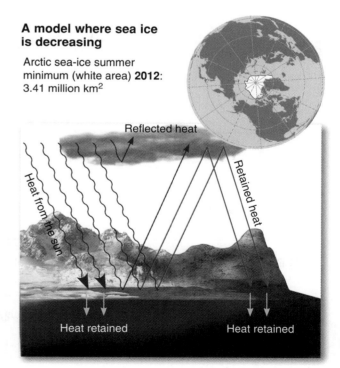

As sea-ice retreats, more non-reflective surface is exposed. Heat is absorbed instead of reflected, warming the air and water and causing sea-ice to form later in the fall than usual. Thinner and less reflective ice forms, perpetuating the cycle.

©2018 **BIOZONE** International
ISBN: 978-1-927309-55-1
Photocopying Prohibited

3. Using the sea ice models on the previous page, calculate the difference in summer sea-ice area between 1980 and 2012:

_____

4. How does low sea-ice albedo and volume affect the next year's sea-ice cover? _____

_____

_____

_____

5. What type of feedback system is operating here? _____

EXPLORE: Sea ice melting

▸ The surface temperature of the Earth is partly regulated by the amount of ice on its surface, which reflects a large amount of heat into space. However, the area and thickness of the polar sea-ice is rapidly decreasing. From 1980 to 2008 the Arctic summer sea-ice minimum almost halved, decreasing by more than 3 million km². The 2012 summer saw the greatest reduction in sea-ice since the beginning of satellite recordings.

▸ This melting of sea-ice can trigger a cycle where less heat is reflected into space during summer, warming seawater and reducing the area and thickness of ice forming in the winter. At the current rate of reduction, it is estimated that there may be no summer sea-ice left in the Arctic by 2050.

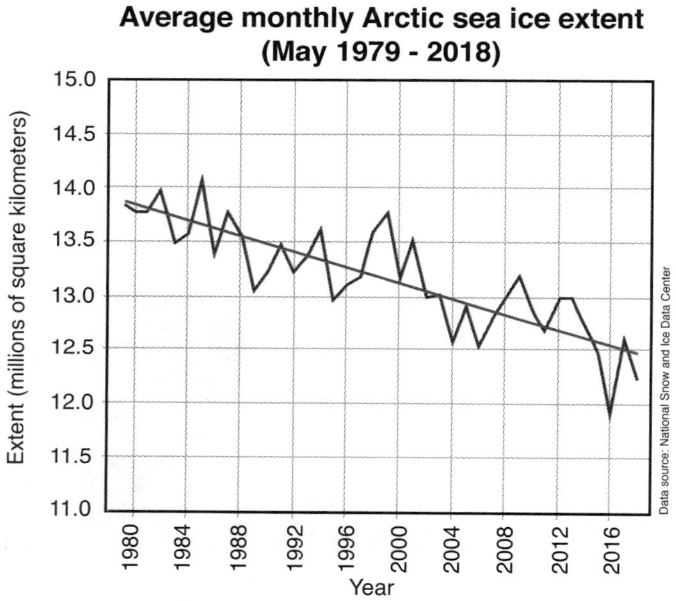

Average monthly Arctic sea ice extent (May 1979 - 2018)

Extent (millions of square kilometers)

Year

Data source: National Snow and Ice Data Center

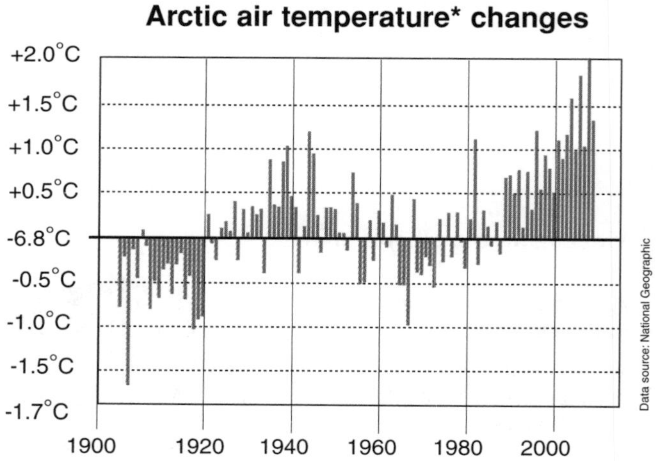

Arctic air temperature* changes

Data source: National Geographic

*Figure shows deviation from the average annual surface air temperature over land. Average calculated on the years 1961-2000.

6. (a) Describe what has happened to the extent of Arctic sea ice since 1980: _____

_____

(b) Describe what has happened to Arctic air temperature since the early 1900s: _____

_____

(c) Based on the data above, do you think that the change is Arctic air temperature is affecting the extent of the Arctic sea ice? Justify your answer:

_____

_____

_____

_____

©2018 **BIOZONE** International
ISBN: 978-1-927309-55-1
Photocopying Prohibited

EXPLORE: Modeling ice sheet melting

The investigation described below provides a model to understand the importance of heat absorbance and reflectivity to ice sheet melting.

Aim

To investigate the effect of albedo on ice sheet melting.

Method

▶ Using two 500 mL Florence or Erlenmeyer flasks, paint one black and coat a second with aluminum foil.

▶ Weigh and record the mass of six ice cubes (~60-90 g) for each flask. The total masses should be equal.

▶ Add 200 mL of 20°C water and the weighed ice cubes to each flask. Seal the flasks and insert a thermometer into each. Record the temperature (time zero).

▶ Leave the flasks in a sunlit area and record the temperature every two minutes for ten minutes. You could also use a 60W tungsten lamp placed 15 cm from the flasks.

▶ After ten minutes remove the ice cubes and weigh them again. Record the values below.

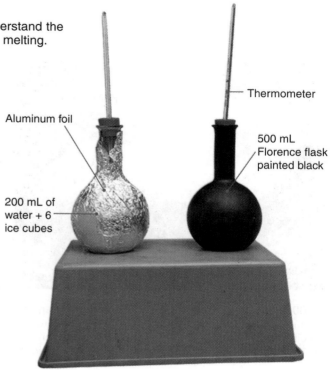

Aluminum foil

Thermometer

500 mL Florence flask painted black

200 mL of water + 6 ice cubes

Data collection

Record the data below and plot it on the grid right:

Time (minutes)	Temperature black flask (°C)	Temperature foil coated flask (°C)
0		
2		
4		
6		
8		
10		
Initial mass of ice (g)		
Final mass of ice (g)		

7. Which flask has the greater albedo? _____

8. Calculate the change in mass of the ice cubes for both the black and foil covered flasks: _____

_____

9. Why is it important to start with the same total mass of ice in each flask? _____

_____

10. What would you change if you wanted to show the effect of more or less sea ice on albedo? _____

_____

11. Write a conclusion for the investigation: _____

_____

_____

_____

_____

©2018 **BIOZONE** International
ISBN: 978-1-927309-55-1
Photocopying Prohibited

81 Models of Climate Change

ENGAGE: Can climate models help prevent tragedies?

Hurricane Katrina over the Gulf of Mexico, 2005

Flooding in New Orleans as a result of Hurricane Katrina

In 2005, Hurricane Katrina caused catastrophic damage along the Gulf coast from central Florida to Texas. New Orleans suffered severe flood damage because large volumes of water from Lake Pontchartrain were pushed against the city's levee system and flood walls causing them to fail in 53 places. Although The National Weather Service and the National Hurricane Center predicted catastrophic damage and evacuation orders were given, 1,464 people died.

We know that one of the crucial factors in the formation of a hurricane is a sea surface temperature of at least 27°C. According to the Environmental Protection Agency (EPA) sea surface temperature has increased since the 1900s, especially in the past three decades. Information like this can be placed into climate models to predict the severity and frequency of storm events. This can be used to help vulnerable communities (such as in New Orleans) design solutions to minimize damage from storm events.

1. Predict the effect of increasing surface sea temperature on the frequency of hurricanes: _____

   _____

   _____

2. Some climate models forecast an increased frequency of damaging hurricanes. Suggest how this information can help the engineers and officials of New Orleans better protect their city and their people.

   _____

   _____

   _____

   _____

EXPLORE: What is a climate model?

▶ **Climate** refers to the statistics of weather conditions (wind, precipitation, temperature, etc) at a location over a long period of time. Climate is the result of interactions among the Earth's systems. The complexity of these interactions makes climate difficult to understand and predict.

▶ Scientists use models to break the Earth's climate systems into components that can be more easily studied and understood. As the knowledge about a system grows, more components can be added so that it more closely represents the real system.

▶ Models used to understand climate are called **climate models**. They are mathematical representations and are very complex because the many different factors affecting the climate must be accounted for.

▶ The factors affecting the climate are interconnected. A change in one factor has an effect on another. Scientists manipulate the various components of the model and see what the outcome is.

▶ Climate models can be developed on different scales, i.e. for a particular region or globally.

   LS4.C ESS3.D ETS1.B CE SSM SC

©2018 **BIOZONE** International
ISBN: 978-1-927309-55-1
Photocopying Prohibited

What are climate models used for?

Climate models have several purposes:

▸ To understand the present climate, including the factors contributing to it.

▸ To project climatic conditions into the future.

▸ To investigate how natural processes or human activity may affect climate.

▸ To predict the effect of certain activities so that changes can be recommended to help prevent or slow down further climate change.

Climate data is utilized by a wide range of people.

▸ Farmers may use the data to plan ahead for changing rainfall patterns by changing crop types or adjusting their livestock management practices.

▸ Cities can plan and implement strategies to manage water supplies and protect infrastructure through periods of low rainfall and flooding.

▸ The energy sector uses the data to forecast energy consumption. For example, energy consumption increases in a heat wave as people use fans and air conditioning to keep cool. Energy production can be increased to meet demand. In the long term, if global warming continues, more electricity capacity may be needed to meet demand.

The hydroelectric energy produced by the Hoover dam (above) varies depending on energy demands.

How are climate models tested?

To see how well models work, scientists enter past data and see how accurately they predict the climate changes that have already occurred. If the models recreate historical trends accurately, we can have confidence that they will also accurately predict future trends in climate change.

The graph on the right shows an example of how climate models are tested. The orange-yellow lines represent data from 14 models and 58 different simulations. The red line represents the average of all 58 simulations. The black line represents the average actual (observed) data for the same period. The gray vertical lines represent large volcanic eruptions during the period.

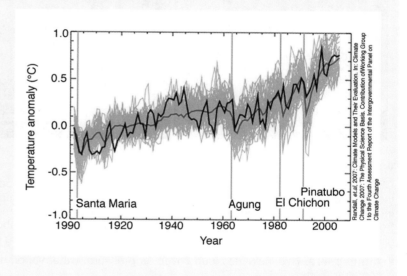

3. (a) How do scientists test the accuracy of their climate models? _____

_____

_____

(b) How do accurate models help manage resources effectively? _____

_____

_____

_____

_____

(c) Study the testing results on the graph above. Do you think the average data from the models accurately reflects the historical data? Why or why not?

_____

_____

_____

©2018 BIOZONE International
ISBN: 978-1-927309-55-1
Photocopying Prohibited

EXPLORE: Building climate models

▶ Climate models have been in use since the 1950s, but these very early versions really only modeled the weather in a particular region.

▶ The sophistication and accuracy of climate models has increased over time. This is because our knowledge about factors contributing to climate has increased and also because developments in computing and mathematics have allowed us to predict more complicated scenarios more accurately.

▶ In 1988, the Intergovernmental Panel on Climate Change (IPCC) was established. Its role is to analyze published climate data and inform the international community about their findings.

▶ The diagram below shows how the sophistication of climate models have changed over time.

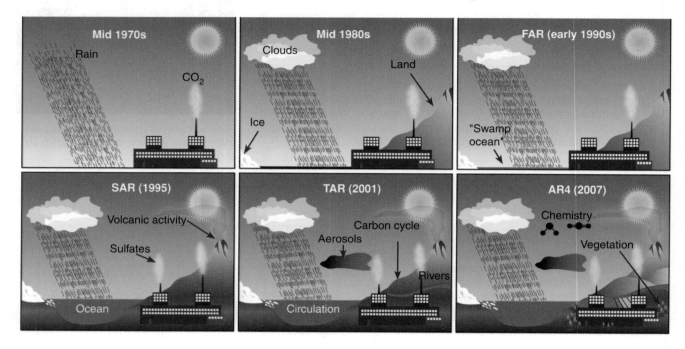

What should a climate model include?

Climate models predict climate change more accurately when the model incorporates all the factors contributing to climate change. Some components influencing climate (e.g. the ocean and atmosphere) have their own models. Data from these separate models can provide more detailed information about the climate model as a whole. Most models now incorporate the following components:

Ocean

- **Atmosphere**: This includes cloud cover, air pressure, water vapor, and gas concentrations.

- **Oceans**: Oceans have a key role in climate regulation. They help to buffer (neutralize) the effects of increasing levels of greenhouse gases in the atmosphere by acting as a carbon sink. They act as a heat store, preventing rapid rises in global atmospheric temperature.

- **Ice sheets and sea ice (the cryosphere)**: These factors influence how much of the Sun's heat is reflected or retained. Increased ice levels reflect more heat away from Earth. Less ice allows more heat to be retained.

Carbon emissions

- **Biogeochemical cycles**: Levels of some atmospheric compounds can greatly influence climate change. Carbon is the most significant, but others such as nitrogen, phosphorus, and sulfur can also influence climate.

- **Biosphere**: The level of plant cover on Earth has a significant impact on the amount of atmospheric carbon. During photosynthesis, plants utilize carbon dioxide from the atmosphere to produce carbohydrates, effectively removing a major greenhouse gas from the atmosphere.

Deforestation

- **Human activity**: Human activity has increased the rate of global warming, especially through the actions of deforestation (large scale tree removal) and carbon emissions into the atmosphere. The addition of greenhouse gases into the atmosphere through human activity is driving current climate change.

- **External influences**: These include energy variations from the Sun (e.g. through sunspot cycles) and levels of carbon dioxide and other aerosols released during volcanic eruptions.

©2018 **BIOZONE** International
ISBN: 978-1-927309-55-1
Photocopying Prohibited

4. (a) How has the complexity of climate models changed over time? _____

_____

_____

_____

_____

   (b) Do you think a complex model will provide more accurate forecasts than a simple model? Explain your answer:

_____

_____

_____

_____

_____

5. (a) Working in pairs or small groups, select one component of a climate model (opposite page) and research its significance to climate change. Summarize your findings and report back to the class.

   (b) Once all the presentations have been made, determine if any factor(s) has a larger influence than any other.

EXPLAIN: Using climate models to predict change

▶ There are elements of uncertainty, even in well tested models. The major source is human activity and, in particular, how will consumption of fossil fuels change in the future? The level of greenhouse gases in the atmosphere will have a significant impact on future climate change.

▶ The IPCC often run a number of different scenarios to predict climate change. Between them, the results provide a best-case and worst-case scenario.

▶ The major scenarios are presented below, but there are subcategories (e.g. A1B) to help make them more accurate:

• **A1** assumes rapid economic and technological growth, a low rate of population growth, and a very high level of energy use. Differences between "rich" and "poor" countries narrow.

• **A2** assumes high population growth, slower technological change and economic growth, and a larger difference between countries and regions than in other scenarios. Energy use is high.

• **B1** assumes a high level of environmental and social consciousness and sustainable development. There is low population growth, high economic and technological advancement, and low energy use. The area devoted to agriculture decreases and reforestation increases.

• **B2** has similar assumptions to B1. However, there are more disparities between industrialized and developing nations, technological and economic growth is slower than in B1, and population growth is greater (but less than A2). Energy use is midway between B1 and A2. Changes in land use are less dramatic than in B1.

Predictions of 2001 models

Climate scenario
A2 ————
A1B ----------
B1 —·—·—·—

Temperature change (°C) vs Year (1900–2100)

Predictions of 2012 models

Climate scenario
RCP 8.5 (~A2) ————
RCP 6.0 (~A1B) ----------
RCP 4.5 (~B1) —·—·—·—

Temperature change (°C) vs Year (1900–2100)

6. Why do you think scientists simulate a number of different scenarios when they run a climate model? _____

_____

_____

_____

_____

©2018 **BIOZONE** International
ISBN: 978-1-927309-55-1
Photocopying Prohibited

7. Study the 2001 and 2012 models of climate change predictions on the previous page.

(a) In the 2001 model, identify the scenario predicted to produce the highest temperature change by 2100:

_____

(b) What factors are likely to contribute to this? _____

_____

(c) Why would scenario B1 produce the lowest temperature increase? _____

_____

_____

(d) How do the predictions between the 2001 and 2012 models differ? _____

_____

_____

_____

EXPLAIN: Using climate models to predict sea level rise

What causes sea level rise?

The increase in the Earth's surface temperature is linked to a rise in global sea level. Sea level rise occurs because of two main factors, thermal expansion and melting ice. When water heats up, it expands and takes up more space. Around 24 million km^3 of water is stored in permanent snow, glaciers, and ice caps. When these melt, they add to the volume of water in the oceans. Sea level rise will not only affect people living in coastal communities, but also Earth's systems. Many models have been developed to predict sea level rise under different scenarios in order to determine its effect.

What effect will sea level rise have in the US?

In 2010, around 39% of the US population (around 123 million people) lived in counties directly on the shoreline. This population is expected to increase by 8% by 2020. Rising sea levels therefore represent a significant hazard to the US. Large cities such as New York (right) are in danger of becoming inundated (flooded) as sea levels rise. Other large cities, such as San Francisco and Los Angeles, are at sea level, or close to it. In New York, a sea level rise of only a few meters would inundate thousands of hectares of highly developed land. Airports, ports, railroads, housing and infrastructure, highways, and industry would be damaged.

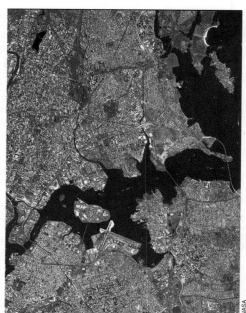

The New York city coastline, seen from space

8. Why is the US so vulnerable to sea level rise?

_____

_____

9. (a) Study the graph on the right. What does it show?

_____

_____

_____

_____

(b) What is the worst case scenario? _____

_____

(c) What is the best case scenario? _____

_____

Predicted global sea level rise under a number of scenarios

Observed | Scenarios

Global sea level rise (cm above 1992)

200 · 160 · 120 · 80 · 40 · 0 · -40

1900 · 1950 · 2000 · 2050 · 2100

Year

Global Sea Level Rise Scenarios for the United States National Climate Assessment, NOAA (2012)

©2018 **BIOZONE** International
ISBN: 978-1-927309-55-1
Photocopying Prohibited

EXPLAIN: How are climate models used to predict distribution changes?

▶ Global warming is changing patterns of temperature distribution, precipitation (rain and snowfall), and vegetation. These changes alter the habitats of organisms and may have serious effects on regional and global biodiversity. As temperatures rise, the distribution of organisms may change as they shift to stay within the range of their temperature tolerance. Alpine plants and animals are particularly vulnerable. As temperatures increase, the snow line increases in altitude pushing alpine organisms to higher elevations (see example of the pika below).

▶ Climate models can be used to determine how climate change might affect the distribution and survival of living organisms. For some organisms their habitat will expand while for others it will shrink.

Case study: The American pika

The North American pika (*Ochotona princeps*) lives in the mountains of western North America. In the US, their range extends from Washington to California and New Mexico. Their typical habitat is the boulder fields at or above the tree level, they use the rocks to make their homes.

In the northern extent of their range pika may be found near sea level, but in the southern areas of their range they are usually found above 2500 meters.

Pika are very sensitive to high temperatures and will die in six hours if exposed to temperatures above 25.5°C. For this reason, they tend to be active in the cooler morning and evening part of the day, and shelter in their burrows during the hot parts of the day. For many populations, a northward migration is not possible because the valleys they would need to pass through are too hot for their survival.

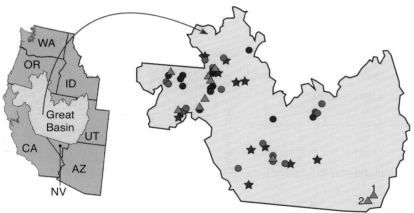

KEY

● Pikas present (2014-2015 survey)

● Pikas present (historical surveys)

★ Pikas locally extinct (historical surveys)

▲ Pikas locally extinct (2014-2015 survey)

1 ▲ Cedar Breaks National Monument

2 ▲ Zion National Park

Modified from Beever, E. *et.al* (2016) Pika (Ochotona princeps) losses from two isolated regions reflect temperature and water balance, but reflect habitat area in a mainland region. J. of mamolgy, 6(5)

A survey in the Great Basin (within the pika range) in 2014-2015 compared current pika populations with historical records (above). If researchers sighted or heard pika, or found fresh hay piles (piles of vegetation) it was recorded as a positive result. Searches were made in the cooler part of the day. The researchers revisited sites on two different days if they thought the pika were no longer present. The results found widespread evidence of a decrease in range, up slope relocation (shifting to higher altitude), and local extinction. For example, pika were detected in 2011 and 2012 in Zion National Park and Cedar Banks National Monument respectively. They were not detected in 2014-2015.

10. Why did researchers look for pika in the cooler parts of the day? _____

_____

11. The Cal-adapt website provides climate model data for a variety of parameters including temperature. Visit their website (see **BIOZONE's Resource Hub** for the URL) to view predicted maximum temperatures data. Select the CanESM2 climate model. This model shows historical and predicted annual average values from 32 different climate models. Use the data and state the predicted average maximum temperature for:

(a) 2017: _____ (b) 2020: _____ (c) 2040: _____

(d) Predict the likely effects of these temperatures on pika: _____

_____

_____

_____

©2018 **BIOZONE** International
ISBN: 978-1-927309-55-1
Photocopying Prohibited

12. Why might up slope relocation be the only way pika will be able to respond to global warming? _____

_____

_____

_____

13. Pika living at lower altitudes are more at risk from climate change than those living at higher elevation. Explain why:

_____

_____

_____

EXPLAIN: Will climate change affect malaria distribution?

Malaria is a serious, potentially fatal, disease affecting 300 million people a year. The parasite (*Plasmodium*) that causes malaria is carried by *Anopheles* mosquitoes. When an infected mosquito bites a human, the parasite infects the red blood cells and multiplies inside them. The parasites then burst out, destroying the blood cells, and can then reinfect another mosquito.

Malaria is a major health problem in tropical regions where the climate is warm and wet enough for the mosquito to breed in large numbers. Larger numbers of mosquitoes create greater opportunity for malaria to be spread to people. With global temperatures predicted to increase, there is potential for *Anopheles* to expand its range, and malaria could spread into areas that are currently malaria-free. Models can be used to predict this.

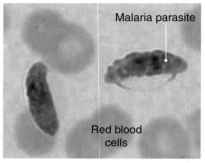
Malaria parasite
Red blood cells

The graph (right) shows the estimated doubling time of the malaria parasite (*P. falciparum*) over a range of temperatures for different mosquito densities. Density values are given for high numbers of mosquitoes relative to humans and for low numbers of mosquitoes relative to humans.

14. What happens to the doubling time with change in temperature?

_____

_____

15. (a) Which mosquito density poses the greatest malaria risk to humans?

_____

(b) Suggest why: _____

_____

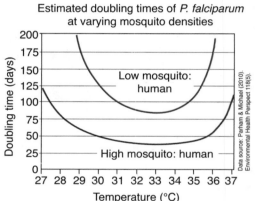
Estimated doubling times of *P. falciparum* at varying mosquito densities

Doubling time (days): 200, 175, 150, 125, 100, 75, 50, 25, 0
Low mosquito: human
High mosquito: human
Temperature (°C): 27 28 29 30 31 32 33 34 35 36 37

Data source: Parham & Michael (2010). Environmental Health Perspect 118(5).

16. The map below shows the current global distribution of malaria (latitude lines included). Malaria does not currently occur in the US, but it does occur in Central and South America. Carry out your own research to find models predicting how climate change might affect the distribution of malaria. Color the map to show any changes. Comment on any changes to the US and suggest what the health impacts of this could be. Summarize your results and attach them to this page.

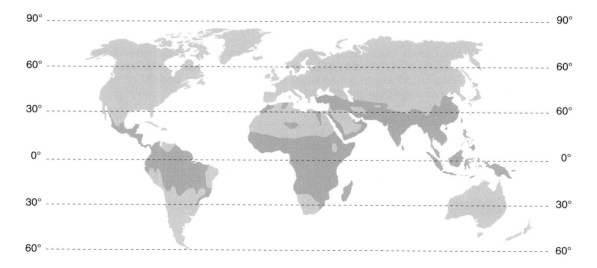

90°   90°
60°   60°
30°   60°
0°   0°
30°   30°
60°   60°

©2018 **BIOZONE** International
ISBN: 978-1-927309-55-1
Photocopying Prohibited

EXPLAIN: How could climate change affect agriculture?

▶ The impacts of climate change on agriculture and horticulture will vary around the globe because of local climate and geography. In some regions, temperature changes will increase the growing season for existing crops or enable a wider variety of crops to be grown.

▶ Changes in temperature or precipitation patterns may benefit some crops, but have negative effects on others. Increasing atmospheric CO_2 levels will enhance the growth of some crops, but rising nighttime temperatures may affect seed set and fruit production.

Effects of increases in temperature on crop yields

Studies on grain production in rice have shown that higher maximum daytime temperatures have little effect on crop yield. However, higher minimum night time temperatures lower crop yield by as much as 5% for every 0.5°C increase. The results are shown in the two graphs below.

The effects of temperature on grain yield

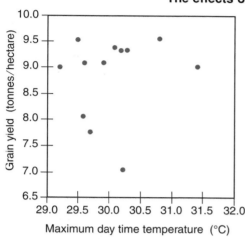

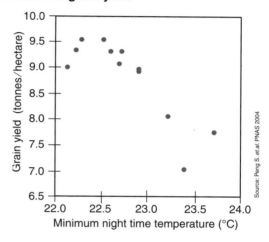

Source: Peng S. et.al. PNAS 2004

Possible effects of increases in temperature on crop damage

The fossil record shows that global temperatures rose sharply around 56 million years ago (the Paleocene-Eocene Thermal Maximum (PETM)). Studies of fossil leaves with insect browse damage indicate that leaf damage peaked at the same time. This gives some historical evidence that, as temperatures rise, plant damage caused by insects also rises. This could have implications for crops.

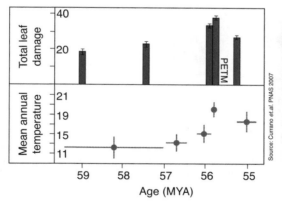

Source: Currano et.al. PNAS 2007

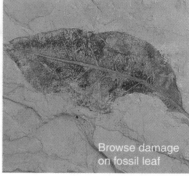

Browse damage on fossil leaf

17. Suggest why global warming might benefit some agricultural crops, while disadvantaging others: _____

_____

_____

_____

_____

18. What evidence is there that global warming might influence the distribution or number of crop pests and so affect agriculture?

_____

_____

_____

©2018 **BIOZONE** International
ISBN: 978-1-927309-55-1
Photocopying Prohibited

82 Solutions to Climate Change

ELABORATE: How can new technologies be used to slow climate change?

If greenhouse gas production was to completely stop today, the Earth's temperature would still continue to rise slightly. In order to prevent continued global warming, most climate scientists agree that we need to reduce our level of greenhouse gas emissions. This may slow the most damaging effects of climate change.

There are many possible ways to achieve this. They include improving energy efficiency, increasing the use of renewable energy sources, and using carbon capture technologies.

We can also promote carbon capture naturally by planting more trees and reducing the rate of deforestation.

Tree plantation, USA

USDA Natural Resources Conservation Service cc2.0

Large-scale carbon producers

▸ Burning fossil fuels in power stations for electricity accounts for about 40% of global carbon dioxide emissions. The transport industry accounts for at least another 30%. Even power stations using high quality coal and oil release huge volumes of CO_2. Systems that capture the CO_2 produced so that it can be stored or used for other purposes are beginning address this problem (below).

▸ Another important source of CO_2 emissions is often overlooked, probably because it is so common we don't stop to consider its effects. The manufacture of cement and concrete account for 5-10% of all CO_2 emissions. New types of cement and techniques for manufacture are aiming to reduce this amount.

Schematics of three possible carbon capture systems

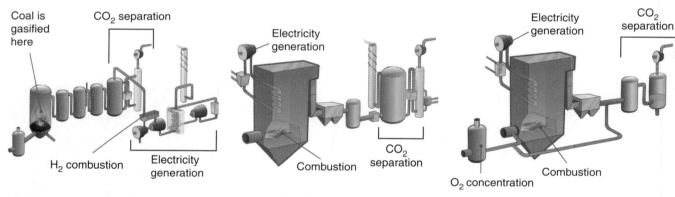

Pre-combustion capture: The coal is converted to CO_2 and H_2 using a gasification process. The CO_2 is recovered while the H_2 gas is combusted.

Post combustion capture: CO_2 is washed from the flue gas after combustion. It is then passed to a desorber to re-gasify the CO_2, where it is then compressed for storage.

Oxyfuel combustion: Concentrated O_2 is used in the furnace, producing only CO_2 gas in the flue gas. This is then compressed for storage. Compressed CO_2 is useful as a inexpensive, nonflammable pressurized gas, e.g. for inflation and for carbonated water.

1. Describe the differences and similarities of the three types of carbon dioxide capture systems described above:

_____

_____

_____

_____

_____

_____

_____

_____

   LS2.C LS4.D ESS3.C ESS3.D ETS1.A ETS1.B ETS1.C SSM SC

©2018 **BIOZONE** International
ISBN: 978-1-927309-55-1
Photocopying Prohibited

Storing captured CO_2

Captured CO_2 can be injected into porous strata between non-porous layers. Power stations near to injection sites can pipe the recovered CO_2 to the injected well. Other stations will need to transport the CO_2 to the site. The transportation of the CO_2 will produce less CO_2 than is captured by the power station, making the option viable.

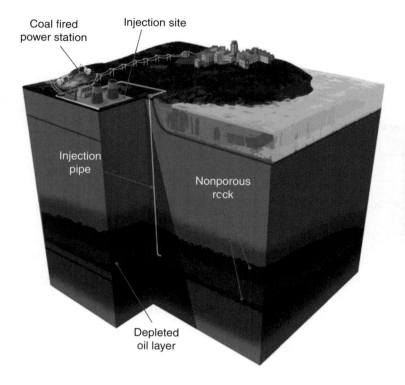

Coal fired power station

Injection site

Injection pipe

Nonporous rcck

Depleted oil layer

CO_2 can be stored by injecting it into depleted oil wells or other deep geological formations, releasing it into deep ocean waters (above), or reacting it with minerals to form solid carbonates. The CO_2 can also be used as a starting point for the production of synthetic fuels.

Deep ocean storage of CO_2 risks lowering ocean pH. Storing CO_2 in geological formations risks sudden release of large quantities of CO_2 if the rock proves unstable. The sudden release of CO_2 can kill animal life in the area (above).

Case study: Lowering emissions in the cement industry

▶ Cement and concrete (the final cured product of cement mixed with water and gravel) are essential to the building industry and the global economy. 4.1 billion tonnes of cement were produced in 2015. This is expected to increase to 4.8 billion tonnes by 2030. This is important to the global climate because producing one tonne of cement also produces about one tonne of CO_2 (and uses the equivalent of 200 kg of coal).

▶ The most common cement used is called Portland cement, which is very strong when set. Its manufacture is highly energy intensive. 40% of the CO_2 emissions come from the burning of fossil fuels to heat limestone ($CaCO_3$) and other minerals to around 1400 °C. At this temperature, the limestone degrades and releases CO_2. This step accounts for about 50% of the CO_2 emissions. Portland cement reabsorbs about half of this CO_2 as it hardens over the life time of the cement.

▶ Reducing the CO_2 emissions for cement manufacture can be done at three steps in the process: Reducing the amount of fossil fuels needed to heat the raw materials, reducing the amount of CO_2 released by the raw materials, and increasing the amount of CO_2 absorbed when setting.

▶ New types of cement using magnesium silicates instead of limestone are under trial. These do not need to be heated to such high temperatures and they do not release CO_2 when heated. This leads to total CO_2 emissions of up to 0.5 tonnes of CO_2 per tonne of cement produced (half that of Portland cement). Additionally, during setting, CO_2 is absorbed at a greater rate than in traditional cement (about 1.1 tonnes per tonne of cement produced). This type of cement is often called carbon negative cement as it actually absorbs more CO_2 than is produced in its manufacture (about 0.6 tonnes of CO_2 absorbed for every tonne of cement produced).

Cement factory

©2018 **BIOZONE** International
ISBN: 978-1-927309-55-1
Photocopying Prohibited

2. Describe how captured carbon dioxide might be used or stored: _____

_____

_____

_____

3. Discuss some of the potential problems with capturing and storing carbon dioxide: _____

_____

_____

_____

4. (a) Approximately how many tonnes of CO_2 were produced by the cement industry in 2015? _____

   (b) Where is this CO_2 produced in the manufacture of cement? _____

   _____

   _____

   (c) Explain why carbon negative cement is carbon negative: _____

   _____

   _____

   (d) Based on the 2015 figures, how much carbon would carbon negative cement absorb?_____

   _____

EVALUATE: Designing and evaluating climate change solutions

Climate change is a major global challenge. The problem is complex, with many contributing factors. It will take collaboration between many countries, and will require many different approaches and solutions to tackle the problem of climate change successfully.

Many researchers suggest that greenhouse gas concentrations must be stabilized around 450-550 parts per million (ppm) to avoid the most damaging impacts of climate change in the future. Current concentrations are about 380 ppm. To reach this target, greenhouse gas emissions need to be reduced by 50% to 80% of what they're on track to be in the next century (IPCC).

5. Your challenge is to analyze and design a solution to tackle anthropogenic climate change. Divide the class into groups.

   (a) Each group should analyze how the greenhouse gases produced from human activity contribute to global warming. Specify qualitative and quantitative criteria for measuring the scope of the problem. Once you have completed your research, set the qualitative and quantitative targets you think are required to resolve the problem.

   (b) Present your group's findings to the class. The class must then decide collectively what goal and targets are required to mitigate the effects of climate change. You may find that there are a number of different opinions expressed, and compromises may have to be made to reach a consensus.

   (c) Now your target is set, work in your groups and evaluate a solution to reach the agreed target. Some groups should investigate the use of carbon reduction techniques, while others can look at how carbon sequestering and storage techniques can be used to achieve your goal. Consider and prioritize a range of criteria when evaluating your solution. This includes cost, safety, reliability, and aesthetics. You should also take into account any social, cultural, and environmental impacts.

   (d) Finish with a "climate change summit" in your classroom. Each group should present their findings and recommendations to the class as a whole.

CA EP&Cs V: The spectrum of what is considered in making decisions about resources and natural systems and how those factors influence decisions (Va).

Many factors must be taken into consideration when trying to solve the complex problem of climate change.

▸ Human need for resources and energy must be balanced with environmental considerations to mitigate climate change.

▸ Solutions must be prioritized against a range of criteria including cost, safety, reliability, aesthetics, cultural, societal, and environmental needs.

©2018 **BIOZONE** International
ISBN: 978-1-927309-55-1
Photocopying Prohibited

83 Human Impact on Ecosystems

ENGAGE: Flood!

Sea levels are expected to rise by up to 58 cm by the year 2090. This is the result of the thermal expansion of ocean water and melting of glaciers and ice shelves. Many of North America's largest cities are near the coast. The predicted rises in sea levels could result in inundation of these cities and entry of salt water into agricultural lands.

According to a recent report, 90 coastal communities in the US are already experiencing chronic flooding. This is defined as 10% or more of a community's usable land flooding 26 times a year. Using an intermediate-high prediction of sea level rise (1.2 m by 2100) more than 70 new communities will face inundation by 2035. A greater number of communities are predicted to be affected if models with higher sea level rise projections are used (2.0 m by 2100). Under both models, the number of affected communities increases with time. The East and Gulf states are particularly at risk, and they include some of the most populated cities in the US.

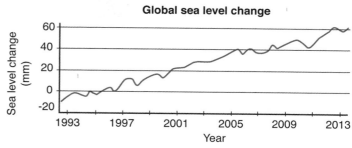

1. How has global sea level changed over time since 1993? _____

_____

_____

2. (a) Use an interactive map to study the predicted degree of inundation in San Francisco. One such interactive model is available for use via **BIOZONE's Resource Hub**. What is the effect of sea level rise over time on San Francisco?

_____

_____

_____

(b) If you have time, find your own neighborhood on the map. Is it affected by the predicted sea level rise? _____

EXPLORE: Case study: The effect of sea level rise on the Everglades

The Florida Everglades are a natural region of tropical wetlands in the southern region of Florida. The Everglades are a subtropical wetland and provide habitat for a wide diversity of species.

Low lying wet lands such as the Everglades risk being inundated by seawater. The Everglades (above) contains a variety of ecosystems including mangrove forests, sawgrass marshes, cypress swamps, and pine lands.

Studies have found that the small fish that make up the foraging base of many coastal species do better under less saline conditions. As the saltwater-freshwater interface moves inland, the production of the Everglades will decline.

Peat collapse is another effect of saltwater intrusion. The intrusion of saltwater removes the peat soil. Peat bogs support a variety of freshwater species, and the ecosystems they support helps filter the freshwater that humans depend on.

©2018 **BIOZONE** International
ISBN: 978-1-927309-55-1
Photocopying Prohibited

This map shows the amount of land that would be inundated around Florida by a 1 m sea level rise, as predicted by 2100.

Florida

Everglades

Land lost

The Everglades that make up a large part of southern Florida are extremely flat. In fact, moving inland they increase just 5 cm in altitude per kilometer. This means that even minor increases in sea level will have major effects on the wetland ecosystems found there.

▶ The rise in the mean sea level and inundation of the land are only part of the effect of sea level rise. As the sea level rises, salt water intrusion into ground water destroys freshwater habitat from underneath. In addition, storm surge becomes more important. As outlying islands, reefs, and other barriers to storm surges become closer to sea level, they are less effective at preventing flooding during storms. Storm surges during severe hurricanes can be up to 15 m above the high tide level. The storm surge can therefore cause severe erosion and damage to the land far beyond the mean high tide mark.

▶ Examples of ecosystem change in the Everglades due to sea level change include the movement of mangrove forests approximately 3 km inland since the 1940s and therefore the equivalent reduction in freshwater ecosystems behind the mangroves.

3. (a) What feature of the Florida Everglades make it susceptible to flooding? _____

_____

(b) Describe the effects of sea level rise on the ecosystems of the Everglades: _____

_____

_____

_____

_____

4. In the US, the American crocodile (right) is only found in southern Florida. They can tolerate high temperatures and brackish water (water that is a mix of sea water and fresh water). The American crocodile is less tolerant of cold than the American alligator. The alligator is more widespread and is found throughout the southeastern US and coexists with the American crocodile in southern Florida.

American crocodile

Tomás Castelazo CC2.5

(a) What could happen to the American crocodile range in Florida if sea levels rise?

_____

_____

_____

_____

(b) How might global warming affect its range overlap with the American alligator?

_____

_____

_____

(c) How could a range overlap affect competition between the two species?

_____

_____

_____

Southern Florida

The areas in orange is the current range of the American crocodile. The gray regions are protected areas.

©2018 **BIOZONE** International
ISBN: 978-1-927309-55-1
Photocopying Prohibited

EXPLORE: How does human activity affect ecosystems?

Human activity can have major effects on ecosystems. These include polluting the air, water, or soil of ecosystems, habitat destruction including deforestation, and removal of important species and introduction of invasive species. There are few places remaining on Earth that are not directly affected by human activities. For most of human civilization these impacts on the environment have had detrimental effects.

Detrimental impacts of human activities on ecosystems

Fragmentation of habitats
Development of natural landscapes separates habitats and makes it difficult for organisms to move between them.

Introduction of new species
Some introductions may have a limited impact, but most are harmful. Introduced species can often become pests as they often have few predators in their new environment. They can quickly spread and cause huge environmental damage.

Simplifying natural ecosystems
Plowing diverse grasslands and replanting them with a low diversity pasture affects the interrelationships of thousands of species.

Overharvesting
Overgrazing of native grasslands by livestock or overharvesting of trees from forests or fish from the sea can alter the balance of species in an ecosystem and have unpredictable consequences.

Pollution
Pollution can include the deliberate dumping of waste into the environment or unintended side effects of chemical use, e.g. estrogens in the environment or emission of ozone-depleting gases.

Reduced biodiversity
Native species may compete for the same resources as livestock. Natives are often actively excluded by fences or removal, reducing the biodiversity of the region.

5. Divide into small groups. Choose one of the categories of human impact from the graphic above. As a group, discuss its current detrimental effects and then discuss the likely effects of climate change on your chosen category.

(a) Summarize your findings in the space below and share them with your class: _____

_____

_____

_____

_____

_____

_____

(b) Do you think any one activity has more impact than the others? Explain why or why not: _____

_____

_____

_____

_____

©2018 **BIOZONE** International
ISBN: 978-1-927309-55-1
Photocopying Prohibited

EXPLORE: Future proofing California's water supply

Rivers are dammed for a variety of reasons including to produce electricity, for flood control, and to provide water reservoirs for irrigation, industry, and domestic use. The reservoirs also provide opportunities for recreational activities such as fishing and boating. Dams are crucial for managing and regulating California's water supply, so that water is available even in times of drought. Recall from IS3 that California has a history of flood and drought cycles.

As climate change progresses, more extreme weather conditions are predicted. For California, it is highly likely that large flood events followed by widespread droughts will become more common and more severe than currently experienced. The implications for managing California's water supply are huge. If California's water management systems are not properly controlled, hydroelectric power generation, flood management, and water supply for irrigation and domestic use will be affected.

The image (right) is of the Exchequer Dam at Lake McClure during drought conditions in 2015. Water levels fell so low that the old dam became exposed.

Exchequer Dam at Lake McClure, California

Paul Hames/California Department of Water Resources

6. Water management at dams can be very complicated as there are often conflicting goals to be considered. Trade-offs are often made when coming to a solution. Illustrate this using flood control as an example.

_____

_____

_____

_____

_____

_____

CA EP&Cs V: The spectrum of what is considered in making decisions about resources and natural systems and how those factors influence decisions (Va)

Careful consideration of many factors (environmental, recreational, and supply) is required when deciding how to operate a dam to satisfy the needs of many groups and prevent environmental harm.

▸ Over summer, recreational demands on water are high but water reserves are often low as water reserves are used to meet high water use (e.g. irrigation) and hydroelectricity demand (e.g. to cool houses).

The table below shows the actual water volumes of four major California water reservoirs between 2008-2018.

	Shasta capacity (AF*)	% of total capacity	Oroville capacity (AF*)	% of total capacity	San Luis capacity (AF*)	% of total capacity	Don Pedro capacity (AF*)	% of total capacity
Total capacity	4,552,000		3,537,577		2,041,000		2,030,000	
Sampling date								
01 June 2018	3,801,325	83.5	2,407,616		1,516,408		1,902,345	
01 June 2017	4,348,164		2,456,354		1,911,116		1,777,488	
01 June 2016 (D)	4,160,600		3,301,763		670,024		1,533,984	
01 June 2015 (D)	2,397,120		1,559,875		1,080,671		825,580	
01 June 2014 (D)	2,167,232		1,729,124		845,241		1,074,740	
01 June 2013 (D)	3,350,588		2,803,391		771,654		1,470,210	
01 June 2012 (D)	4,289,799		3,498,687		1,295,771		1,670,118	
01 June 2011	4,491,268		3,404,584		1,826,881		1,724,316	
01 June 2010	4,460,923		2,505,086		1,442,207		1,908,538	
01 June 2009 (D)	3,110,352		2,281,337		697,887		1,441,066	
01 June 2008 (D)	2,798,626		1,762,033		1,096,109		1,050,858	

*AF = acre foot (the volume of one acre of surface area to a depth of one foot),   D = drought

©2018 **BIOZONE** International
ISBN: 978-1-927309-55-1
Photocopying Prohibited

7. (a) Calculate the percentage of total capacity for each reservoir on each date and record it on the table. The first calculation has been done for you.

(b) Plot the changes in percentage of total capacity over time for each reservoir on the grid below:

(c) Study the graph you have drawn above. In general, what happens to reservoir levels when there is a drought?

_____

_____

(d) Extreme weather events (including droughts) are predicted to become more common in California as a result of climate change. Predict the effect of more frequent, longer lasting droughts on California's reservoirs:

_____

_____

_____

_____

(e) What problems could occur as a result of this? _____

_____

_____

_____

8. Climate change may also increase the occurrence of severe rainfall events. Use your knowledge from IS3 to predict how this could affect California's reservoirs:

_____

_____

_____

_____

©2018 **BIOZONE** International
ISBN: 978-1-927309-55-1
Photocopying Prohibited

EXPLAIN: How can climate change benefit invasive species?

An introduced species is one that has evolved in one place and has been transported by humans, either intentionally or accidentally, to another region. If the introduced species has a negative effect on their new ecosystem, they are called **invasive species**. Their new ecosystem often lacks the natural predators and diseases that would usually control their population growth, so their numbers can expand very rapidly. Native species may suffer as a result.

Case study: California's native mussels

Climate change may alter habitats by affecting carbon dioxide levels, temperature, and precipitation rates. As habitats change, this could enable invasive species to spread into new areas.

The native mussel *Mytilus trossulus* (a shellfish) used to be abundant along the coast of California. The Mediterranean blue mussel (*M. galloprovincialis*) was introduced into Californian waters in the early 1990s. It spread quickly along the Californian coastline and is rapidly out-competing and displacing *M. trossulus*. *M. trossulus* is now restricted to the northern Californian coastline (right). Genetic analysis has shown that *M. galloprovincialis* forms hybrids with the native *M. trossulus*.

The physiological tolerance of the two species is different. *M. galloprovincialis* has a higher tolerance for warmer water temperatures and higher salinities than the native *M. trossulus* which prefers lower temperatures and salinity.

An experiment measured heart rate to determine the effect of temperature on the two species. The data in the graph (right) shows how resting heart rate is affected by temperature for animals acclimated to different salinities. Heart rate is a measure of stress in mussels. If animals become too stressed they could die or would have to find a suitable habitat to reduce their stress.

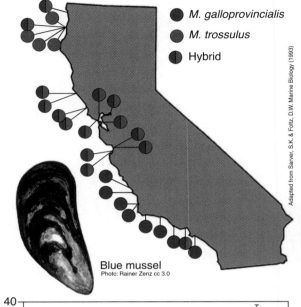

Blue mussel
Photo: Rainer Zenz cc 3.0

Adapted from Sarver, S.K. & Foltz, D.W. Marine Biology (1993)

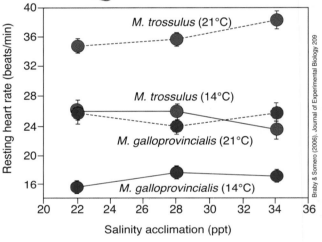

Braby & Somero (2006). Journal of Experimental Biology 209

9. Explain how the introduction of *M. galloprovincialis* affected the distribution of *M. trossulus* in California:

_____

_____

_____

_____

_____

_____

10. (a) Heart rate is a measure of animal stress. Explain how each species of *Mytilus* responds to an increase in temperature (14°C compared to 21°C):

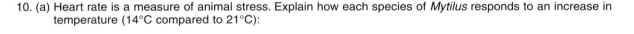

_____

_____

(b) What do you think would happen to the distribution of M. *trossulus and M. galloprovincialis* in response to the rising sea temperatures predicted to occur as a result of climate change?

_____

_____

_____

_____

(c) Predict the effect of increasing temperature on the heart rate of the hybrid species? _____

_____

_____

©2018 **BIOZONE** International
ISBN: 978-1-927309-55-1
Photocopying Prohibited

ELABORATE: Climate change and disease

Sea star wasting disease affects many species of starfish. White lesions first appear on the body of infected individuals and spread rapidly. Not long after, the tissue around the lesions begins to decay (right), ultimately causing the limb to fall off. Death can occur three days after the lesions first appear and large numbers of individuals can be affected, resulting in rapid population decline. A mass outbreak occurred along the West coast of the US in 2013.

The cause of the disease is unknown, but may be caused by a viral pathogen (*Densovirus*) or a marine species of *Vibrio*, a bacterium. Increased coastal water temperatures may also be an important factor. Temperature affects the generation (doubling) time of *Vibrio* bacteria. This is illustrated in the graph (below). Temperature also affects how long infected individuals survive after infection (below, right).

Elizabeth Cerny-Chipman Oregon State University CC2.0.

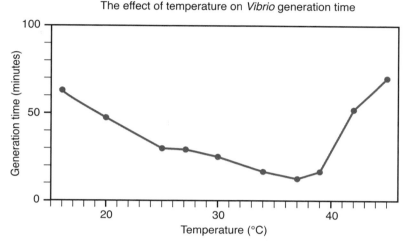

The effect of temperature on *Vibrio* generation time

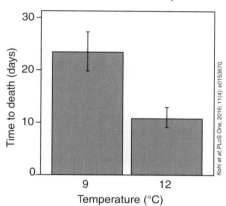

Average time to death of the ochre sea star, *Pisaster ochraceus*, at two temperatures

Kohl et al. PLoS One. 2016; 11(4): e0153670.

Scientists are concerned about how the predicted increase in sea temperature due to global warming will affect the incidence of sea star wasting disease. Current summer sea temperatures off the Californian coast are on average 20°C for more southern regions. Temperatures have a lower summer average in the northern regions (11°C).

Use the simulation on **BIOZONE's Resource Hub** to predict the effect of temperature change on *Vibrio* generation time.

11. From the simulation, determine the doubling time of *Vibrio* at:

(a) 20.0°C: _____   (b) 20.5°C: _____   (c) 21.0°C: _____

12. (a) Follow the simulation instructions on **BIOZONE's Resource Hub** and plot a graph to show the growth curves at each temperature. Print out your graph and attach it to this page.

(b) Use your plots to describe how temperature changes affect the generation time of *Vibrio* bacteria: _____

_____

_____

13. Assuming *Vibrio* infection causes sea star wasting disease:

(a) Predict the effect of warming sea temperatures on the incidence and effect of the disease: _____

_____

_____

(b) Which sea star populations in California (northern or southern) will be most affected by sea star wasting. Explain:

_____

_____

_____

_____

You can help monitor this disease by visiting **BIOZONE's Resource Hub** to access the Oregon State University's sea star wasting syndrome observation log.

©2018 **BIOZONE** International
ISBN: 978-1-927309-55-1
Photocopying Prohibited

EVALUATE: Designing a solution to a real world problem

▶ Deciding on a course of action for a complex situation (such as preserving biodiversity in the face of climate change) is not always simple. Environmental, cultural, recreational, and economic impacts must be taken into account. This wide range of criteria must be prioritized to help create an effective solution.

▶ The map below shows a hypothetical coastal area of 9,300 ha (93 km^2) in which two separate populations of an endangered bird species exist within a forested area of public land. A proposal to turn part of the area into a wildlife reserve to protect the endangered birds has been put forward by local conservation groups. The proposal would allow a single area of up to 1,500 ha (15 km^2) to be reserved exclusively for conservation efforts.

▶ In addition to homing the endangered birds, the area is known to have large deposits of coal. Trampers regularly use the tramping tracks in the area and hunters also spend time in the area because part of it has an established population of introduced game animals. Climate change models predict a sea level rise in the area. The new coastline predicted from a moderate rise in sea level is shown in red on the map below.

14. Study the map below and draw on to the map where you would place the proposed reserve, taking into account economic, cultural, and environmental values. There are a few lines for extra notes at the bottom of the page.
On a separate sheet, write a report justifying your decision as to where you placed the proposed reserve.

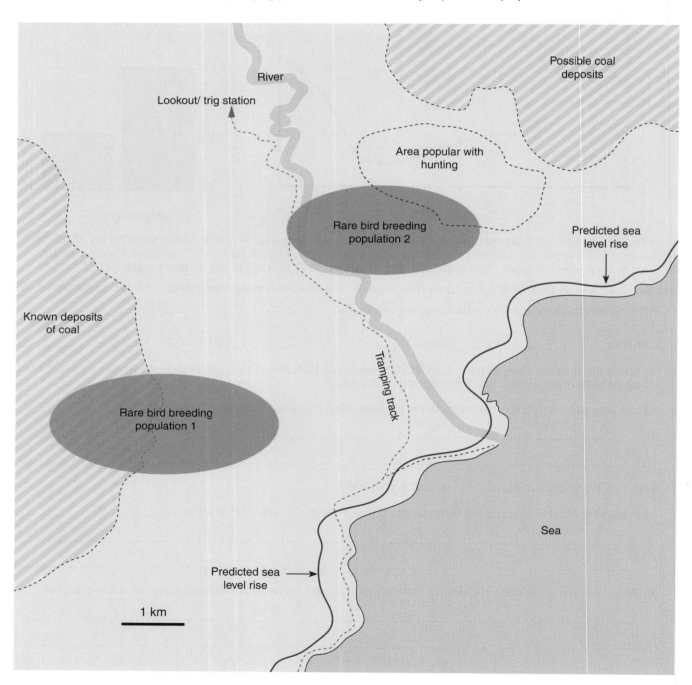

84 Weather Whiplash Revisited

Drought affected land

Three cyclones over the Western Pacific

Flash flooding in Death Valley

California's climate typically features cyclical seasonal weather events (dry to wet and dry again). These cycles are normal and important to sustaining life and agriculture in the region. However, in recent times, the swing between seasonal events has become more extreme (e.g. drought to flood). These events are called weather whiplash.

1. Using what you have learned in this chapter, explain how climate change can contribute to the frequency and severity of such events. Discuss the potential effects of these events and how they may affect ecosystems, biodiversity, and people:

85 Summative Assessment

1. The graphs below show data for past and current CO_2 concentrations and temperature, and models for CO_2 and temperature for the future.

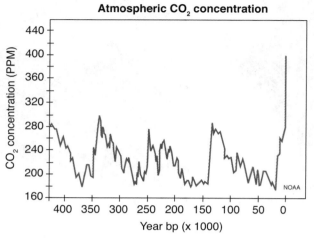

Atmospheric CO_2 concentration

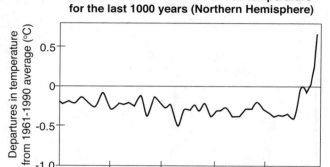

Variations of the Earth's surface temperature for the last 1000 years (Northern Hemisphere)

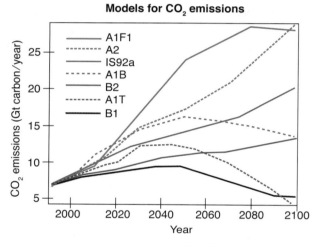

Models for CO_2 emissions

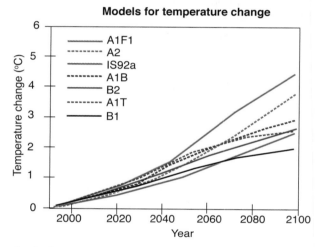

Models for temperature change

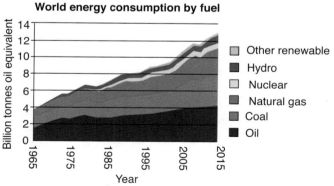

World energy consumption by fuel

Other renewable
Hydro
Nuclear
Natural gas
Coal
Oil

Use the graphs above to make an evidence-based forecast of the future rate of climate change, including an analysis of whether or not the rate will increase or decrease. You may use extra sources of information to further support your analysis. Use extra paper if you need it and attach it to this page:

_____

_____

_____

_____

_____

_____

    LS2.C ESS4.C ESS3.D CE SC

©2018 **BIOZONE** International
ISBN: 978-1-927309-55-1
Photocopying Prohibited

2. Table 1 below shows the expected temperature, precipitation changes, and grain yield for wheat grown in the Eastern Washington area over the next 65 years. The data was produced using the CCSM3 global climate model (which predicts more warming and less precipitation globally, although not necessarily locally).

Table 1

	Baseline	2020	2040	2080
Precipitation (mm)	535.8	549.9	543.9	588.3
Mean temperature °C	8.5	10.2	11.2	12.0
Yield (kg) No CO_2 effect	5713	6022	5116	5209
Yield (kg) CO_2 effect	5713	6546	6034	7033

Table 2 below shows the percentage crop yield response of wheat and other crop plants to changes in the environment including temperature and CO_2 changes.

Table 2

Crop	(A) Temperature (+1.2°C)	(B) CO_2 increase (380 to 440 ppm)	Temperature (A), and CO_2 (B), and irrigation
Wheat	-6.7	+6.8	+0.1
Corn (midwest)	-4.0	+1.0	-3.0
Soybean	-3.5	+7.4	+3.9
Cotton	-5.7	+9.2	+3.5

With reference to Table 1:

(a) Describe the change in rainfall expected in the Eastern Washington area over the next 65 years: _____

_____

(b) Describe the change in temperature expected in the Eastern Washington area over the next 65 years: _____

_____

(c) Describe the effect of climate change on grain yield, including effects of increased CO_2: _____

_____

_____

_____

With reference to Table 2:

(d) How will a change in temperature affect other crops grown in the USA? _____

_____

_____

(e) How will a change in CO_2 affect other crops grown in the USA? _____

_____

_____

3. (a) Plants carry out photosynthesis, producing organic molecules. The chemical equation for photosynthesis is:

$$6CO_2 + 6H_2O \rightarrow C_6H_{12}O_6 + 6O_2$$

Why does the grain yield increase with more atmospheric CO_2? _____

_____

_____

(b) What overall effect might climate change have on crop yield? How might this affect farmers and consumers?

_____

_____

_____

©2018 **BIOZONE** International
ISBN: 978-1-927309-55-1
Photocopying Prohibited

Change in winter destination, 17 US bird species

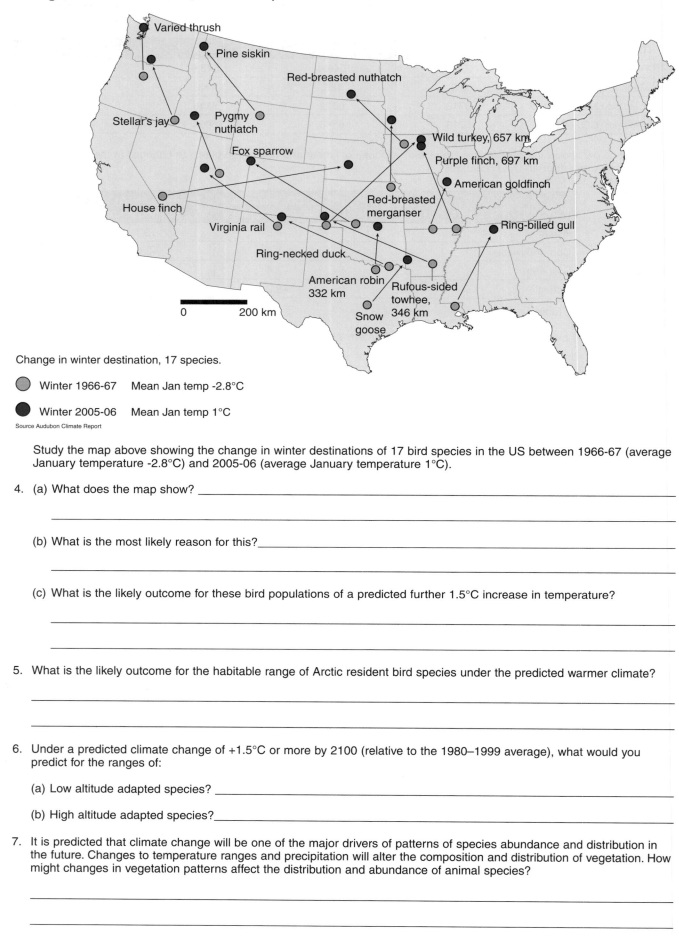

Change in winter destination, 17 species.

◯   Winter 1966-67   Mean Jan temp -2.8°C

●   Winter 2005-06   Mean Jan temp 1°C

Source Audubon Climate Report

Study the map above showing the change in winter destinations of 17 bird species in the US between 1966-67 (average January temperature -2.8°C) and 2005-06 (average January temperature 1°C).

4. (a) What does the map show? _____

_____

(b) What is the most likely reason for this?_____

_____

(c) What is the likely outcome for these bird populations of a predicted further 1.5°C increase in temperature?

_____

_____

5. What is the likely outcome for the habitable range of Arctic resident bird species under the predicted warmer climate?

_____

_____

6. Under a predicted climate change of +1.5°C or more by 2100 (relative to the 1980–1999 average), what would you predict for the ranges of:

(a) Low altitude adapted species? _____

(b) High altitude adapted species?_____

7. It is predicted that climate change will be one of the major drivers of patterns of species abundance and distribution in the future. Changes to temperature ranges and precipitation will alter the composition and distribution of vegetation. How might changes in vegetation patterns affect the distribution and abundance of animal species?

_____

_____

_____

©2018 **BIOZONE** International
ISBN: 978-1-927309-55-1
Photocopying Prohibited

8. Species do not respond to climate change equally. Discuss some of the likely consequences of climate change to the composition and stability of ecosystems. What ecosystems do you think are most at risk and why?

_____

_____

_____

_____

_____

_____

_____

_____

9. According to the USGS, California uses more water than any other US state. On average in California, 50% of water use is environmental, 40% is used in agriculture, and urban consumption makes up the last 10%. Environmental water usage includes water required to maintain stream and wetland habitat and also for maintaining the quality of water used in agricultural and urban applications.

(a) Study the graphic (right) and comment on the relative amounts of water use throughout California:

_____

_____

_____

_____

_____

(b) In dry years, state authorities can reduce water allocation for the environment so farms and urban areas are adequately supplied. Droughts have been more common in recent years and could be longer and more severe in the future as a consequence of climate change.

Discuss the problems associated with supplying adequate water resources for competing demands. Decide how you would prioritize water supply in California and discuss the effect of your choice on the three main water uses.

_____

_____

_____

_____

_____

_____

_____

_____

_____

_____

_____

_____

Average annual applied water use (1998-2010)

Department of Water Resources (2013). California Water Plan Update (Bulletin 160-13).

Statewide applied water use (MAF)

Wet year (2006)
104 MAF

62%

8%

29%

Dry year (2001)
61 MAF

13%

50%

36%

Environment   Urban   Agriculture

©2018 **BIOZONE** International
ISBN: 978-1-927309-55-1
Photocopying Prohibited

SEP support

Basic Skills for Students in Life Science

Science and engineering practices
Supported as noted and throughout other chapters in context

Asking questions and defining problems

☐ 1 Demonstrate an understanding of science as inquiry. Appreciate that unexpected results may lead to new questions and to new discoveries. — **86**

☐ 2 Formulate and evaluate questions that you can feasibly investigate. Ask questions that arise from observation or examining models or theories, or to find out more information, determine relationships, or refine a model. — **86 88**

Developing and using models

☐ 3 Develop and use models to describe systems or their components and how they work, to explain or make predictions about phenomena, or to generate data to support explanations or solve problems. — **87**

☐ 4 Develop a complex model that allows you to manipulate and test a system or process. — **87**

Planning and carrying out investigations

☐ 5 Plan and conduct investigations to provide data to test a hypothesis, support an explanation, or test a solution to a problem. Identify and evaluate the importance of any assumptions in the design of your investigation. — **88 95 96**

☐ 6 Consider and evaluate the accuracy and precision of the data that you collect. — **89**

☐ 7 Use appropriate tools to collect and record data. Data may be quantitative (continuous or discontinuous), qualitative, or ranked. — **94 95**

☐ 8 Make and test hypotheses about the effect on a dependent variable when an independent variable is manipulated. Understand and use controls appropriately. — **96**

Analyzing and interpreting data

☐ 9 Analyze data in order to make valid and reliable scientific claims. Consider limitations of data analysis (e.g. measurement error, bias) when analyzing and interpreting data. — **97-101**

☐ 10 Apply concepts of statistics and probability to answer questions and solve problems. Summarize data and describe its features using descriptive statistics. — **99-101**

Use mathematics and computational thinking

☐ 11 Demonstrate an ability to use mathematics and computational tools to analyze, represent, and model data. Recognize and use appropriate units in calculations. — **90-93 97-101**

☐ 12 Create and use simple computational simulations based on mathematical models. — **87**

☐ 13 Demonstrate an ability to apply ratios, rates, percentages, and unit conversions. — **91 92**

Construct explanations and design solutions

☐ 14 Apply scientific evidence, ideas, and principles to explain phenomena and solve problems. — **86**

Engage in argument from evidence

☐ 15 Use evidence to defend and evaluate claims and explanations about science. — **86**

☐ 16 Provide and/or receive critiques on scientific arguments by using scientific methodology. — **86**

Obtain, evaluate, and communicate information

☐ 17 Evaluate the validity and reliability of designs, methods, claims, and/or evidence. Communicate scientific and/or technical information in multiple formats. — **86**

☐ 18 Demonstrate an ability to read critically and compare, integrate, and evaluate sources of information in different media and formats. — **86**

86 How Do We Do Science?

▶ Science is a way of understanding the world we live in: how it formed, the rules it obeys and how it changes over time. It is based on a rigorous, dynamic process of observation, investigation, and analysis. Science distinguishes itself from other ways of understanding by using empirical standards, logical arguments, and skeptical review. Science allows what we understand to change over time as the body of knowledge increases.

▶ It is important to realize that science is a human endeavor and requires creativity and imagination. New research and ways of thinking can be based on the well-argued idea of a single person.

▶ Science influences and is influenced by society and technology. As society's beliefs and needs change, and technology advances, what is or can be researched is also affected. Scientific discoveries advance technology and can change society's beliefs.

▶ Science can never answer questions with absolute certainty. It can be confident of certain outcomes, but only within the limits of the data. Science might help us predict with 99.9% certainty a system will behave a certain way, but that still means there's one chance in a thousand it won't.

Exploration and discovery
Questioning, observing, and sharing information

Investigating and testing ideas
Carrying out investigations, comparing results to predictions and developing models that explain the patterns seen.

Benefits and outcomes
Using results to develop technology, solve problems, answer questions and educate.

Analysis and feedback
Review and discussion of results. Publication and repeat investigations.

Science is a process through which we can understand what we see

Science has application and relevance in the modern world

Science is ongoing: it moves in a direction of greater understanding

Science is a global human endeavour

Science is exciting, dynamic, creative, and collaborative

▶ **Jigsaw activities:** The following question relates to a class activity in which the class splits into groups, which then reconvene to present their ideas to the entire class.

1. The buttons above make five statements about science. Divide your class into five groups, with each group addressing one statement. Discuss the statement as a group and present a brief written summary of what the statement means, what evidence there is to support it, and whether you agree with it. Have one person in each group present the group's views to the class a whole.

©2018 **BIOZONE** International
ISBN: 978-1-927309-55-1
Photocopying Prohibited

SSM

87 Systems and System Models

▶ A **system** is a set of interrelated components that work together. Energy flow in ecosystems (such as the Alaskan ecosystem on the right), gene regulation, interactions between organ systems, and feedback mechanisms are all examples of systems studied in biology.

▶ Scientists often used models to learn about biological systems. A **model** is a representation of a system and is useful for breaking a complex system down into smaller parts that can be studied more easily. Often only part of a system is modeled. As scientists gather more information about a system, more data can be put into the model so that eventually it represents the real system more closely.

Modeling systems

There are many different ways to model systems or their components. Often seeing data presented in different ways can help to understand it better. Some common examples of models are shown here.

Visual models

Visual models can include drawings, such as these plant cells (below right) or three dimensional physical or computer generated models. Three dimensional models can be made out of materials such as modeling clay and ice-cream sticks, like the model of a water molecule (below left).

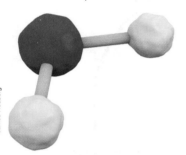

James Hedberg

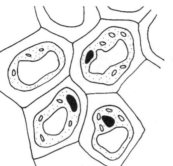

Mathematical models

Displaying data in a graph or as a mathematical equation, as shown below for logistic growth, often helps us to see relationships between different parts of a system.

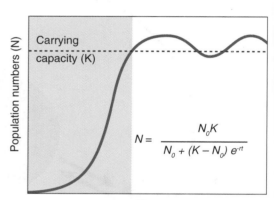

Population numbers (N)

Carrying capacity (K)

$$N = \frac{N_0 K}{N_0 + (K - N_0)\, e^{-rt}}$$

Time

Analogy

An analogy is a comparison between two things. Comparing a biological system to an everyday object can sometimes help us to understand it better. For example, the heart pumps blood in blood vessels in much the same way a fire truck pumps water from a fire hydrant through a hose. Similarly, the DNA in chromosomes is like a library. Extending that analogy further, the steps in baking a cake from a recipe book provide an analogy for how the instructions in DNA (the recipe) are translated into a specific protein (the cake).

The DNA in chromosomes is like ...... a library of books

1. What is a system? _____

_____

2. (a) What is a model? _____

_____

(b) Why do scientists often study one part of a system rather than the whole system? _____

_____

_____

_____

  SSM

©2018 **BIOZONE** International
ISBN: 978-1-927309-55-1
Photocopying Prohibited

88 Observations, Hypotheses, and Assumptions

Observations and hypotheses

▸ An observation is watching or recording what is happening. Observation is the basis for forming hypotheses and making predictions. An observation may generate a number of hypotheses (tentative explanations for what we see). Each hypothesis will lead to one or more predictions, which can be tested by investigation.

▸ A hypothesis is often written as a statement to include the prediction: "If X is true, then if I do Y (the experiment), I expect Z (the prediction)". Hypotheses are accepted, changed, or rejected on the basis of investigations. A hypothesis should have a sound theoretical basis and should be testable.

Observation 1:

▸ Some caterpillar species are brightly colored and appear to be highly visible to predators such as insectivorous birds. Predators appear to avoid these caterpillars.

▸ These caterpillars are often found in groups, rather than as solitary animals.

Observation 2:

▸ Some caterpillar species have excellent camouflage. When alerted to danger they are difficult to see because they blend into the background.

▸ These caterpillars are usually found alone.

Assumptions

Any investigation requires you to make **assumptions** about the system you are working with. Assumptions are features of the system you are studying that you assume to be true but that you do not (or cannot) test. Some assumptions about the examples above include:

• Insect eating birds have color vision.
• Caterpillars that look bright to us, also appear bright to insectivorous birds.
• Birds can learn about the taste of prey by eating them.

Read the two observations about the caterpillars above and then answer the following questions:

1. Generate a hypothesis to explain the observation that some caterpillars are brightly colored and highly visible while others are camouflaged and blend into their surroundings:

_____

_____

_____

_____

2. Describe two assumptions being made in your hypothesis: _____

_____

_____

_____

3. Generate a prediction about the behavior of insect eating birds towards caterpillars: _____

_____

_____

_____

©2018 **BIOZONE** International
ISBN: 978-1-927309-55-1
Photocopying Prohibited

89 Accuracy and Precision

The terms accuracy and precision are often used when talking about measurements.

▶ **Accuracy** refers to how close a measured value is to its true value, i.e. the correctness of the measurement.

▶ **Precision** refers to the closeness of repeated measurements to each other, i.e. the ability to be exact. For example, a digital device, such as a pH meter (right) will give very precise measurements, but its accuracy depends on correct calibration.

Using the analogy of a target, repeated measurements are compared to arrows shot at a target. This analogy is useful when distinguishing between accuracy and precision.

Accurate but imprecise	**Inaccurate and imprecise**	**Precise but inaccurate**	**Accurate and precise**

The measurements are all close to the true value but quite spread apart.

Analogy: The arrows are all close to the bullseye.

The measurements are all far apart and not close to the true value.

Analogy: The arrows are spread around the target.

The measurements are all clustered close together but not close to the true value.

Analogy: The arrows are all clustered close together but not near the bullseye.

The measurements are all close to the true value and also clustered close together.

Analogy: The arrows are clustered close together near the bullseye.

Significant figures

Significant figures (sf) are the digits of a number that carry meaning contributing to its precision. They communicate how well you could actually measure the data.

For example, you might measure the height of 100 people to the nearest cm. When you calculate their mean height, the answer is 175.0215 cm. If you reported this number, it implies that your measurement technique was accurate to 4 decimal places. You would have to round the result to the number of significant figures you had accurately measured. In this instance the answer is 175 cm.

Non-zero numbers (1-9) are always **significant**.

All zeros between non-zero numbers are always **significant**.

$$0.005704510$$

Zeros to the left of the first non-zero digit after a decimal point are **not significant**.

Zeros at the end of number where there is a decimal place are **significant** (e.g. 4600.0 has five sf).
BUT
Zeros at the end of a number where there is no decimal point are **not significant** (e.g. 4600 has two sf).

1. (a) Why are precise but inaccurate measurements not helpful in a biological investigation? _____

_____

_____

(b) Experimental work often relies on instrumentation to collect data. How can you best ensure the accuracy of your recordings in these situations?

_____

2. State the number of significant figures in the following examples:

(a) 3.15985 _____

(b) 0.0012 _____

(c) 1000 _____

(d) 1000.0 _____

(e) 42.3006 _____

(f) 120 _____

©2018 **BIOZONE** International
ISBN: 978-1-927309-55-1
Photocopying Prohibited

90 Working With Numbers

▸ Using correct mathematical notation and being able to carry out simple calculations and conversions are fundamental skills in biology. This activity will help you to practice your skills in this area.

Commonly used mathematical symbols

In mathematics, universal symbols are used to represent mathematical concepts. They save time and space when writing. Some commonly used symbols are shown below.

=   Equal to

<   The value on the left is **less than** the value on the right

<<   The value on the left is **much less than** the value on the right

>   The value on the left is **greater than** the value on the right

>>   The value on the left is **much greater than** the value on the right

∝   Proportional to. A ∝ B means that A = a constant X B

~   Approximately equal to

Decimal and standard form

▸ **Decimal form** (also called ordinary form) is the longhand way of writing a number (e.g. 15,000,000). Very large or very small numbers can take up too much space if written in decimal form and are often expressed in a condensed **standard form**. For example, 15,000,000 is written as 1.5×10^7 in standard form.

▸ In standard form a number is always written as A x 10^n, where A is a number between 1 and 10, and n (the exponent) indicates how many places to move the decimal point. n can be positive or negative.

▸ For the example above, A = 1.5 and n = 7 because the decimal point moved seven places (see below).

$$1\,5\,0\,0\,0\,\,0\,0\,0 = 1.5 \times 10^7$$

▸ Small numbers can also be written in standard form. The exponent (n) will be negative. For example, 0.00101 is written as 1.01×10^{-3}.

$$0.\,0\,0\,1\,0\,1 = 1.01 \times 10^{-3}$$

▸ Converting can make calculations easier. Work through the following example to solve $4.5 \times 10^4 + 6.45 \times 10^5$.

1. Convert $4.5 \times 10^4 + 6.45 \times 10^5$ to decimal form:

   _____

2. Add the two numbers together: _____

3. Convert to standard form: _____

Estimates

▸ When carrying out calculations, typing the wrong number into your calculator can put your answer out by several orders of magnitude. An **estimate** is a way of roughly calculating what answer you should get, and helps you decide if your final calculation is correct.

▸ Numbers are often rounded to help make estimation easier. The rounding rule is, if the next digit is 5 or more, round up. If the next digit is 4 or less, it stays as it is.

▸ For example, to estimate 6.8 x 704 you would round the numbers to 7 x 700 = 4900. The actual answer is 4787, so the estimate tells us the answer (4787) is probably right.

Use the following examples to practise estimating:

4. 43.2 x 1044: _____

5. 3.4 x 72 ÷ 15: _____

6. 658 ÷ 22: _____

Conversion factors and expressing units

▸ Measurements can be converted from one set of units to another by the use of a **conversion factor**. A conversion factor is a numerical factor that multiplies or divides one unit to convert it into another.

▸ Conversion factors are commonly used to convert non-SI units to SI units (e.g. converting pounds to kilograms). Note that mL and cm^3 are equivalent, as are L and dm^3.

In the space below, convert 5.6 cm^3 to mm^3 (1 cm^3 = 1000 mm^3):

7. _____

▸ The value of a variable must be written with its units where possible. SI units or their derivations should be used in recording measurements: volume in cm^3 (mL) or dm^3 (L), mass in kilograms (kg) or grams (g), length in meters (m), time in seconds (s). To denote 'per', you can use a solidus (/) or a negative exponent, e.g. per second is written as /s or s^{-1} and per meter squared is written as /m^2 or m^{-2}.

▸ For example the rate of oxygen consumption could be expressed as:

Oxygen consumption (mL/g/s) *or*
Oxygen consumption mL g^{-1} s^{-1})

©2018 **BIOZONE** International
ISBN: 978-1-927309-55-1
Photocopying Prohibited

91 Tallies, Percentages, and Rates

▶ The data collected by measuring or counting in the field or laboratory is called **raw data**. Raw data often needs to be processed into a form that makes it easier to identify its important features (e.g. trends) and make meaningful comparisons between samples or treatments.

▶ Basic calculations, such as totals (the sum of all data values for a variable), are commonly used to compare treatments. Some common methods of processing data include creating tally charts, and calculating percentages and rates. These are explained below.

Tally Chart
Records the number of times a value occurs in a data set

HEIGHT (cm)	TALLY	TOTAL
0-0.99	111	3
1-1.99	++++ 1	6
2-2.99	++++ ++++	10
3-3.99	++++ ++++ 11	12
4-4.99	111	3
5-5.99	11	2

Percentages
Expressed as a fraction of 100

Men	Body mass (Kg)	Lean body mass (Kg)	% lean body mass
Athlete	70	60	85.7
Lean	68	56	82.3
Normal weight	83	65	78.3
Overweight	96	62	64.6
Obese	125	65	52.0

Rates
Expressed as a measure per unit time

Time (minutes)	Cumulative sweat loss (mL)	Rate of sweat loss (mL/min)
0	0	0
10	50	5
20	130	8
30	220	9
60	560	11.3

- A useful first step in analysis; a neatly constructed tally chart doubles as a simple histogram.

- Cross out each value on the list as you tally it to prevent double entries.

- Percentages express the proportion of data falling into any one category, for example, in pie graphs.

- Allows meaningful comparison between different samples.

- Useful to monitor change (e.g. percentage increase from one year to the next).

- Rates show how a variable changes over a standard time period (e.g. one second, one minute, or one hour).

- Rates allow meaningful comparison of data that may have been recorded over different time periods.

Example: Height of 6 day old seedlings

Example: Percentage of lean body mass in men

Example: Rate of sweat loss during exercise in cyclists

1. What is raw data? _____

_____

2. Why is it useful to process raw data and express it differently, e.g. as a rate or a percentage? _____

_____

_____

_____

3. Identify the best data transformation in each of the following examples:

(a) Comparing harvest (in kg) of different grain crops from a farm: _____

(b) Comparing amount of water loss from different plant species over time: _____

(c) Initial analysis of the heights of individuals at a school: _____

©2018 **BIOZONE** International
ISBN: 978-1-927309-55-1
Photocopying Prohibited

92 Fractions and Ratios

▶ Fractions and ratios are widely used in biology and are often used to provide a meaningful comparison of sample data where the sample sizes are different.

Fractions

▶ Fractions express how many parts of a whole are present.

▶ Fractions are expressed as two numbers separated by a solidus (/). For example 1/2.

▶ The top number is the numerator. The bottom number is the denominator. The denominator can not be zero.

Simplifying fractions

▶ Fractions are often written in their simplest form (the top and bottom numbers cannot be any smaller, while still being whole numbers). Simplifying makes working with fractions easier.

▶ To simplify a fraction, the numerator and denominator are divided by the highest common number that divides into both numbers equally.

▶ For example, in a class of 20 students, five had blue eyes. This fraction is 5/20. To simplify this fraction 5 and 20 are divided by the highest common factor (5).

$$5 \div 5 = 1 \text{ and } 20 \div 5 = 4$$

▶ The simplified fraction is 1/4.

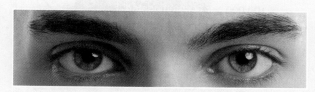

Adding fractions

▶ To add fractions the denominators must be the same. If the denominators are the same the numerators are simply added. E.g. 5/12 + 3/12 = 8/12

▶ When the denominators are different one (or both) fractions must be multiplied to give a common denominator, e.g. 4/10 + 1/2. By multiplying 1/2 by 5 the fraction becomes 5/10. The fractions can now be added together (4/10 + 5/10 = 9/10).

Ratios

▶ Ratios give the relative amount of two or more quantities (it shows how much of one thing there is relative to another).

▶ Ratios provide an easy way to identify patterns.

▶ Ratios do not require units.

▶ Ratios are usually expressed as $a : b$.

▶ In the example below, there are 3 blue squares and 1 gray square. The ratio would be written as 3:1.

Calculating ratios

▶ Ratios are calculated by dividing all the values by the smallest number.

▶ Ratios are often used in Mendelian genetics to calculate phenotype (appearance) ratios. Some examples for pea plants are given below.

882 inflated pod   *299 constricted pod*

To obtain the ratio divide both numbers by 299.
299 ÷ 299 =1
882 ÷ 299 = 2.95
The ratio = 2.95 : 1

495   *152*   *158*   *55*
round yellow *wrinkled yellow* *round green* *wrinkled green*

For the example above of pea seed shape and color, all of the values were divided by 55. The ratio obtained was:
9 : 2.8 : 2.9 : 1

---

1. (a) A student prepared a slide of the cells of an onion root tip and counted the cells at various stages in the cell cycle. The results are presented in the table (right). Calculate the ratio of cells in each stage (show your working):

   _____

   _____

   (b) Assuming the same ratio applies in all the slides examined in the class, calculate the number of cells in each phase for a cell total count of 4800.

Cell cycle stage	No. of cells counted	No. of cells calculated
Interphase	140	
Prophase	70	
Telophase	15	
Metaphase	10	
Anaphase	5	
Total	**240**	**4800**

2. Simplify the following fractions:

   (a) 3/9 : _____   (b) 84/90: _____   (c) 11/121: _____

3. In the fraction example pictured above, 5/20 students had blue eyes. In another class, 5/12 students had blue eyes. What fraction of students had blue eyes in both classes combined?

   _____

©2018 **BIOZONE** International
ISBN: 978-1-927309-55-1
Photocopying Prohibited

93 Dealing with Large Numbers

▶ In biology, numerical data indicating scale can often decrease or increase exponentially. Examples include the exponential growth of populations, exponential decay of radioisotopes, and the pH scale.

▶ Exponential changes in numbers are defined by a function. A function is simply a rule that allows us to calculate an output for any given input. Exponential functions are common in biology and may involve very large numbers.

▶ Log transformations of exponential numbers can make them easier to handle.

Exponential function

▶ Exponential growth occurs at an increasingly rapid rate in proportion to the growing total number or size.

▶ In an exponential function, the base number is fixed (constant) and the exponent is variable.

▶ The equation for an exponential function is $y = c^x$.

▶ Exponential growth and decay (reduction) are possible.

▶ Exponential changes in numbers are easy to identify because the curve has a J-shape appearance due to its increasing steepness over time.

▶ An example of exponential growth is the growth of a microbial population in an unlimiting, optimal growth environment.

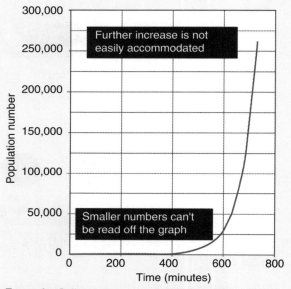

Example: Cell growth in a yeast culture where growth is not limited by lack of nutrients or build up of toxins.

Log transformations

▶ A log transformation makes very large numbers easier to work with. The log of a number is the exponent to which a fixed value (the base) is raised to get that number. So $\log_{10} (1000) = 3$ because $10^3 = 1000$.

▶ Both $\log_{10}$ (common logs) and $\log_e$ (natural logs or *ln*) are commonly used.

▶ Log transformations are useful for data where there is an exponential increase or decrease in numbers. In this case, the transformation will produce a straight line plot.

▶ To find the log10 of a number, e.g. 32, using a calculator, key in log 32 = . The answer should be 1.51.

▶ Alternatively, the untransformed data can be plotted directly on a log-linear scale (as below). This is not difficult. You just need to remember that the log axis runs in exponential cycles. The paper makes the log for you.

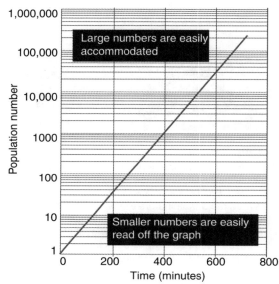

Example: The same yeast cell growth plotted on a log-linear scale. The y axis present 6 exponential cycles

1. Why is it useful to plot exponential growth using semi-log paper? _____

_____

_____

2. What would you do to show yeast exponential growth (left plot above) as a straight line plot on normal graph paper?

_____

3. Log transformations are often used when a value of interest ranges over several orders of magnitude. Can you think of other examples of data from the natural world where the data collected might show this behavior?

_____

_____

_____

©2018 **BIOZONE** International
ISBN: 978-1-927309-55-1
Photocopying Prohibited

94 Apparatus and Measurement

▸ The apparatus used in experimental work must be appropriate for the experiment or analysis and it must be used correctly to eliminate experimental errors.

Selecting the correct equipment

It is important that you choose equipment that is appropriate for the type of measurement you want to take. For example, if you wanted to accurately weigh out 5.65 g of sucrose, you need a balance that accurately weighs to two decimal places. A balance that weighs to only one decimal place would not allow you to make an accurate enough measurement.

Study the glassware (right). Which would you use if you wanted to measure 225 mL? The graduated cylinder has graduations every 10 mL whereas the beaker has graduations every 50 mL. It would be more accurate to measure 225 mL in a graduated cylinder.

Percentage errors

Percentage error is a way of mathematically expressing how far out your result is from the ideal result. The equation for measuring percentage error is:

$$\frac{\text{experimental value - ideal value}}{\text{ideal value}} \times 100$$

For example, to determine the accuracy of a 5 mL pipette, dispense 5 mL of water from the pipette and weigh the dispensed volume on a balance. The mass (g) = volume (mL). The volume is 4.98 mL.

$$\frac{\text{experimental value (4.98) - ideal value (5.0)}}{\text{ideal value (5.0)}} \times 100$$

The percentage error = –0.4% (the negative sign tells you the pipette is dispensing **less** than it should).

Recognizing potential sources of error

It is important to know how to use equipment correctly to reduce errors. A spectrophotometer measures the amount of light absorbed by a solution at a certain wavelength. This information can be used to determine the concentration of the absorbing molecule (e.g. density of bacteria in a culture). The more concentrated the solution, the more light is absorbed. Incorrect use of the spectrophotometer can alter the results. Common mistakes include incorrect calibration, errors in sample preparation, and errors in sample measurement.

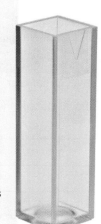

A cuvette (left) is a small clear tube designed to hold spectrophotometer samples. Inaccurate readings occur when:

- The cuvette is dirty or scratched (light is absorbed giving a falsely high reading).

- Some cuvettes have a frosted side to aid alignment. If the cuvette is aligned incorrectly, the frosted side absorbs light, giving a false reading.

- Not enough sample is in the cuvette and the beam passes over, rather than through the sample, giving a lower absorbance reading.

1. Assume that you have the following measuring devices available: 50 mL beaker, 50 mL graduated cylinder, 25 mL graduated cylinder, 10 mL pipette, 10 mL beaker. What would you use to accurately measure:

   (a) 21 mL: _____ (b) 48 mL: _____ (c) 9 mL: _____

2. Calculate the percentage error for the following situations (show your working):

   (a) A 1 mL pipette delivers a measured volume of 0.98 mL: _____

   _____

   (b) A 10 mL pipette delivers a measured volume of 9.98 mL: _____

   _____

©2018 **BIOZONE** International
ISBN: 978-1-927309-55-1
Photocopying Prohibited

95 Types of Data

▸ Data is information collected during an investigation. Data may be quantitative, qualitative, or ranked.

▸ Quantitative data can be continuous or discontinuous and is more easily analyzed than qualitative data.

When planning a biological investigation, it is important to consider the type of data that will be collected. It is best to collect quantitative or numerical data, because it is easier to analyze it objectively (without bias).

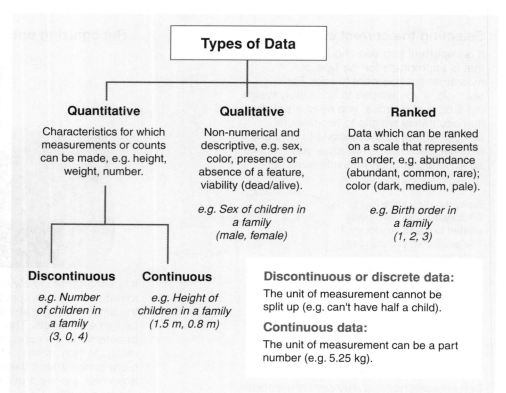

Types of Data

Quantitative
Characteristics for which measurements or counts can be made, e.g. height, weight, number.

Qualitative
Non-numerical and descriptive, e.g. sex, color, presence or absence of a feature, viability (dead/alive).

e.g. Sex of children in a family (male, female)

Ranked
Data which can be ranked on a scale that represents an order, e.g. abundance (abundant, common, rare); color (dark, medium, pale).

e.g. Birth order in a family (1, 2, 3)

Discontinuous
e.g. Number of children in a family (3, 0, 4)

Continuous
e.g. Height of children in a family (1.5 m, 0.8 m)

Discontinuous or discrete data:
The unit of measurement cannot be split up (e.g. can't have half a child).

Continuous data:
The unit of measurement can be a part number (e.g. 5.25 kg).

A: Skin color

B: Eggs per nest

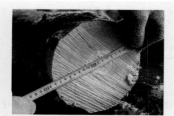

C: Tree trunk diameter

1. For each of the photographic examples A-C above, classify the data as quantitative, ranked, or qualitative:

   (a) Skin color: _____

   (b) Number of eggs per nest: _____

   (c) Tree trunk diameter: _____

2. Why is it best to collect quantitative data where possible in biological studies? _____

   _____

   _____

   _____

   _____

3. Give an example of data that could not be collected quantitatively and explain your answer: _____

   _____

   _____

   _____

©2018 **BIOZONE** International
ISBN: 978-1-927309-55-1
Photocopying Prohibited

96 Variables and Controls

▸ Variables may be dependent, independent, or controlled. A control in an experiment allows you to determine the effect of the independent variable.

Types of variables

A **variable** is a factor that can be changed during an experiment (e.g. temperature). Investigations often look at how changing one variable affects another.

There are several types of variables:

- Independent
- Dependent
- Controlled

Only one variable should be changed at a time. Any changes seen are a result of the changed variable.

Remember! The dependent variable is 'dependent' on the independent variable.

Example: *When heating water, the temperature of the water depends on the time it is heated for. Temperature (dependent variable) depends on time (independent variable).*

Dependent variable
- Measured during the investigation.
- Recorded on the y axis of the graph.

Controlled variable
- Factors that are kept the same.

Independent variable
- Set by the experimenter, it is the variable that is changed.
- Recorded on the graph's x axis.

Experimental controls

▸ A **control** is the standard or reference treatment in an experiment. Controls make sure that the results of an experiment are due to the variable being tested (e.g. nutrient level) and not due to another factor (e.g. equipment not working correctly).

▸ A control is identical to the original experiment except it lacks the altered variable. The control undergoes the same preparation, experimental conditions, observations, measurements, and analysis as the test group.

▸ If the control works as expected, it means the experiment has run correctly, and the results are due to the effect of the variable being tested.

Test plant (nutrient added)

Control plant (no nutrient added)

An experiment was designed to test the effect of a nutrient on plant growth. The control plant had no nutrient added. Its growth sets the baseline for the experiment. Any growth in the test plant beyond what is seen in the control plant is due to the presence of the nutrient.

1. What is the difference between a dependent variable and an independent variable? _____

_____

_____

2. Why do we control the variables we are not investigating? _____

_____

_____

3. What is the purpose of the experimental control? _____

_____

_____

©2018 **BIOZONE** International
ISBN: 978-1-927309-55-1
Photocopying Prohibited

97 Drawing Graphs

▶ Graphs are useful for visually displaying numerical data, trends, and relationships between variables. Different types of graph are appropriate for different types of data.

▶ Scatter plots and line graphs are appropriate where data are continuous for both variables. Bar charts are appropriate where data are categorical for one variable. For frequency distributions, histograms are appropriate. Pie charts are useful if you want to compare proportions in different categories.

▶ For continuous data with calculated means, points can be joined. On scatter plots, a line of best fit is often drawn.

▶ Presenting graphs properly demands attention to a few basic details, including correct orientation and labeling of the axes, and accurate plotting of points.

Guidelines for line graphs

WHEN TO USE: Use a line graph when both variables are continuous and one variable (the independent variable) affects another, the dependent variable. Important features include:

▶ The data must be continuous for both variables. The independent variable is often time or experimental treatment. The dependent variable is generally the biological response.

▶ The relationship between two variables can be represented as a continuum and the data points are plotted accurately and connected directly (point to point).

▶ Line graphs may be drawn with measure of error (right). The data are presented as points (which are calculated means), with error bars above and below, indicating the variability in the data (e.g. standard deviation).

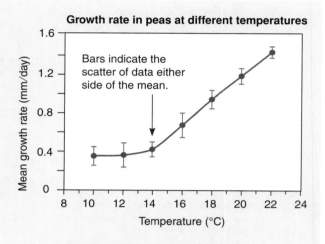

Growth rate in peas at different temperatures

Bars indicate the scatter of data either side of the mean.

Plotting multiple data sets

A single figure (graph) can be used to show two or more data sets, i.e. more than one curve can be plotted per set of axes. This type of presentation is useful when comparing the trends for two or more treatments, or the response of one species against the response of another. Important points regarding this format are:

▶ If the two data sets use the same measurement units and a similar range of values for the dependent variable, one scale on the y axis is used.

▶ If the two data sets use different units and/or have a very different range of values for the dependent variable, two scales for the y axis are used (see right). The scales can be adjusted if necessary to avoid overlapping plots

▶ The two curves are distinguished with a key.

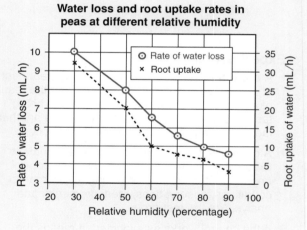

Water loss and root uptake rates in peas at different relative humidity

○ Rate of water loss
× Root uptake

Guidelines for scatter graphs

WHEN TO USE: Use a scatter graph to display continuous data where there are two interdependent variables.

▶ The data must be continuous for both variables.
▶ There is no independent variable, but the variables are often correlated, i.e. they vary together in a predictable way.
▶ Useful to determine the relationship between two variables.
▶ The points on the graph are not connected, but a line of best fit is often drawn through the points to show the relationship between the variables (this may be computer generated with a value assigned to the goodness of the fit).
▶ Obvious outliers (points that lie well outside most of the scatter) are usually disregarded from analyses.

Body length vs brood size in *Daphnia*

Line of best fit

Outlier

Interpolation: For both line and scatter graphs, the fitted line can be used to find an unknown value inside the set of data points. This is called interpolation.

©2018 **BIOZONE** International
ISBN: 978-1-927309-55-1
Photocopying Prohibited

Guidelines for histograms

WHEN TO USE: Use a histogram when one variable is continuous and the other is a frequency (counts). These plots produce a frequency distribution, because the y-axis shows the number of times a measurement or value was obtained. Important features of histograms include:

▶ The data are numerical and continuous (e.g. height or weight) so the bars touch.

▶ The x-axis usually records the class interval. The y-axis usually records the number of individuals in each class interval.

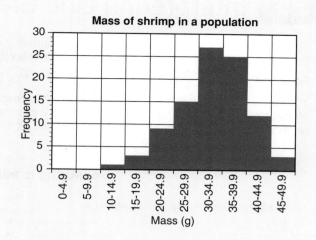

Guidelines for bar and column graphs

WHEN TO USE: Use a bar or column graph for data that are non-numerical and discrete (categorical) for one variable. There are no dependent or independent variables. Important features include:

▶ Data for one variable are discontinuous, non-numerical categories (e.g. place, color, species), so the bars do not touch.

▶ Data values may be entered by the bars if you wish.

▶ Multiple sets of data can be displayed side by side to compare (e.g. males and females in the same age group).

▶ Axes may be reversed so that the categories are on the x axis, i.e. the bars can be vertical or horizontal. When they are vertical, these graphs are called column graphs.

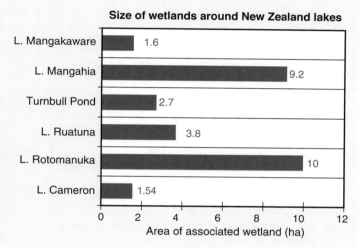

1. Determine what type of graph is appropriate for each of the following examples:

    (a) Arm span vs height in humans: _____

    (b) Daily energy requirement for different species of deer: _____

    (c) Number of fish of each size in a population: _____

    (d) Volume of water used per person per day in different North American cites: _____

    (e) Mean catalase reaction rate at different temperatures: _____

    (f) Number of eggs per brood in different breeds of chickens: _____

    (g) Mean monthly rainfall vs mean monthly temperature: _____

2. For the plots on the previous page:

    (a) Use interpolation to determine the mean growth rate of pea seedlings at 17°C: _____

    (b) Use interpolation to determine the number of eggs per brood in a 1.5 mm long *Daphnia*: _____

    (c) Use interpolation to determine the rate of water loss in peas at 40% relative humidity:_____

3. Extrapolation, i.e. predicting a data value that lies outside the range of available data, is not recommended practice.

    (a) Suggest why you should not extrapolate to find data values?_____

    _____

    (b) Can you think of an example to illustrate your decision?_____

    _____

©2018 **BIOZONE** International
ISBN: 978-1-927309-55-1
Photocopying Prohibited

98 Interpreting Line Graphs

▶ The equation for a linear (straight) line on a graph is y = mx + c. The equation can be used to calculate the gradient (slope) of a straight line and tells us about the relationship between x and y (how fast y is changing relative to x).

▶ For a straight line, the rate of change of y relative to x is always constant.

▶ You can use the equation to find a value for y (the response variable) for any value of x (the manipulated variable or time) providing you know the gradient and the y intercept. However, this makes the assumption that the straight line relationship does not change, which may not always be true (extrapolation is not usually a good idea).

Measuring gradients and intercepts

The equation for a straight line is written as:

y = mx + c

Where :

y = the y-axis value

m = the slope (or gradient)

x = the x-axis value

c = the y intercept (where the line crosses the y-axis).

Determining "m" and "c"

To find "c" just find where the line crosses the y-axis.

To find m:

1. Choose any two points on the line.

2. Draw a right-angled triangle between the two points on the line.

3. Use the scale on each axis to find the triangle's vertical length and horizontal length.

4. Calculate the gradient of the line using the following equation:

$$\frac{\text{change in y}}{\text{change in x}}$$

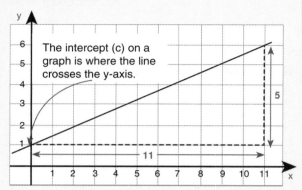

For the example above:

c = 1

m = 0.45 (5 ÷11)

Once c and m have been determined you can choose any value for x and find the corresponding value for y.

For example, when x = 9, the equation would be:

y = 9 x 0.45 + 1

y = 5.05

A line may have a positive, negative, or zero slope

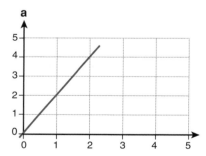

a

Positive gradients: the line slopes upward to the right (y is increasing as x increases).

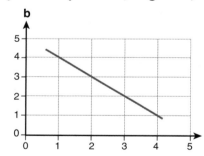

b

Negative gradients: the line slopes downward to the right (y is decreasing as x increases).

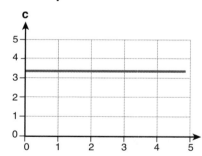

c

Zero gradients: the line is horizontal (y does not change as x increases).

1. For the graph (right):

(a) Identify the value of c: _____

(b) Calculate the value of m: _____

(c) Determine y if x = 2: _____

_____

(d) Describe the slope of the line: _____

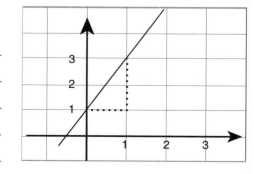

©2018 **BIOZONE** International
ISBN: 978-1-927309-55-1
Photocopying Prohibited

99 Mean, Median, and Mode

▶ Descriptive statistics are used to summarize a data set and describe its basic features. The type of statistic calculated depends on the type of data and its distribution.

Descriptive statistics

▶ When we describe a set of data, it is usual to give a measure of **central tendency**. This is a single value identifying the central position within that set of data.

▶ **Descriptive statistics**, such as mean, median, and mode, are all valid measures of central tendency depending of the type of data and its distribution. They help to summarize features of the data, so are often called summary statistics.

▶ The appropriate statistic for different types of data variables and their distributions is described below.

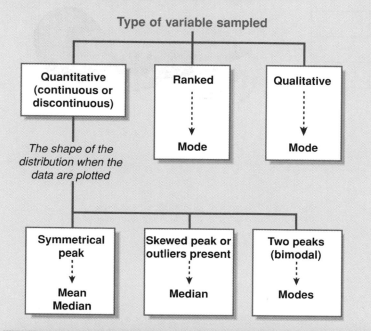

Distribution of data

Variability in continuous data is often displayed as a frequency distribution. There are several types of distribution.

- Normal distribution (A): Data has a symmetrical spread about the mean. It has a classical bell shape when plotted.
- Skewed data (B): Data is not centered around the middle but has a "tail" to the left or right.
- Bimodal data (C): Data which has two peaks.

The shape of the distribution will determine which statistic (mean, median, or mode) should be used to describe the central tendency of the sample data.

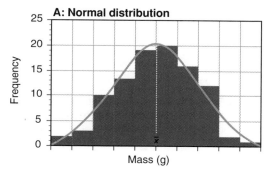

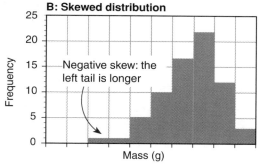

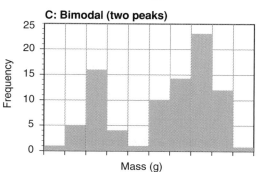

Statistic	Definition and when to use it	How to calculate it
Mean	• The average of all data entries. • Measure of central tendency for normally distributed data.	• Add up all the data entries. • Divide by the total number of data entries.
Median	• The middle value when data entries are placed in rank order. • A good measure of central tendency for skewed distributions.	• Arrange the data in increasing rank order. • Identify the middle value. • For an even number of entries, find the mid point of the two middle values.
Mode	• The most common data value. • Suitable for bimodal distributions and qualitative (categorical) data.	• Identify the category with the highest number of data entries using a tally chart or a bar graph.

©2018 **BIOZONE** International
ISBN: 978-1-927309-55-1
Photocopying Prohibited

1. The birth weights of 60 newborn babies are provided right. Create a tally chart (frequency table) of the weights in the table provided below. Choose an appropriate grouping of weights.

Weight (kg)	Tally	Total

Birth weights (kg)

NEED HELP?
See Activity 98

3.740	3.380	4.510	3.135	3.260
3.830	2.660	3.800	3.090	3.430
3.530	3.375	4.170	3.830	3.510
3.095	3.840	4.400	3.970	3.230
3.630	3.630	3.770	3.840	3.570
1.560	3.810	3.400	4.710	3.620
3.910	2.640	3.825	4.050	3.260
4.180	3.955	3.130	4.560	3.315
3.570	2.980	3.400	3.350	3.230
2.660	3.350	3.260	3.380	3.790
3.150	3.780	4.100	3.690	2.620
3.400	3.260	3.220	1.495	3.030

2. (a) On the graph paper (right) draw a frequency histogram for the birth weight data.

(b) What type of distribution does the data have?

_____

_____

_____

(c) Predict whether mean, median, or mode would be the best measure of central tendency for the data:

_____

(d) Explain your reason for your answer in (c):

_____

_____

_____

(e) Calculate the mean, median, and mode for the birth weight data:

Mean: _____

Median: _____

Mode: _____

(f) What do you notice about the results in (e)? _____

_____

_____

(g) Explain the reason for this: _____

_____

_____

©2018 **BIOZONE** International
ISBN: 978-1-927309-55-1
Photocopying Prohibited

100 What is Standard Deviation?

▸ While it is important to know the mean of a data set, it is also important to know how well the mean represents the data set as a whole. This is evaluated using a simple measure of the spread in the data called **standard deviation**.

▸ Standard deviation can be used to evaluate how reliably the mean represents the data. In general, if the standard deviation is small, the mean will more accurately represent the data than if it is large.

Standard deviation

▸ Standard deviation is usually presented as $\bar{x} \pm s$. In normally distributed data, 68% of all data values will lie within one standard deviation (s) of the mean ($\bar{x}$) and 95% of all data values will lie within two standard deviations of the mean (right).

▸ Different sets of data can have the same mean and range, yet a different data distribution. In both the data sets below, 68% of the values lie within the range $\bar{x} \pm 1s$ and 95% of the values lie within $\bar{x} \pm 2s$. However, in B, the data values are more tightly clustered around the mean.

▸ Standard deviation is easily calculated using an spreadsheet. Data should be entered as columns. In a free cell, type the formula for standard deviation (this varies depending on the program) and select the cells containing the data values, enclosing them in parentheses.

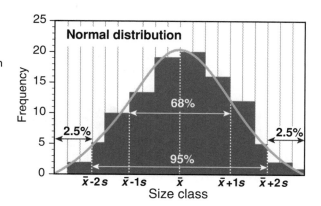

Normal distribution

Histogram A has a larger standard deviation; the values are spread widely around the mean.

Both plots show a normal distribution with a symmetrical spread of values about the mean.

Histogram B has a smaller standard deviation; the values are clustered more tightly around the mean.

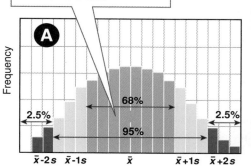

A

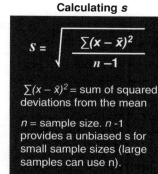

Calculating s

$$s = \sqrt{\frac{\sum(x - \bar{x})^2}{n - 1}}$$

$\sum(x - \bar{x})^2$ = sum of squared deviations from the mean

n = sample size. n -1 provides a unbiased s for small sample sizes (large samples can use n).

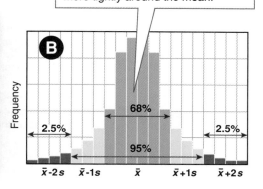

B

1. Two sample data sets of rat body length have the same mean. The first data set has a much larger standard deviation than the second data set. What does this tell you about the spread of data around the mean in each case? Which data set is likely to provide the most reliable estimate of body length in the rat population being sampled and why?

_____

_____

_____

2. The data on the right shows the heights for 29 male swimmers.

   (a) Calculate the mean for the data: _____

   (b) Use manual calculation, a calculator, or a spreadsheet to calculate the standard deviation (s) for the data:

   _____

   (c) State the mean ± 1s: _____

   (d) What percentage of values are within 1s of the mean? _____

   (e) What does this tell you about the spread of the data? _____

   _____

   _____

Raw data: Height (cm)					
178	177	188	176	186	175
180	181	178	178	176	175
180	185	185	175	189	174
178	186	176	185	177	176
176	188	180	186	177	

©2018 **BIOZONE** International
ISBN: 978-1-927309-55-1
Photocopying Prohibited

101 Detecting Bias in Samples

▸ Bias is the selection for or against one particular group and can influence the findings of an investigation. Bias can occur when sampling is not random and certain members of a population are under or over represented.

▸ Small sample sizes can bias results. Bias can be reduced by random sampling (sampling in which all members of the population have an equal chance of being selected). Using appropriate data collection methods will also reduce bias. This exercise shows how random sampling, large sample size, and sampling bias affect our statistical assessment of variation in a population.

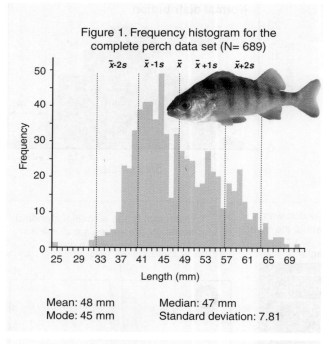

Figure 1. Frequency histogram for the complete perch data set (N= 689)

Mean: 48 mm   Median: 47 mm
Mode: 45 mm   Standard deviation: 7.81

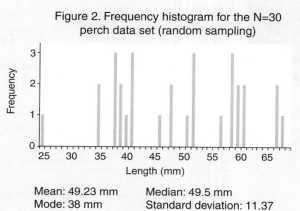

Figure 2. Frequency histogram for the N=30 perch data set (random sampling)

Mean: 49.23 mm   Median: 49.5 mm
Mode: 38 mm   Standard deviation: 11.37

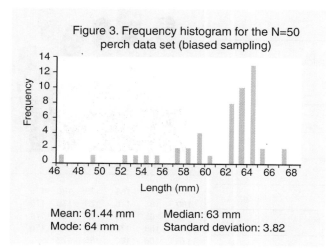

Figure 3. Frequency histogram for the N=50 perch data set (biased sampling)

Mean: 61.44 mm   Median: 63 mm
Mode: 64 mm   Standard deviation: 3.82

▸ In this exercise, perch were collected and their body lengths (mm) were measured. Data are presented as a frequency histogram and with descriptive statistics (mean, median, mode and standard deviation).

▸ Figure 1 shows the results for the complete data set. The sample set was large (N= 689) and the perch were randomly sampled. The data are close to having a normal distribution.

▸ Figures 2 and 3 show results for two smaller sample sets drawn from the same population. The data collected in Figure 2 were obtained by random sampling but the sample was relatively small (N = 30). The person gathering the data displayed in Figure 3 used a net with a large mesh size to collect the perch.

1. (a) Compare the results for the two small data sets (Figures 2 and 3). How close are the mean and median to each other in each sample set?

_____

_____

_____

_____

(b) Compare the standard deviation for each sample set:

_____

_____

_____

(c) Describe how each of the smaller sample sets compares to the large sample set (Figure 1):

_____

_____

_____

_____

_____

(d) Why do you think the two smaller sample sets look so different to each other?

_____

_____

_____

_____

_____

©2018 **BIOZONE** International
ISBN: 978-1-927309-55-1
Photocopying Prohibited

Image credits

We acknowledge the generosity of those who have provided photographs for this edition: • J Podos for photos of the Galápagos finches • D. Dibenski for the photo of the flocking auklets • Dartmouth College Electron Microscope Facility • • Watson and Crick DNA photo: A. Barrington-Brown, © Gonville and Caius College, Cambridge / Coloured by Science Photo Library • Photos of chicken combs courtesy Marc King • Biston photos: Olaf Leillinger • NY State Department of Health • Hominin skullimages by Bone clones • Rove beetle mimic ©Taku Shimada (permission with credit)

We also acknowledge the photographers that have made their images available through Wikimedia Commons under Creative Commons Licences 2.0, 2.5. or 3.0:
• Peter D. Tillman • VIUDeepBay_ • Michaelmaggs • JesseW900 • Cgoodwin • Grant Singleton CSIRO • Antelope_aka_Pronghorns.jpg: Larry Lamsa • Brocken Inaglory • Luc Viatour www.Lucnix.be • Scott McD1 • Adbar • 350z33 • Mike baird-WP • Nigel Wedge • TR Shankar • I, Kenpei • Takahashi • ATamari • Pjt56 • Alex Wild • Wojsyl • Martin D, Thompson A, Stewart I, Gilbert E, Hope K, Kawai G, Griffiths A - Extrem Physiol Med • Kristian Peters • Igor Liberti • CSIRO • BenAveling • Eugen Zibiso • Keisotyo • Jmpost • thisisbossi • Hcrepin • Charles Robert Knight • Jobjabramon • Muntuwandi • H Zell • David Blaiki • Andreas Eichler • OpenStax college • Thomas Splettstoesser • Diacritica • All images: Drosophila sciencecourseware.org • Christopher Michel • Cindy McCravey • Jeffmock • Evan Baldonado • Image; Google satellite • Emmanuelm • Gregory Antipa • Zforgacs • Suseno • Y tambe • Lusb • Lexicon Genetics/HGRI • Joanne H. Hsu • Jpbarrass • Matthias Zepper • Ansgar Walk • Bjørn Christian Tørrissen • Photaro • California High-Speed Rail Authority • Scott Ehardt • Jerald E. Dewey, USDA Forest Service, Bugwood.org • Joshua Stevens • Poppy • Steve Jurvetson • Mircea Madau • Tomás Castelazo • Paul Hames/California Department of Water Resources • Elizabeth Cerny-Chipman Oregon State University • Tuxyso •
• Derdadort • Radim Holiš • Matt Affolter (QFL247) • katorisi • G310Luke • Till Niermann • Yosemite • Rick Kimpel • JeremyaGreene • California Department of Water Resources • William Croyle, California Dept of Water Resources, Pubic Domain • Hans Hillewaert • Nobu Tamura • Pavel Gol'din, Dmitry Startsev, and Tatiana Krakhmalnaya • Kevin Guertin • NoraSmb • Rocky Mountain Laboratories, NIAID, NIH • Xiangyux • Changehali • Lip Kee Yap• Lylambda • Frank Wouters • Alastair Rae • A C Moraes • Ellen Levy Finch • • jim gifford • jiel • Bobby 111 • Dr David Midgley • Kevin Cole • Alastair Rae • Joe Schneid, Louisville, Kentucky • Qing Hai Fan MAF • Wilson44691 • Moussa Direct Ltd.• Heinz-Josef Lücking • Jen • Obersachse • Didier Descouen • Radim Holiš • Connie Ma • Decumanus • Tillman • Dr David Midgley Kevin • Bauman • Haplochromis • Karl Magnacca • Tiago Fioreze • Rainer Zenz

Contributors identified by coded credits are: BF: Brian Finerran (University of Canterbury), BH: Brendan Hicks (Uni. of Waikato), CDC: Center for Disease Control and Prevention, Atlanta, USA, EII: Education Interactive Imaging, NASA: National Aeronautics and Space Administration, NIH: National Institutes of Health, RA: Richard Allan, USCG: U.S. Coast Guard, U.S. USDA: United States Department of Agriculture, USFW: U.S. Fish and Wildlife Service, USGS: United States Geological Survey

Royalty free images, purchased by BIOZONE International Ltd, are used throughout this workbook and have been obtained from the following sources: Corel Corporation from their Professional Photos CD-ROM collection; IMSI (Intl Microcomputer Software Inc.) images from IMSI's MasterClips® and MasterPhotos™ Collection, 1895 Francisco Blvd. East, San Rafael, CA 94901-5506, USA; ©1996 Digital Stock, Medicine and Health Care collection; © 2005 JupiterImages Corporation www.clipart.com; ©Hemera Technologies Inc, 1997-2001; ©Click Art, ©T/Maker Company; ©1994., ©Digital Vision; Gazelle Technologies Inc.; PhotoDisc®, Inc. USA, www.photodisc.com. • TechPool Studios, for their clipart collection of human anatomy: Copyright ©1994, TechPool Studios Corp. USA (some of these images were modified by Biozone) • Totem Graphics, for their clipart collection • Corel Corporation, for use of their clipart from the Corel MEGAGALLERY collection • 3D images created using Bryce, Vue 6, Poser, and Pymol • iStock images • Art Today.

Index

©2018 BIOZONE International
Photocopying Prohibited